ELECTROTECHNICAL SYSTEMS

Simulation with Simulink® and SimPowerSystems™

ELECTROTECHNICAL SYSTEMS

Simulation with Simulink® and SimPowerSystems™

Viktor M. Perelmuter

CRC Press
Taylor & Francis Group
Boca Raton   London   New York

CRC Press is an imprint of the
Taylor & Francis Group, an **informa** business

CRC Press
Taylor & Francis Group
6000 Broken Sound Parkway NW, Suite 300
Boca Raton, FL 33487-2742

© 2013 by Taylor & Francis Group, LLC
CRC Press is an imprint of Taylor & Francis Group, an Informa business

No claim to original U.S. Government works

Version Date: 20120803

International Standard Book Number: 978-1-4665-1402-7 (Hardback)

To my wife

Contents

Preface...xi
Author xv

Chapter 1   Special Features of SimPowerSystems™ Models

      1.1   General Characteristics 1
      1.2   Graphical User Interface Powergui 7
      References 16

Chapter 2   Models of Power Circuit Devices... 17

      2.1   Electrical Sources.. 17
      2.2   Impedances and Loads 22
      2.3   Transformers...36
      2.4   Transmission Line Models 51
      2.5   Miscellaneous 59
      References 63

Chapter 3   Measuring and Control Blocks...65

      3.1   Measurement of Main Circuit Quantities.............................65
      3.2   Meters with Employment of Simulink® Blocks 68
      3.3   Control Blocks...77
      References 83

Chapter 4   Simulation of Power Electronics Devices 85

      4.1   Models of Power Semiconductor Devices.............................85
      4.2   Control Blocks for Power Electronics 92
      4.3   Simulation of Converter with Thyristors............................ 101
      4.4   Simulation of a High-Voltage Direct Current Electric
              Power Transmission System 108
      4.5   Simulation of Converters with Forced-Commutated
              Devices 115
      4.6   Cascaded H-Bridge Multilevel Inverter Simulation.............. 122
      4.7   Four-Level Inverter with "Flying" Capacitor Simulation...... 126
      4.8   Simulation of Z-Source Converters....................................134
      4.9   Simulation of Resonant Inverters 143
      4.10   Simulation of Modular Multilevel Converters......................152
      4.11   Simulation of Matrix Converters..157
      References 166

Chapter 5 Electric Machine and Electric Drive Simulation 167

    5.1 Direct Current (DC) Motors and Drives 167
        5.1.1 DC Drives with Chopper Control........................... 167
        5.1.2 Saturation Consideration 179
        5.1.3 Continuous Models of DC Electrical Drives
              in SimPowerSystems™ 187
    5.2 Induction Motors and Electric Drives 190
        5.2.1 Model Description.. 190
        5.2.2 Simulation of IM with Two-Level
              Voltage-Source Inverter (VSI) and DTC.................. 196
        5.2.3 Models of the Standard IM Drives
              in SimPowerSystems™ 204
        5.2.4 IM with Two-Level VSI and an Active Front-End
              Rectifier 209
        5.2.5 IM with Three-Level VSI....................................... 211
            5.2.5.1 IM with Three-Level VSI and DTC 211
            5.2.5.2 IM with Three-Level Inverter and *L-C*
                     Filter.. 214
        5.2.6 Simulation of IM Supplied from CHB Inverter 216
        5.2.7 IM Supplied from the Four-Level Inverter
              with "Flying" Capacitors... 221
        5.2.8 Simulation of the Five-Level H-Bridge
              Neutral-Point Clamped Inverter (5L-HNPC)
              Supplying IM..223
        5.2.9 Simulation of the IM with Phase-Wound Rotor 236
        5.2.10 IM with Current Source Inverter.............................240
        5.2.11 Simulation of IM Soft-Start 242
        5.2.12 IM Model with Six Terminals.................................244
        5.2.13 Model of Six-Phase IM 247
        5.2.14 Simulation of the Special Operation Modes
              of the Line-Fed IM...254
    5.3 Synchronous Motors (SM) and Electric Drives 256
        5.3.1 SM Model..256
        5.3.2 Simulation of the Electrical Drive with SM
              and Load-Commutated Converters 264
        5.3.3 Model of Six-Phase SM... 271
        5.3.4 Cycloconverter Simulation 277
        5.3.5 SM with VSI Simulation 284
            5.3.5.1 Standard Model.....................................284
            5.3.5.2 Power Electrical Drive with
                     Three-Level VSI 287
            5.3.5.3 Power Electrical Drive with CHB
                     Inverter.. 291
        5.3.6 Simplified SM Model 297

5.4   Synchronous Motor with Permanent Magnets 299
5.5   Switched Reluctance Motor Simulation 305
5.6   Mechanical Coupling Simulation 311
References 315

Chapter 6   Electric Power Production and Transmission Simulation 317
6.1   Computation of Transmission Line Parameters 317
6.2   Use of the Simplified SM Model 324
6.3   Simulation of Systems with Hydraulic-Turbine Generators 326
6.4   Simulation of Systems with Steam Turbine-Synchronous
       Generator 338
6.5   Simulation of Wind Generation Systems (WG) 349
       6.5.1   WG with an Induction Generator (IG) 349
       6.5.2   WG with a Synchronous Generator
                with Permanent Magnets (SGPM) 356
       6.5.3   WG with SGPM and Diesel-Generator 359
       6.5.4   Simulation of a Stand-Alone WG 363
6.6   Simulation of the Unit: Diesel—Squirrel-Cage IG 369
6.7   FACTS Simulation 372
       6.7.1   Static Synchronous Compensator Simulation 372
       6.7.2   STATCOM Simulation 384
                6.7.2.1   Models of Standard STATCOM
                           Systems 384
                6.7.2.2   DSTATCOM Simulation 390
                6.7.2.3   STATCOM with Cascaded H-Bridge
                           Multilevel Inverter Simulation 392
       6.7.3   Active Filter Simulation 397
       6.7.4   Static Synchronous Series Compensator
                Simulation 403
       6.7.5   Unified Power Flow Controller Simulation 407
       6.7.6   Phase-Shifting Transformer Simulation 415
References 418

List of the Models on CD 419

Index 423

Preface

MATLAB® is the high-level programming language developed for solving technical problems. It is used widely in academic courses and in scientific and engineering activities. The graphical programming language Simulink® is included in MATLAB. Simulink is used for simulating dynamical systems. It is supposed that the system under investigation is developed as a functional diagram consisting of blocks that are equivalent, by their functions, to the program blocks that are included in the Simulink library. Under development of the system model, the developer transfers the demanded blocks from the Simulink library in his scheme of the model and connects them in accordance to the system functional diagram.

This feature of Simulink made it very popular among engineers and scientific workers. There are several publications in the various fields of Simulink employment. One can find a full list of publications on MATLAB and Simulink use in the MathWorks website, http://www.mathworks.com/support/books/. The list contains more than 1000 books and shows widespread employment of MATLAB and Simulink. These books expound a theory, describe real-life examples, and provide exercises using MATLAB, Simulink, and other products of MathWorks. They provide the fundamental material for university instructors of the technical and natural sciences and mathematics and are the reliable reference for researchers in academy and in industry.

Electrical engineering, electrical power engineering, and power electronics are continually developing technical branches that have specific areas of study. Models using standard Simulink blocks can be developed to investigate these objects. However, such models often prove to be very complicated and demand that their investigator or developer know the detailed features and properties of the object under investigation. In order to simplify and to precipitate investigations in the aforementioned areas, a set of SimPowerSystems™ blocks is developed. SimPowerSystems uses the Simulink environment that allows it to create the system model with the help of the simple procedure "click and drag."

This book provides a description of the simulation methods of the various electrical systems that use SimPowerSystems. There are very few books on this subject.

This book tries to fill this gap and pursues the following aims:

1. It provides the necessary descriptions and explanations of blocks and models. Electrical objects are rather complicated systems and are described by a set of differential (and often integral) and algebraic equations. When the models of these objects are created, some simplifications are carried out and certain assumptions are made. Under using of these models it is necessary to understand clearly, which simplifications are made, in order to estimate, if they are admissible in the considered problem.

2. It organizes the material in a logical sequence, from simple to complex, to make it easier to study SimPowerSystems as well as to enable this book to be used in the educational process covering various electrical fields.

3. It provides an opportunity to use SimPowerSystems for simulation of very different and complex electrical systems under development and allows researchers and engineers to investigate developed electrical systems.

This book also provides the reader with the following opportunities:

1. To study the toolbox SimPowerSystems comprehensively

2. To understand the principles, features, and detailed functions of various electrical systems, such as electrical drives, power electronics, and systems for production and distribution of the electrical energy, better and to acquire knowledge about the functioning of electrical systems developed recently in various fields

3. To precipitate the development of and to investigate electrical systems by employing the models that are elaborated in this book

To aid in achieving these goals, the book has a companion compact disc (CD) that contains nearly 100 models of the electrotechnical systems created by using SimPowerSystems. One part of these models are the adaptations, one way or another, of the demonstrational models from SimPowerSystems, the others are developed by the author under fulfilling of the research works and designing in fields of the electrical drives and power electronics or specially for this book. In the elaborated models, I have tried not only to show how the models can be used, but also to provide an opportunity to study the corresponding theoretical material. The models have been created and described in the same logical consistency with which this book is written, following the principle "from the simple to the complicated." SimPowerSystems User's Guide contains several demonstrational models that give much interesting and useful information, but the theoretical background is usually absent. Besides, the demonstrational models do not include a number of the important electrotechnical applications, in particular those that are related to power electronics and electrical drives. Some of the demonstrational models do not depict specific features of the modeled devices completely. The models in these areas are developed and placed on the appended CD. I have also tried to avoid the intricate models that are interesting mainly for the specialists in this technical branch in order to cater to a wider audience. The reader can place the models on CD in his folder, and afterwards to investigate their operation under given or selected by the reader parameters, to use them as a base for the analogous models developed by him and so on.

The books that deal with Simulink models placed on the appended CD contain often the drawings of these model diagrams in the book text; it makes reading of the book lighter. In our case, this is practically impossible because the diagrams are very complicated, and the incoming subsystems and their parameters have to be specified too. Thus, the main principles of the models are described in the book; a computer will, however, be required to study these further.

The contents of this book can be summarized as follows:

Chapter 1 covers some common characteristics of SimPowerSystems and describes the graphical user interface, Powergui, which makes analysis and simulation much easier. Chapter 2 describes the models of the main circuit elements, with the help of which the full model of the investigated system is created: supply sources, resistors and loads, transformers, transmission line models. Some of the models are developed, for instance, *transmission line with regulated transformer* and *rectifier with four phase-shifting transformers*.

A number of the measuring and control blocks are included in SimPowerSystems. These blocks use the Simulink blocks and are employed together with the power elements contained in SimPowerSystems. They are described in Chapter 3 and essentially decrease the time for model development. Besides, the blocks intended for connection of SimPowerSystems with the Simulink blocks are described here.

Chapter 4 describes the models of semiconductor devices that are used in power electronics: diodes, thyristors and GTO thyristors, power transistors, and control blocks used with these devices. The models of the various thyristor converters, high-voltage DC transmission lines, two- and three-level voltage source inverters (VSI), cascaded inverters, inverters with "flying" capacitors, Z-source inverters, resonance and modular inverters, and matrix converters are elaborated using of the aforementioned power and control blocks.

Models of DC and AC motors are described in Chapter 5. Since the DC motor model in SimPowerSystems does not take into account saturation of its magnetic system, the modified model with saturation is given. The possibility of DC electrical drive simulation is demonstrated by the chopper control of the series and shunt DC motors and by four-quadrant DC thyristor drives with field weakening and saturation. The induction motor (IM) is the most widespread type of motor; therefore, its simulation methods are considered important. Simulation of the IM that is supplied by two- or three-level VSI with the direct torque control (DTC) or with the vector control, of the IM that is supplied by the current source inverter (CSI), of the wound-rotor IM are considered in this chapter. The model of the IM soft-start, IM simulation under supplying by the cascaded H-bridge multilevel (CHB) inverter, by the four-level inverter with the "flying" capacitors, and by the five-level H-bridge neutral-point clamped inverter are described. The models of IM with disconnected windings (six terminals) and of six-phase IM are also covered.

Models of electrical drives with the load-commutated thyristor inverter and the synchronous motor (SM) and those with SM and cycloconverter are developed. The six-phase SM model and the models of electrical drives with the SM with permanent magnets as well as with switch reluctance motor are described.

Chapter 6 deals with simulation of power production and transmission systems. The command of SimPowerSystems is explained, which gives the possibility, knowing the transmission line construction, to determine the parameters of its electrical model. Simulation of the production and transportation of electrical energy by using hydraulic turbine generators, steam turbine generators, wind generators, and diesel generators are considered. In conclusion, the models of the up-do-date systems that improve the quality of electrical energy—active filters, static compensators of the different types—are developed and described.

It is worth noting that blocks SimPowerSystems have certain limitations. This is mainly because they are intended to simulate systems, not circuits and separate devices. This means that, for instance, the models of semiconductor devices cannot be used to calculate their switching processes, overvoltages, delays, currents during switch-over at these conditions, and so on. Other programs have to be used for this.

This book can be used by engineers to study new electrical systems and to investigate existing ones. It should also prove useful for graduate and undergraduate students undergoing courses in electrical fields in higher educational institutions.

MATLAB® is a registered trademark of The MathWorks, Inc. For product information, please contact:

The MathWorks, Inc.
3 Apple Hill Drive
Natick, MA, 01760-2098 USA
Tel: 508-647-7000
Fax: 508-647-7001
E-mail: info@mathworks.com
Web: www.mathworks.com

Author

Viktor M. Perelmuter, PhD, received his diploma in electrical engineering (with honors) from the National Technical University "Kharkov Polytechnic Institute" in 1958. He has worked in the Research Electrotechnical Institute, Kharkov, Ukraine, thyristor drive department, until 2000, including as department chief from 1988 to 2000. He was at the head of elaboration of the power electrical drives for the heavy industry, in particular, for the metallurgy, repeatedly took part in putting these drives into operation at the metallurgical plants. This work was commended with a number of honorary diplomas. From 1993 to 2000, he served as director of the Joint venture "Elpriv" at Kharkov. From 1965 to 1998, he served as supervisor under the graduation works in Technical University, Kharkov, and from 1975 to 1985, he served as chairman of the State Examination Committee in the Ukrainian Correspondence Polytechnic Institute.

Along with his engineering activities, Dr. Perelmuter headed the scientific work in the fields of electrical drives, power electronics, and control systems. He is the author or coauthor of 9 books and about 75 articles and holds 19 patents in USSR and Ukraine. He received his PhD (candidate of the technical sciences) from the Electromechanical Institute Moscow, SU, in 1967, his diploma as senior scientific worker (confirmation of the Supreme Promoting Committee by USSR Council of Ministers) in 1981, and his DScTech from the Electrical Power Institute Moscow, SU, in 1991.

Since 2001, Dr. Perelmuter has been working as scientific advisor in the National Technical University "Kharkov Polytechnic Institute" and in the Research Electrotechnical Institute, Kharkov, Ukraine. He is also a member of the IEEE.

1 Special Features of SimPowerSystems™ Models

1.1 GENERAL CHARACTERISTICS

MATLAB® is a high-performance programming language developed for solving technical problems. It is used widely in the educational process, particularly in scientific and engineering activities. The significant advantage of MATLAB is the possibility of its extension with add-on toolboxes (collections of special-purpose MATLAB functions, available separately) that are orientated toward the particular branches of the science and technique.

MATLAB programming language belongs to the text languages: The commands of the program are written as a sequence of the lines of symbols and are executed in the order to be written (except the special cases of conditional and unconditional transfers and cycles). This gives to the language universality, but is not always convenient in practice where the initial problem is usually graphically represented as drawings, electrical diagrams, block diagrams, and so on. In order to write the device operation program, it is necessary to express the operation algorithm of every block of this device and their interaction in the mathematical form the chosen programming language can "understand." This fact makes an increased demand on personnel's skill who are involved in the development and investigation of new systems. The employment of text languages often makes it more difficult to divide the problem into smaller components, which is done to delegate subsections of the problem to be solved by more people in order to solve the main problem.

That is why, together with the text languages, the graphical languages are developed, in which the executed operations are depicted by the certain blocks, and the connections between blocks determine the interaction between certain operations. Usually, these languages are intended for a specific technical branch, and the blocks correspond to the elements, units, and devices that are typical to that branch. At that, the connections between the program blocks correspond to the links between the aforementioned technical components. When the developer of the technical device uses the program block for simulation of the technical block of a complex system, he or she may not know the functional program of this block; it is sufficient for him or her to know that the functioning program of the block corresponds fairly precisely to the operation of the real block.

The graphical programming language Simulink® is included in MATLAB; this language is intended for dynamic system investigation. It is supposed that the system under investigation is developed as a functional diagram (block diagram) that

1

consists of the blocks that are equivalent, by their functions, to the program blocks included in Simulink library. This library is diverse enough and expands constantly. When the system model is elaborated, the developer "clicks and drags" the demanded blocks from Simulink library into his model diagram and connects them according to the system functional diagram.

Electrotechnical systems play an important place in our lives and have a wide range of usage. Scientists and engineers are involved in their research and development. Therefore, the development of investigation methods of these systems is a very important task. In most cases, these systems turn out to be very difficult for the full analysis of steady-state and transient processes, and a method of simulation proves to be the only possibility. A lot of elements and devices that form an electrotechnical system have a complex intrinsic structure, and employment of the standard Simulink blocks for their simulation turns out to be difficult and demands from the staff, who are developing such systems, in-depth knowledge about the device structure and the equations describing the processes proceeding in it.

Therefore, models of more complicated blocks and devices are developed. The set of SimPowerSystems™ blocks contains the models of the complex enough but standard devices and units, whose fields of application are production, transmission, transformation, and utilization of the electric power, electrical drives, and power electronics. SimPowerSystems operates in the Simulink environment. Since Simulink uses MATLAB as the computational engine, the developer can use MATLAB toolboxes and Simulink blocksets.

The first step in simulation is development of the model scheme consisting of the blocks of Simulink and SimPowerSystems. Under the block connection, it is necessary to take into consideration the following. The inputs and outputs of Simulink blocks are denoted by ">." The lines connecting them transmit directional signals, as is usual for block diagrams. The inputs and outputs of SimPowerSystems blocks are the electrical wires; they transfer electric energy in both directions and are denoted as "□." The lines of these two kinds cannot be connected with each other. The SimPowerSystems blocks that require interaction with the Simulink blocks have special inputs and outputs for that purpose, which are also denoted as ">." Besides, there are the special blocks in SimPowerSystems that convert quantities of SimPowerSystems in the signals that are accessible to Simulink blocks and, conversely, convert Simulink signals into voltages and currents of SimPowerSystems.

After the blocks are connected, it is necessary to specify their parameters. For that purpose, the block dialog box (or, the parameter window) opens, in which the block parameters are entered according to block function. It is possible also to define the alphabetical symbols for the requested parameters and their numerical values to be fed into the special program MATLAB (SP) with extension .m that is executed before simulation. If the command sim('name Simulink program') is executed at the end of SP, the block parameters that are identified by their symbols are carried to the corresponding blocks, and thereafter can be used without SP. For that purpose, the command to be given for workspace storage is save <'name'> where name is the name under which the workspace is stored in the file with an extension mat.

Then it is possible to call the option of the menu *File/Model Properties/Callbacks/ Model initialization function* in the window with the given Simulink model and to write in the opening window: `load` (`'name'`) where `name` is the assigned-above name of the workspace. Also it is possible to write in the same window the command: `run` (`'Spname'`). Then, SP will be executed automatically under every Simulink program start.

After making the model scheme, it is necessary to decide if this model will be considered as a continuous or discrete one; that is, with variable or constant step the differential equations will be integrated.

The general recommendations of the SimPowerSystems authors are: The integration with a variable step is used for small size systems (the system of equations is fewer than the 30th order and has fewer than six electronic switches) because it is faster and more precise. However, for large systems, in particular those with nonlinear blocks, simulation can be slow. In such cases, it is reasonable to discretize the simulated system and to use simulation with the constant step. For discretization, the option *Discretize electrical model* is selected with indication of the sample time in the block of the graphical user interface **Powergui**. While using the integration method with a constant step, it is better to set the step value of 5–10 μs for models without fully controlled electronic components and 0.5–3 μs for models with IGBT, GTO, and MOSFET. The so-called stiff method ode 23 tb is commonly used for the models given in this book. For computation of the models given in this book, cases might be observed, when, under simulation with the variable step, the computation stopped practically ("hanged up"), and, under simulation with the constant step, ended successfully. The relative tolerance of 0.001 is set as default; however, there were cases when, for making computations faster or for prevention of "hanging up," a relative tolerance of 0.01 or 0.0001 was set. Thus, for important tasks, it is reasonable to experiment with various methods and with different values of tolerance and step.

The Phasor Simulation Method allows to make calculation much faster. Phasor is a time (unlike space) vector of the sinusoidal voltage or current having the definite frequency. For the sinusoidal voltage $u(t) = U_m \sin(\omega t + \alpha)$, this vector is $\mathbf{U} = U_m e^{j\alpha} = U_m \cos \alpha + j U_m \sin \alpha$, the same for the current vector. All system phasors have the same frequency. Phasors are summed up according to the rules of vector algebra. Two main laws of the electrical circuit theory are valid for them: The sum of all voltage drop phasors round a close circuit is zero and the sum of all current phasors flowing into a junction of a circuit is zero as well. If the voltage phasors at the beginning and at the end of the series circuit with parameters R, L, C are known, the current phasor is

$$\mathbf{I} = \frac{\mathbf{V_a} - \mathbf{V_b}}{\mathbf{Z}} = \frac{\mathbf{V_a} - \mathbf{V_b}}{(R + j\omega L - j/\omega C)} = (\mathbf{V_a} - \mathbf{V_b})\mathbf{Y} \qquad (1.1)$$

where
  $\mathbf{Y} = G + jB$ is an admittance (full conductance)
  G is the conductance
  B is the susceptance

The full circuit power is

$$\mathbf{S} = 0.5\mathbf{VI}^*\tag{1.2}$$

If not peak, but rms values are used for phasors, the factor 0.5 falls out.

In computations using the Phasor Simulation Method, it is not the instantaneous values of the voltages and currents, but their phasors that are related with equation (1.1) that are found. At that, the differential equations change into algebraic ones and fast running processes are neglected. This method is used mainly for studying the electromechanical oscillations in the electrical power system containing many generators and motors, but it can be applied to any linear system. It is necessary to keep in mind that this method gives a solution only at one particular frequency. For transition to the given simulation mode, the option *Phasor* is selected in **Powergui** with the indication of frequency.

Use of the *Accelerator* mode makes simulation much faster. This mode is selected in the model window in the line Toolbar, in the field for choice of simulation mode, instead of the usually used *Normal* mode. It can be chosen also in the falling menu in the item *Simulation*. A C-compiled code of the model is created in this case. It is worth noting that in this mode, some time elapses between start of simulation and start of computations, which depends on many factors and sometimes can be rather big.

When the simulation process is started, a special engine of initialization that makes the model in the state space starts and builds the equivalent system that can be simulated by Simulink. This model is stored in the block **Powergui**. The blocks are divided into linear and nonlinear ones, the space–state matrixes **A**, **B**, **C**, **D** of the system linear part are computed, and system discretization is carried out at this stage if the corresponding modeling mode is selected; if the Phasor mode is chosen, the state–space model is replaced with the complex transfer matrix $\mathbf{H}(j\omega)$ that relates voltage and current phasors. The S-Function block is used for simulation of the linear part.

The nonlinear blocks are modeled as controlled current sources. Their inputs are the corresponding voltages (the subspace of the outputs of the linear part), and their outputs come to the inputs of the linear part as the input submatrix $\mathbf{u}_2$; another submatrix $\mathbf{u}_1$ is formed by the outputs of the independent voltage and current sources. The proper Simulink blocks are used for modeling of the nonlinear blocks and of the independent sources (Figure 1.1) [1].

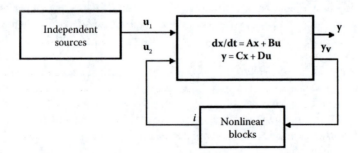

FIGURE 1.1   Block diagram of the system Simulink® model.

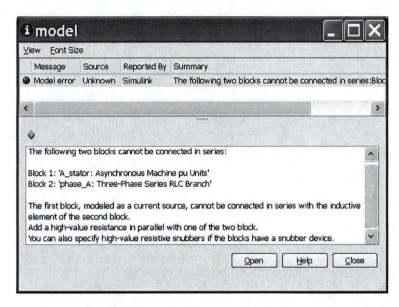

FIGURE 1.2 Example of the report about the error.

SimPowerSystems blocks have the peculiarity that many blocks are modeled as current sources and two such blocks cannot be connected in series; so, for instance, if an inductance in the stator circuit of the motor is set, under model start an error report would appear as shown in Figure 1.2. The mentioned resistor can have very big resistance and not affect the system processes.

The new method is developed in the fifth version of SimPowerSystems for simulation of the power electronic components and switches in the continuous mode. With usual methods, these elements cannot have the zero resistance in the on state and cannot be connected in series with the inductances; so a resistive or resistive-capacitive snubber has to be added. If in reality the snubber is not used, its resistance is chosen large enough, in order to not affect the processes in the model; in such a case, method *ode 23 tb* for the differential equation decision has to be used. However, the speed of simulation can be low. Using the new method, "Ideal Switching Device method," the snubbers are not needed. The simulation method of the switches is chosen with the help of the new **Powergui** version (see further).

A number of commands are provided in SimPowerSystems that are usually executed automatically at the start of simulation; they can also be executed by selecting the command strings provided in the Command window. As a result, the information appears that can be useful for the experienced user to receive more information about the model, but in principle, its knowledge is not obligatory. Therefore, only one version of one command is given here; carrying out this command enables the user to better understand the model structure, and the obtained matrixes of the linear part can be used for future analysis, for example, by using the Control System Toolbox.

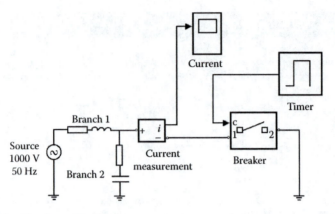

FIGURE 1.3 Simple model scheme.

The command has the following syntax: sps = power_analyze('sys','ss'), where sys is the model name. It computes the equivalent state–space system of the electrical model and estimates matrixes **A**, **B**, **C**, **D** of the model in the standard form:

$$\frac{dx}{dt} = Ax + Bu, \quad y = Cx + Du \tag{1.3}$$

The quantities in vector **x** are currents in the inductances and voltages on the capacitors. The nonlinear devices are modeled as current sources controlled by voltages with the help of nonlinear functions. The system inputs (vector **u**) are sources of the voltage and the current and currents of nonlinear devices. The output vector **y** contains the measured currents and voltages and voltages across nonlinear devices.

As a simple example, Model1_1 is considered (see Figure 1.3). This model has two variable quantities: the inductance current of branch 1 and the voltage across the capacitor of branch 2. The Breaker is modeled as a current source in SimPowerSystems. As described previously, the model has two inputs: the source voltage and the current output of the Breaker. The outputs are the measured current and the voltage at the Break input. Indeed, by executing the command

```
sps = power_analyze('model1_1','ss'),

we receive:

a =
            Il_Branch1   Uc_Branch2
Il_Branch1     -1417        -8.333
Uc_Branch2  6.25e+004         0

b =
             I_Breaker    U_Source
Il_Branch1      1000        8.333
Uc_Branch2  -6.25e+004        0
```

```
c =
            Il_Branch1   Uc_Branch2
U_Breaker        120           1
I_Current Me      0            0

d =
            I_Breaker     U_Source
U_Breaker       -120          0
I_Current Me      1           0
```

1.2   GRAPHICAL USER INTERFACE POWERGUI

The graphical user interface **Powergui** gives a number of possibilities for model tuning and understanding its operation. The function and the view of this block are different in various SimPowerSystems versions. In the third one, it is auxiliary and one can dispense with it, and in the fourth and the fifth versions, the use of this block is mandatory because it is used for the storage of the equivalent model scheme. The model can have only one block, and its name is not changeable. In some previous versions, if this block was not incorporated into the model scheme while designing the model, it appeared automatically at the start of model operation; however, beginning from the 5.1 version, it is necessary to put it into the model scheme from the library or from the other model manually.

When the block icon is double-clicked, its window opens, as shown in Figure 1.4, which shows the fifth version. By selecting the item "Configure parameters," the dialog box opens and shows two options: "Solver" and "Preferences." The configuration of the window for the first option depends on the selected simulation method (continuous, discrete, or phasor). Its view for the first choice is shown in Figure 1.5.

If the first option, "Enable use of ideal switching devices" (see Section 1.1), is marked, the following four lines appear, in which the details of the used method are made precise: to switch off the snubbers, to set the inner resistances of the switches and the power electronic devices to zero, and to set the voltage drop across these devices to zero. In the later **Powergui** version, a new option appears: "Display circuit differential equations." If this option is selected, the differential equations of the model are displayed in the command window at the start of simulation. If "Discrete" or "Phasor" methods are chosen, the sample time (s) or the frequency (Hz) are indicated, respectively.

The window "Preferences" has four fields. If "Display SimPowerSystems warning and messages" is marked, various reports can appear on the screen about this model during simulation. The second field is used for selecting any special function in the mode *Accelerator*; however, this field is not considered here. The third field defines the initial conditions for simulation. There are three opportunities: "Blocks," "Steady," and "Zero." In the first case, the initial values are given in the model blocks; in the second one, they are equal to the steady-state values; and in the third, they are zero. The fourth field concerns the cases to be considered when creating the model; changes are made in library blocks and at the same time the links between the model and the library are deactivated. By selecting "Warning," the deactivated links are

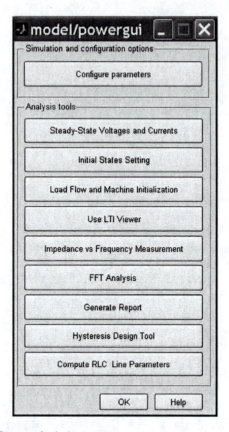

FIGURE 1.4   Main **Powergui** window.

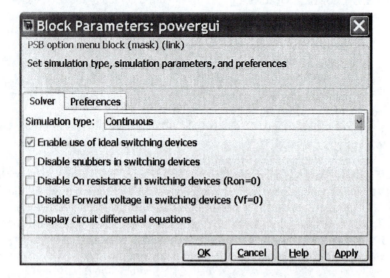

FIGURE 1.5   **Powergui** dialog box under continuous method of simulation.

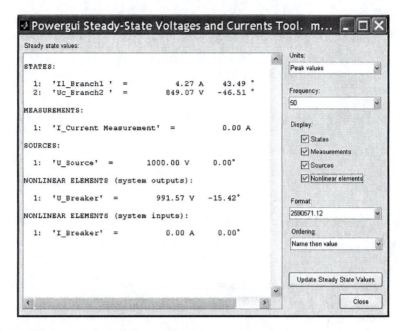

FIGURE 1.6   **Powergui** window under steady-state estimation.

found, and the report appears in the command window. By selecting "Yes," these links are restored automatically, and by selecting "No," checking is not executed and the links are not restored.

In the **Powergui** window in Figure 1.4, the available analysis tools are shown.

The item "Steady-State Voltages and Currents" gives the opportunity to determine the initial steady states of the quantities that are included in vectors **x**, **u**, **y** as mentioned in Section 1.1. Figure 1.6 shows the window given for the model as shown in Figure 1.3. In the right part of the window, there is a menu that shows only part of the quantities: states, measurements, sources, and inputs and outputs of nonlinear elements.

The item "Initial States Setting" sets the initial values of the inductance currents and capacitor voltages for simulation. Usually, they are determined by the vectors that are received using the previous item, but they can be taken as equal to zero or to the selected values.

The next item "Load Flow and Machine Initialization" gives the opportunity to execute simulation, beginning from the steady state. Analysis of voltages and currents, of the requested power for the load supplying and for electrical energy transmission, and of emergency operations in the steady state takes up a lot of time in the case of electrical power systems; since the computation of a steady state is fairly difficult, one begins simulation simply with the arbitrary initial values and waits for getting the steady state, which often requires more time. This item makes simulation much quicker.

Suppose that the system consists of n nodes or collecting buses. To each ith node, the generator lines that draw currents $\mathbf{I}_{gi}$ and have the complex powers $\mathbf{S}_{gi}$, the load lines consuming currents $\mathbf{I}_{li}$ and having the complex powers $\mathbf{S}_{li}$ and the lines that transfer in transit the complex powers $\mathbf{S}_{ti}$ with currents $\mathbf{I}_{ti}$ are connected, so that for the vectors

$$\mathbf{I}_{gi} = \mathbf{I}_{li} + \mathbf{I}_{ti} \tag{1.4}$$

If the node voltage vector is $\mathbf{V}_i$, then

$$\mathbf{V}_i \mathbf{I}_{gi}^* = \mathbf{V}_i \mathbf{I}_{li}^* + \mathbf{V}_i \mathbf{I}_{ti}^* \tag{1.5}$$

or

$$\mathbf{S}_{gi} = \mathbf{S}_{li} + \mathbf{S}_{ti} \tag{1.6}$$

Because

$$\mathbf{S}_{gi} = P_{gi} + jQ_{gi} \tag{1.7}$$

and remains the same for the other power, for the active and reactive powers, respectively

$$P_{gi} = P_{li} + P_{ti} \tag{1.8}$$

$$Q_{gi} = Q_{li} + Q_{ti} \tag{1.9}$$

Thus, knowing the vectors of the in-flow and out-flow currents, and also voltage vectors, the active and reactive powers can be computed.

It can be written for the currents $\mathbf{I}_{ti}$:

$$\mathbf{I}_{ti} = \sum_{j=1}^{n} \mathbf{y}_{ij} \mathbf{V}_j, \quad i = 1, 2, \dots n \tag{1.10}$$

$\mathbf{y}_{ij}$ are complex admittances between the nodes i and j, or

$$\mathbf{I}_{ti} = \sum_{j=1}^{n} y_{ij} V_j \langle -\gamma_{ij} - \delta_j \rangle, \quad i = 1, 2, \dots n \tag{1.11}$$

where
   expression in the braces means the vector angle
   y_{ij} is the admittance module
   γ_{ij} is its angle
   V_j is the voltage vector module
   δ_j is its angle

After substitution in (1.5) and taking into account (1.6) through (1.9), it follows for each node (bus):

$$P_{gi} = P_{li} + \sum_{j=1}^{n} y_{ij} V_i V_j \cos(\delta_i - \gamma_{ij} - \delta_j) \tag{1.12}$$

$$Q_{gi} = Q_{li} + \sum_{j=1}^{n} y_{ij} V_i V_j \sin(\delta_i - \gamma_{ij} - \delta_j) \tag{1.13}$$

The load powers P_{li}, Q_{li} are supposed to be known. Thus, for each node, there are two equations with four unknown quantities: P_{gi}, Q_{gi}, V_i, δ_i; therefore, two quantities have to be assigned, but deciding which one to use will depend on the bus type. Three possible types exist: (1) The reference bus in SimPowerSystems called "Swing bus"; in this node (bus), it is accepted $V_i = 1$ $\delta_i = 0$, for unknown quantities P_{gi}, Q_{gi}. Only one such node can exist in the system. Usually, it is the generator of the largest power or the bus of connection with the other system. (2) The load buses, for which P_{gi}, Q_{gi} are known. Usually, they are the buses without generators, so $P_{gi} = 0$, $Q_{gi} = 0$. The unknown quantities are V_i, δ_i. (3) The generator buses; the system generator—prime mover—usually has the power and voltage regulators with fixed references, so it is accepted: P_{gi}, V_i are known and Q_{gi}, δ_i have to be found.

The system of nonlinear equations that is received this way is solved by an approximate method; subsequently, the computed values of voltages, currents, and powers are fed into the model automatically, and simulation starts.

In order to better understand the **Powergui** operation, model1_2 is considered. Explanations provided here are brief, and the reader can return to this model later.

M1 is the synchronous machine (SM) model of 900 MVA power, of 20 kV voltage. On the second page of the dialog box, the item "Initial conditions" has to be [0 0 0 0 0 0 0 0 1]. If this is not the case, this vector has to be fed into the model. The step-up transformer 20 kV/230 kV of the same power is set at the SM output that transfers the electrical energy at a distance of up to 25 km (**Line 1**). The same unit generator-transformer is connected parallel to the line in its end (M2 and T2). There is the *R-L-C* load **L1** 10 km away from this point, and at a distance of 10 km more, the *R-L* load **L2** is connected. The load parameters are given in their dialog boxes. Owing to the importance of **L2**, it is supplied through the parallel line, **Line 4**. The generators have the constant input mechanical power references ($P_{ref1} = P_{ref2} = 1$) and the excitation systems with the voltage controllers and references of $V_{ref1} = V_{ref2} = 1$. **Multimeter** measures the T1 input current. If to select simulation mode *Continuous* in **Powergui** and integration method *ode* 3 with fixed-step 1 μs, the process shown in Figure 1.7 is obtained on the screen of the **Scope** after fulfilling simulation. But it is necessary to check beforehand that in the dialog boxes of **T1**, **T2**, and the loads, the *Winding Currents,* the *Winding Voltages*, and the *Branch Voltages*, respectively, are marked. The transient process can be seen distinctly, and the simulation carried out rather slow. If to select in **Powergui** "Discrete" mode with the sample time of 5 μs, simulation is carried out much faster. In **Powergui**, if we select the "Phasors"

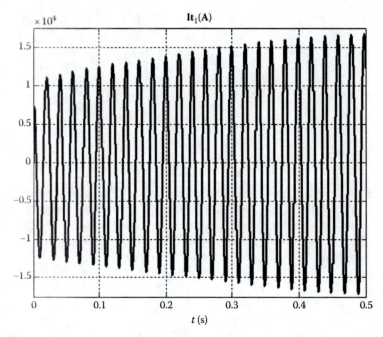

FIGURE 1.7   The process in model1_2 under zero initial conditions in continuous mode.

mode with a frequency of 50 Hz and the integration method *ode* 45 and afterwards start the model, the following error report appears—mismatch of data types. This is because the phasor is a complex quantity, but the Scope recognizes the real data. It is necessary to make sure that "Magnitude" is set in the low field on the right of the dialog box of the block **Multimeter**. Simulation comes to an end almost instantly, and the result is shown in Figure 1.8.

Let us return to the *Continuous* mode and call the item "Load Flow and Machine Initialization." In the right area of the opening window, for M1 "Swing bus" and for M2 "P&V generator" are selected. The rest of the parameters in this area of the window are set automatically. After clicking on the "Update Load Flow," a window appears as shown in Figure 1.9.

For SM, the rated data, the actual values of the voltages and currents and their phase angles about M1 phase voltage, the values of torque and input and output power in SI and pu, and the excitation voltage in pu are given. Pay attention to the changes that took place in the model automatically: The reference SM powers change from 1 to $P_{ref1} = 0.617$ and $P_{ref2} = 0.78$, and in the SM fields "Initial Conditions" nonzero initial values appear. If the item "Steady-State Voltages and Currents" is chosen, a window as shown in Figure 1.10 appears. During simulation, the process shown in Figure 1.11 takes place. It can be seen that transient is absent.

Carry on with the study of **Powergui**. The item "Uses LTI Viewer" gives the opportunity to use methods that are found in the MATLAB toolbox "Control System Toolbox." Since the reader is not expected to be aware of this toolbox, this item is not described here. Interested readers can get to know this toolbox from references [2,3].

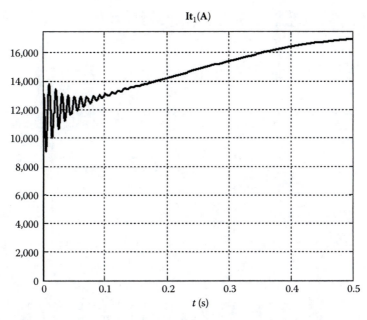

FIGURE 1.8 The process in model1_2 under zero initial conditions in Phasor mode.

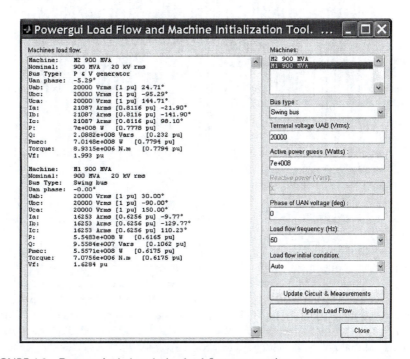

FIGURE 1.9 **Powergui** window during load-flow computation.

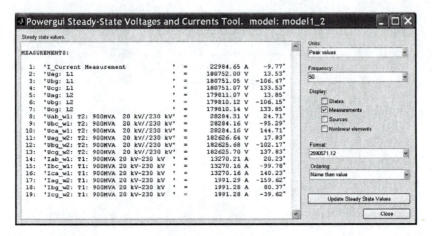

FIGURE 1.10   **Powergui** window during steady-state estimation after load-flow computation.

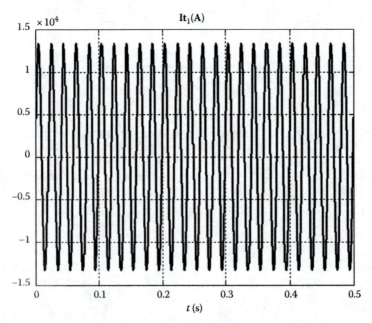

FIGURE 1.11   The process in model1_2 under initial conditions corresponding to load-flow computation.

The item "Impedance vs Frequency Measurement" gives the opportunity to build the frequency characteristic of the complex resistance of any model block. Model1_2a is a copy of model1_2. With the former, two blocks of **Impedance Measurement** from the folder *Measurement* are used. In the dialog box of **L1**, the scheme of null-point connection is changed: Instead of Y(grounded), Y(neutral) is selected. At that point, the neutral point terminal appears on the block picture. One of the blocks, **Impedance Measurement**, measures the impedance of **L1** (phase A),

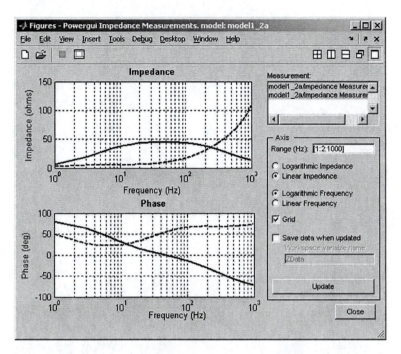

FIGURE 1.12 **Powergui** window during frequency-characteristic computation.

and another measures **Line 4** phase impedance. By selecting the option "Impedance vs Frequency Measurement," a window opens that has the fields for amplitude and phase characteristics in its central area, and in the right area there are the fields that enable the viewer to select the desired representation peculiarity. Linear impedance and logarithmic frequency presentations, grid, and frequency range from 1 to 1000 Hz with step of 2 Hz are chosen. Both impedances are built on the same diagram (Figure 1.12). The load resonance frequency is about 40 Hz.

The item "FFT Analysis" is chosen for the frequency analysis with the help of Fourier transformation of recorded data. Under the data record, the scopes have to be in the following format: "Structure with Time." If this option is selected after simulation, a window appears showing a number of fields (Figure 1.13). In the upper-right field, a list of available structures that are formed in Workspace appears. There is only one structure in the case under consideration that is formed by ScopeData. The structure, the number of inputs, and the number of signals in this input are selected. Then the initial time of the processed signal, the number of cycles of the fundamental frequency and the value of this frequency, the maximal frequency for computation of Fourier transformation, the unit for axis x, and a style of result show—Bar or List—are set.

The processed signal is displayed in the left-upper window; at that point, a part of the signal ("Display FFT window") or all signal ("Display selected signal") can be chosen. If we call the command Display, the result of Fourier transformation appears in the low left field in the selected style of representation. Observe that in the case under consideration, the relative harmonic amplitudes are very small and THD = 0 because the linear system with sinusoidal input signals is considered.

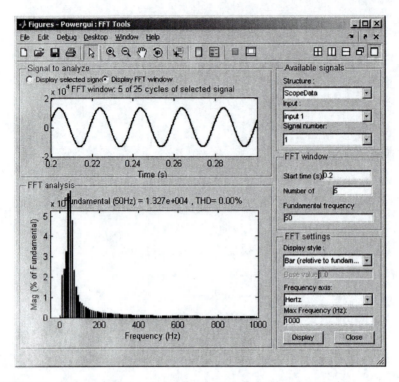

FIGURE 1.13 **Powergui** window during Fourier transformation computation.

The next option, "Generate Report," gives the opportunity to store computation results mentioned previously: the initial states, the steady states, load flow, and machine initialization—as the text file with model name and extension .rep.

The item "Hysteresis Design tool" is used with transformer models while taking saturation into account and will be considered later. The last item, "Compute RLC Line Parameters," is used for computing the active and reactive resistances of electrical power transmission line by using the conductor data, data of the line, and support construction. This procedure is considered in Chapter 6.

From the previous discussion, it follows that **Powergui** is a convenient tool that makes operations with SimPowerSystems models lighter.

REFERENCES

1. Mathworks, *SimPowerSystems™, User's Guide*, 2004–2011.
2. Mathworks, *Control System Toolbox™, User's Guide*, 2004–2011.
3. Perelmuter, V.M., *MATLAB Control System Toolbox and Robust Control Toolbox* (in Russian), Solon-Press, Moscow, Russia, 2008.

2 Models of Power Circuit Devices

2.1 ELECTRICAL SOURCES

The following blocks simulating supply sources are included in SimPowerSystems™ [1]:

1. DC Voltage Source
2. AC Voltage Source
3. AC Current Source
4. Controlled Voltage Source
5. Controlled Current Source
6. Three-Phase Programmable Voltage Source
7. Three-Phase Source
8. Battery

All the blocks are in the folder *Electrical Sources*. Block pictures are shown in Figure 2.1.

The block parameters are put into their dialog boxes. For the block of point 1 (p.1), the direct voltage (V) is given. Blocks of p.2 and p.3 generate the sinusoidal voltage and current quantities, respectively; the voltage (V) or the current (A) amplitude, the phase angle (degree), and the frequency (Hz) are specified. In addition, the sample time given by default is accepted as zero. If it is preferable to measure the output block quantities without employment of the additional voltage or current sensors, the option "Measurement" is marked in the dialog box; the output quantities become available for measurement by the block **Multimeter**, which is described in the next chapter.

Controlled Voltage Source and Controlled Current Source can operate in two modes: as the sources reproducing the Simulink® signal that comes to their inputs or as the original sources. In the first case, the flag "Initialize" is reset. In the second case (flag "Initialize" is set), the DC or AC voltage or current can be chosen in the appearing window, and afterward their parameters are defined.

The **Three-Phase Programmable Voltage Source** simulates a nonstationary and nonsymmetric voltage source. For studying this block, the simple model2_1 is used. The block has several possible configurations of the dialog box. The main one is shown in Figure 2.2. In the first line, the phase-to-phase voltage, phase angle, and voltage frequency are defined. In the line "Time of variation" is indicated, what (or which) parameters of the supply voltage have to vary. In the drop-down menu,

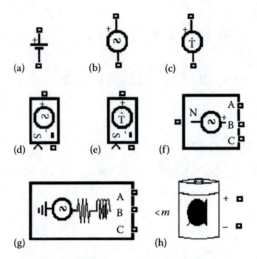

FIGURE 2.1 Pictures of Source models. (a) DC Voltage Source; (b) AC Voltage Source; (c) AC Current Source; (d) Controlled Voltage Source; (e) Controlled Current Source; (f) Three-Phase Programmable Voltage Source; (g) Three-Phase Source; (h) Battery.

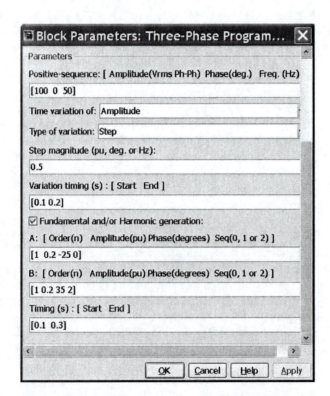

FIGURE 2.2 Dialog box of the block Three-Phase Programmable Voltage Source.

one can choose: "None"—no change, "Amplitude," "Phase," "Frequency." If *None* is selected, the following three lines disappear. If some quantity is selected, in the line "Type of variation" of the drop-down menu, it is possible to choose: "Step," "Ramp" (the linear changing with fixed rate), "Modulation," and "Table of time-amplitude pairs." Depending on the selected character of variations, the meaning of the next line can be changed, and another line can appear. For "Step," the step value is given, moreover, in pu for the voltage, in degree for the phase, and in Hz for frequency. Under "Ramp," the changing speed in the corresponding values is entered. For "Modulation," its amplitude is defined in corresponding units, and the new line, "Frequency," appears in which modulation frequency in Hz is given. Under choice of table representation, two lines appear; in the first one the wanted values of the selected parameter are given, and in the second one, the times when these values are realized.

In addition, the start and end times of the actual changes are defined. As an example, in Figure 2.3 the phase A voltage is shown under step changing of the amplitude by 0.5 at 0.1 and 0.2 s, as is defined in Figure 2.2 (without nonsymmetric consideration).

If you mark the option "Fundamental and/or Harmonic generation" as in Figure 2.2, three lines appear. It is possible to add two harmonics denoted *A* and *B*, and the time interval when these harmonics are active is indicated in the last line. The parameters of each harmonic are given as a vector with four components: the harmonic order about fundamental, the harmonic amplitude in pu, its phase (degree), and type of sequence (1—positive sequence; 2—negative sequence; 0—zero sequence). For example, it is given for harmonic *A* [1 0.2–25 0] and for harmonic *B* [1 0.2 35 2]. It means that zero sequence and negative sequence with the amplitude of 0.2 and

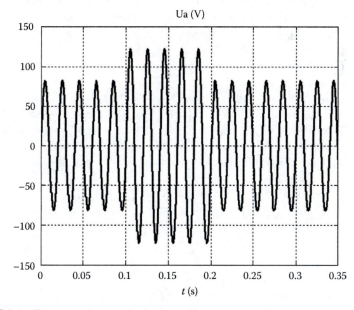

FIGURE 2.3 Phase A voltage under step changing of the amplitude by 0.5.

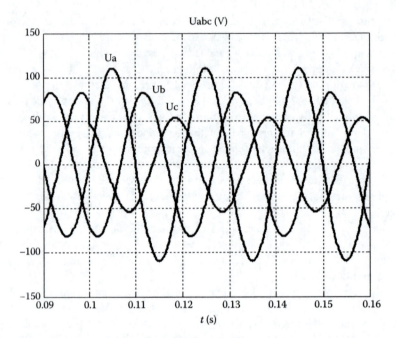

FIGURE 2.4   Phase A voltage with the positive, negative, and zero sequences.

indicated phases are added to the fundamental voltage. If we put "None" in the option "Time variation of," the curves given in Figure 2.4 are formed.

Suppose that instead of the fundamental, the seventh and the fifth harmonics are added: for A [7 0.1 0 1], for B [5 0.2 0 2]. After simulation, the option *Powergui/FFT Analysis* is called, and its window is arranged, as shown in Figure 2.5. In the upper area of the window, four periods of the phase A voltage are shown; in the lower area, the plot of the Fourier coefficients for this voltage is shown.

Block **Three-Phase Source** models a three-phase voltage source that has Y-connection and R-L internal impedance. The neutral can be floating (in the line "Internal connection" of the dialog box, Y is selected), internally grounded (Yg is selected), or accessible (Yn is selected). In the last case, the fourth terminal appears. There are two ways to define R and L. If the option "Specify impedance using short-circuit level" is not marked, the lines are seen, in which values of R (Ω) and L (H) are entered. If this option is marked, the window shown in Figure 2.6 appears.

The source inductance is determined as

$$L = \left(\frac{1}{2\pi f}\right) \frac{(V_{base})^2}{P_{sc}} \tag{2.1}$$

where
  P_{sc} is the three-phase short-circuit power (*VA*)
  V_{base} is the base voltage

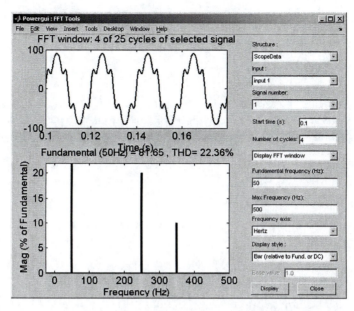

FIGURE 2.5 Fourier transformation window under presence of the fifth and seventh harmonics.

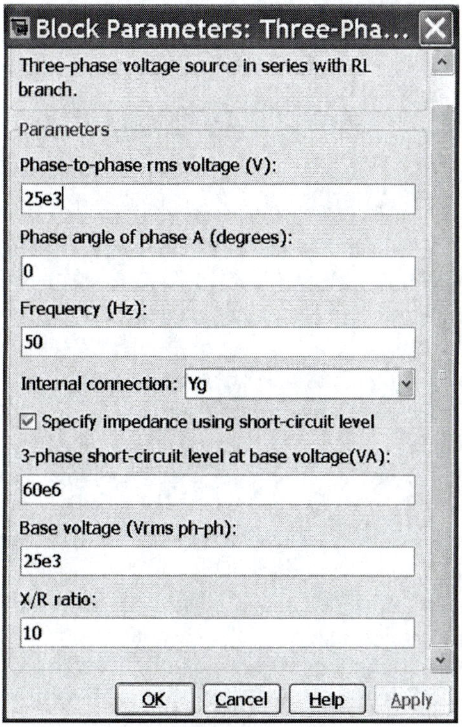

FIGURE 2.6 Dialog box of the block Three-Phase Source.

and the quality factor is

$$\frac{X}{R} = \frac{2\pi f L}{R}$$
(2.2)

One of the parameters, R or L, can be set to zero.

Block **Battery** is not considered in this book, so only one parameter is interesting here: the nominal voltage. Three signals are united at the output m: the charge in pu, the current, and the voltage of the battery.

2.2 IMPEDANCES AND LOADS

The following blocks are included in SimPowerSystems that model various types of resistances and loads:

1. Parallel RLC Branch
2. Parallel RLC Load
3. Series RLC Branch
4. Series RLC Load
5. Three-Phase Parallel RLC Branch
6. Three-Phase Parallel RLC Load
7. Three-Phase Series RLC Branch
8. Three-Phase Series RLC Load
9. Three-Phase Dynamic Load
10. Three-Phase Harmonic Filter
11. Mutual Inductance
12. Three-Phase Mutual Inductance Z1–Z0

The difference between branch models and load models is that for the former the values of $R(\Omega)$, $L(H)$, and $C(F)$ are defined, whereas for the latter the active, reactive inductive, and reactive capacitive (W or var) powers are given under certain voltage and frequency. The load pictures are shown in Figure 2.7.

If some element (R, L, C) is not needed, the corresponding power is set to zero. At that point, the absent element disappears from the block picture. For single-phase loads, the initial values of the capacitor voltage and inductance current can be fixed when the options "Set the initial capacitor voltage" and "Set the initial inductor current" are marked.

If it is necessary to measure the load voltage or (and) the current with the help of the block "Measurement," the quantities to be measured are chosen using the drop-down menu "Measurement": voltage, current, or both. Three-phase powers are defined for three-phase loads, and the voltages and currents are measured in each phase. In the field "Configuration," the scheme of connection can be selected, Y or Delta, and in the first case, the neutral connection can be chosen: ground, floating, accessible neutral.

The pictures of the branch models are shown in Figure 2.8.

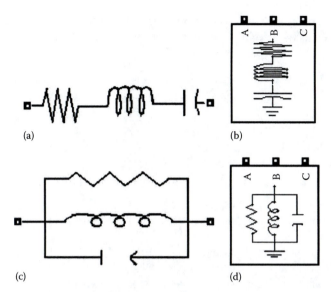

FIGURE 2.7   Pictures of the load models. (a) Series RLC Load; (b) Three-Phase Series RLC Load; (c) Parallel RLC Load; (d) Three-Phase Parallel RLC Load.

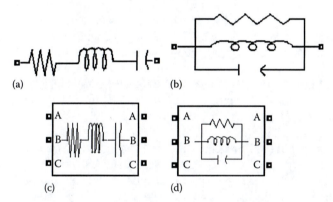

FIGURE 2.8   Pictures of the branch models. (a) Series RLC Branch; (b) Parallel RLC Branch; (c) Three-Phase Series RLC Branch; (d) Three-Phase Parallel RLC Branch.

In the previous versions of SimPowerSystems, all three parameters had to be defined in the block dialog box, and when some element is absent, its value has to be set to 0 or to infinity, depending on the branch scheme. In the last versions, the existing elements are defined at once (i.e., R; R and L; L and C; R, L, and C; and so on). At that point, only existing elements are shown on the block picture, and only corresponding fields appear in the dialog box. The open circuit can be chosen to define "Open circuit" in the line "Branch Type." The initial inductor current and the initial capacitor voltage can be defined for the single-phase branches (if these elements are present). Also for loads, the voltages and currents can be measured.

Block **Three-Phase Dynamic Load** models a three-phase load, whose active P and reactive Q powers vary as a function of the positive sequence of the applied voltage V. The negative and zero sequences are not simulated; it means that even under nonsymmetrical voltage, the current is symmetrical. The load impedance is constant, while $V < V_{min}$. When $V > V_{min}$, P and Q vary:

$$P(s) = P_0 \left(\frac{V}{V_0} \right)^{n_p} \frac{1+T_{p1}s}{1+T_{p2}s} \tag{2.3}$$

$$Q(s) = Q_0 \left(\frac{V}{V_0} \right)^{n_q} \frac{1+T_{q1}s}{1+T_{q2}s} \tag{2.4}$$

where
  P_0, Q_0 are the active and reactive powers under the initial voltage V_0 (the positive sequence)
  T_{p1} and T_{p2} are the time constants that control the dynamic of the active power
  T_{q1} and T_{q2} are the same for the reactive power
  n_p and n_q are the coefficients that usually are in the range 1–3
  s is the Laplace operator

Under the constant current load, $n_p = 1$, $n_q = 1$, because in this case, the power is proportional to the voltage, and under constant load impedance, $n_p = 2$, $n_q = 2$.

The block picture and its dialog box are shown in Figures 2.9 and 2.10. The model can be used in two modes. To mark the option "External control of PQ," the following three lines disappear, and the additional input PQ appears, as in Figure 2.9. The vector signal $[P, Q]$ from the Simulink block enters at this input, defining model powers. If this option is not marked, as in Figure 2.10, the information about n_p, n_q as the vector $[n_p, n_q]$, the time constants as the vector $[T_{p1}, T_{p2}, T_{q1}, T_{q2}]$, and V_{min} in pu have to be entered.

At the output m, the vector signal with three components is formed: the positive-sequence voltage (pu), the active power P (W), and the reactive power Q (var).

Employment of this block gives the opportunity to take into account the dependence of the load power consumption on the supply voltage more precisely. For example, for the resistor furnaces, the impedance is constant, and for a number of melting and galvanic units, the current is constant. For IM and SM, the load torque is defined by the technology of the driven mechanism and does not depend on the

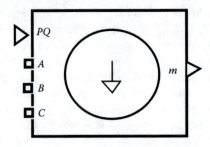

FIGURE 2.9 Picture of the block Three-Phase Dynamic Load.

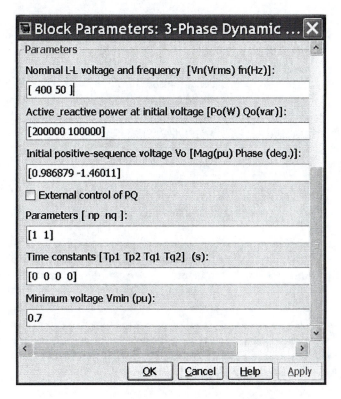

FIGURE 2.10 Dialog box of the block Three-Phase Dynamic Load.

supply voltage, and the speed is defined by the supply frequency or the speed control, so that P does not depend on V ($n_p = 0$). The Q changing depends on the motor type, its parameters, and mode of operation. In Ref. [2], the data of n_p, n_q, cos φ are given for individual loads, but in practice, they are not very useful, because the grid load is determined by the combination of the various loads. More useful are some summarized data as, for instance, those given in Ref. [3] for the different types of consumers (in order cos φ, n_p, n_q): residential consumers—0.87–0.99, 0.9–1.7, 2.4–3.1; commercial consumers—0.85–0.9, 0.5–0.8, 2.4–2.5; industrial consumers—0.85, 0.1, 0.6; steel mills—0.83, 0.6, 2; and primary aluminum—0.9, 1.8, 2.

As an example, the model2_2 is considered. A **Three-Phase Programmable Voltage Source** supplies a **Three-Phase Dynamic Load** through the series circuit that models the inner source and line impedance. The source voltage is $V = 400$ V, the line inductance per phase is 0.064 mH, the load is $V = 400$ V, $P = 200$ kW, and $Q = 100$ kvar. Besides the dynamic load, the constant load of 1 kW, 1 kvar connects to the source. The small active load R is set in parallel. The programmable source operates in the mode of positive-sequence amplitude modulation, the modulation amplitude is 0.7 pu, the modulation frequency is 2 Hz, and the modulation is active in the interval 0.2–0.7 s. The dynamic load has $V_{min} = 0.7$ pu, $n_p = 1$, $n_q = 1$ (constant current), and the time constants are equal to zero.

The block **Multimeter** measures the load voltage and the currents of the source and of the constant load; their difference is the current of the dynamic load. **Scope2** shows the current and the voltage instantaneous values, and **Scope1** fixes the voltage positive sequence on the first axis, pu, the active and reactive powers on the second axis, and the current positive sequence on the third axis. The block **Three-Phase Sequence Analyzer** from the folder *Extra Library/Measurements*, which is considered in Chapter 3, is used for the computation of the last quantity. In order to begin simulation from the steady state, the command `Load Flow and Machine Initialization/Update Load Flow` in **Powergui** is carried out. One can note that the line "Initial positive-sequence voltage" changes from the initial set values of [1 0] to [0.987–1.46] in the load dialog box. After simulation, the plots of the voltage and current positive sequences are obtained, which are shown in Figure 2.11. While $U_1 > 0.7$, the current is constant, and under $U_1 < 0.7$ it changes according to the supply voltage.

A **Three-Phase Harmonic Filter** is intended for simulation of the parallel filters in the power systems that are used for reducing the voltage distortions and for the power factor correction. These filters are capacitive at the fundamental frequency. Usually, the band-pass filters are used that are tuned for low-order harmonic filtering: the 5th, the 7th, the 11th, and so on. The band-pass filters can be tuned for filtering one frequency (single-tuned filter) or two frequencies (double-tuned filter). High-pass filters are used for filtering a wide range of frequencies. A special type of high-pass filter, so-called C-type high-pass filter, is used to provide reactive power and to shun the parallel resonance.

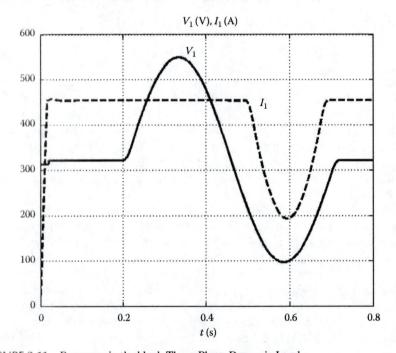

FIGURE 2.11   Processes in the block Three-Phase Dynamic Load.

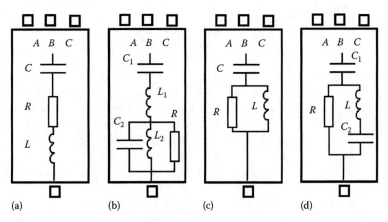

FIGURE 2.12 Pictures of the filter models. (a) single-tuned filter; (b) double-tuned filter; (c) high-pass filter; (d) C-type high-pass filter.

The filters contain R, L, C elements. The values of these elements are determined by the following data: the filter type, the reactive power under the rated voltage, the tuned frequency or frequencies, and quality factor. Four filter types that can be simulated in SimPowerSystems are shown in Figure 2.12.

For the simplest filter—single-tuned filter—the following relationships are valid [1]:

$$X_L = \omega L, \quad X_c = \frac{1}{C\omega} \tag{2.5}$$

The order of the filtered harmonic (here f_1 is the fundamental frequency)

$$n = \frac{f_n}{f_1} = \sqrt{\frac{X_c}{X_L}} \tag{2.6}$$

The quality factor

$$Q = \frac{nX_L}{R} = \frac{X_c}{nR} \tag{2.7}$$

The reactive power at frequency f_1

$$Q_c = \frac{V^2}{X_c} \frac{n^2}{(n^2 - 1)} \tag{2.8}$$

The loss of power at frequency f_1

$$P = \frac{Q_c}{Q} \frac{n}{n^2 - 1}$$ (2.9)

The filter dialog box for tuning as a single-tuned filter is shown in Figure 2.13. In the first line, the wanted filter type is selected from the drop-down menu. In the next line, the neutral connection is chosen: Y (ground), Y (floating), Y with accessible neutral or Delta. The meaning of the following lines is clear from Figure 2.13.

For the double-tuned filter, two tuned frequencies, f_1 and f_2, have to be given. Both the series and the parallel circuits are calculated for mean geometric frequency:

$$f_m = \sqrt{f_1 f_2}$$ (2.10)

The quality factor

$$Q = \frac{R}{2\pi f_m L_2}$$ (2.11)

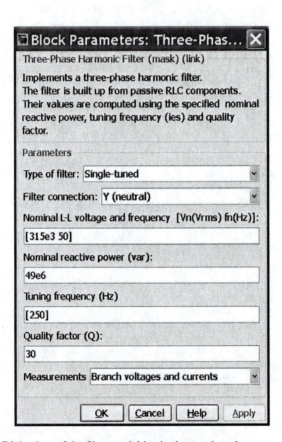

FIGURE 2.13 Dialog box of the filter model in single-tuned mode.

In the high-pass filter, L and R are connected in parallel instead of in series. It results in a wide-band filter, whose impedance at the high frequency is limited by R. The quality factor is

$$Q = \frac{R}{2\pi f_n L} \qquad (2.12)$$

C-type high-pass filter is a modification of the high-pass filter, in which the inductance L is replaced by the series LC circuit that is tuned in resonance at the fundamental frequency. This filter provides a decrease in the loss. The quality factor is defined by (2.12).

As an example, filter computations are given for Ref. [1]: $V = 315\,\text{kV}$, $Q_c = 49\,\text{Mvar}$, $f_1 = 50\,\text{Hz}$.

The single-tuned filter is tuned for suppression of the fifth harmonic with $Q = 30$. From (2.8)

$$49 \times 10^6 = \frac{315^2 \times 10^6}{X_c} \frac{25}{24}, \quad \text{then } X_c = 2109.3\,\Omega$$

From (2.6),

$$25 = \frac{2109.3}{X_L}, \quad \text{then } X_L = 84.38\,\Omega$$

$$L = \frac{84.38}{314} = 0.269\,\text{H}, \quad C = \frac{1}{(314 \times 2109.3)} = 1.51\,\mu\text{F}$$

From (2.9),

$$P = 49 \times 10^6 \frac{5}{(24 \times 30)} = 340\,\text{kW}$$

From (2.7),

$$R = \frac{X_c}{nQ} = \frac{2109.3}{(5 \times 30)} = 14.06\,\Omega$$

In order to have the filter frequency characteristic, the model of four blocks shown in Figure 2.14 has to be made. The used block **Impedance Measurement** has been considered in Chapter 1 already. A multiplication factor of 0.5 has to be set in the block dialog box, because two filter phases are connected in series. After carrying out the command *Impedance vs Frequency Measurement* in **Powergui**, the filter frequency characteristics that are shown in Figure 2.15 are plotted.

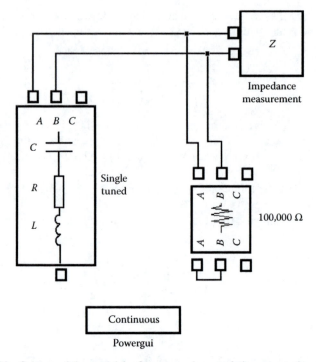

FIGURE 2.14   Scheme of the model for frequency characteristic computation.

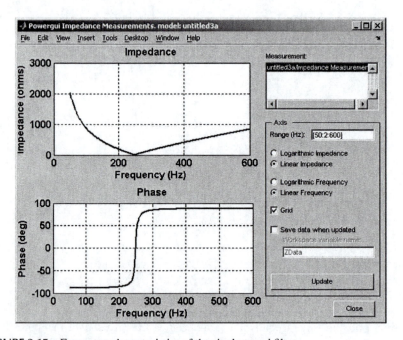

FIGURE 2.15   Frequency characteristics of the single-tuned filter.

The double-tuned filter is tuned for the 11th and the 13th harmonics under $Q = 16$. It means

$$f_m = 50\sqrt{11 \times 13} = 598 \text{ Hz}$$

that is, filter parameters are calculated for the 12th harmonic, then

$$X_{c1} = \frac{315^2}{49} \frac{144}{143} = 2039 \,\Omega, \quad C_1 = 1.56 \,\mu\text{F}$$

$$L_1 = 2039/144/314 = 0.045 \text{ H}.$$

For the parallel circuit, the relationship has to be fulfilled

$$(2\pi f_m)^2 = \frac{1}{L_2 C_2}, \quad \text{i.e., } L_2 C_2 = 0.0704 \times 10^{-6}$$

Accept $C_2 = 50 \,\mu\text{F}$, $L_2 = 1.409 \text{ mH}$.
    The resistance

$$R = 2\pi f_m Q L_2 = 2\pi \times 598 \times 16 \times 1.409 \times 10^{-3} = 84.7 \,\Omega$$

In the dialog box of the model, the necessary change is to be made, and afterward, the command *Powergui/Impedance vs Frequency Measurement* is activated. The frequency range is 250:2:2000 Hz. The frequency characteristics of this filter that are built with the method described earlier are shown in Figure 2.16. One can see two minimums of the filter impedance at frequencies 550 and 650 Hz.
    The high-pass filter is tuned for the twenty-fourth harmonic under $Q = 10$; then

$$X_c = \frac{315^2}{49} \frac{576}{575} = 2028.5 \,\Omega, \quad C_1 = 1.57 \,\mu\text{F}$$

$$L = 2028.5/576/314 = 0.0112 \text{ H}$$

$$P = 49 \times 10^6 \frac{24}{(575 \times 10)} = 204 \text{ kW}$$

From (2.12),

$$R = Q 2\pi f_n L = 10 \times 314 \times 24 \times 0.0112 = 844 \,\Omega.$$

The filter frequency characteristics received by the method described here are shown in Figure 2.17.

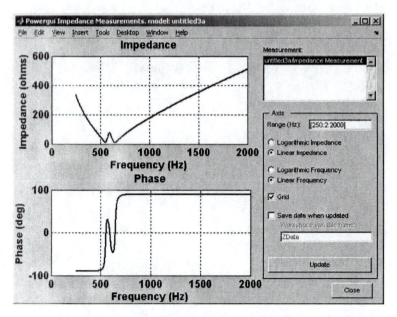

FIGURE 2.16   Frequency characteristics of the double-tuned filter.

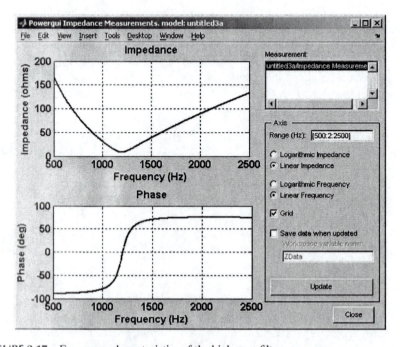

FIGURE 2.17   Frequency characteristics of the high-pass filter.

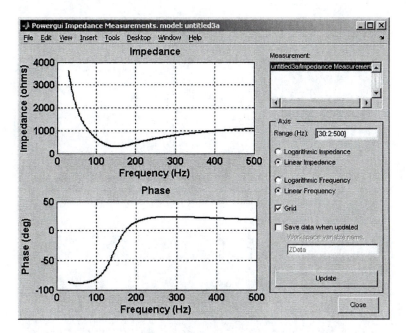

FIGURE 2.18   Frequency characteristics of the C-type filter.

C-type filter is tuned for the third harmonic under $Q = 1.75$, then

$$X_c = 315^2/49 \,(9/8) = 2278\ \Omega, \quad C_1 = 1.4\ \mu F, \quad L = 2278/9/314 = 0.806\ H$$

$$C_2 = 1/0.806/314^2 = 12.58\ \mu F,$$

$$R = 2\pi f_m QL = 314 \times 3 \times 1.75 \times 0.806 = 1328.7\ \Omega.$$

The filter characteristic is given in Figure 2.18.

The model of **Mutual Inductance** can operate in two modes: as a two- or three-winding reactor with equal mutual inductances or as a reactor with any winding number and with different mutual inductances. The choice is fulfilled in the dialog box with the help of the option "Type of mutual inductance"; the third winding appears if the option "Three windings mutual inductance" is marked. In Figure 2.19, the various variants of the block pictures are shown. If the previously mentioned options are marked, the parameter window has a view shown in Figure 2.20. In fact, this block is a model of the two- or three-winding transformer with a transformation ratio of 1. The structure accepted in the block is given in Figure 2.21. So, the resistances and full inductances of the windings are put into two or three lines of the dialog box. In the line *Mutual impedance*, the values R_m, L_m are defined. The inequalities R_1, R_2, $R_3 \neq R_m$, L_1, L_2, $L_3 \neq L_m$ have to be carried out. For the internal and mutual inductances, negative values are permissible. The windings can be left unconnected (floating).

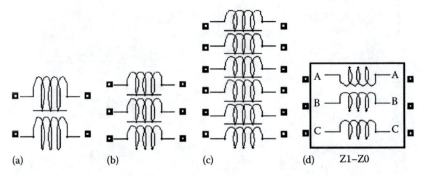

FIGURE 2.19 Mutual inductance models. (a) Two windings; (b) Three windings; (c) Six windings; (d) Three-phase inductance.

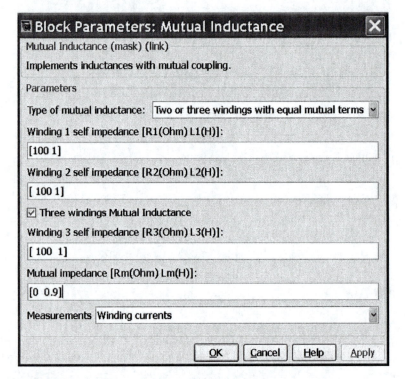

FIGURE 2.20 Dialog box of the model Mutual Inductance.

Under the choice "Generalized mutual inductance," the number of windings N ($N = 6$ in Figure 2.19c) and $N \times N$ matrix of the internal and mutual inductances must be defined. This model is often used for transmission line simulation.

The block operation is displayed in the model2_3, which consists of three parts. The upper part shows an influence of the mutual inductance on the response under voltage changing by step. The **Three-Phase Breaker** shorts out the second and third windings, and at the first one the voltage step of $100\,\text{V}$ is given at $t = 0.1\,\text{s}$.

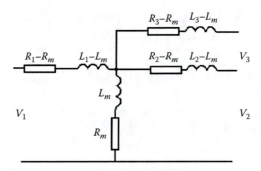

FIGURE 2.21 Scheme of the mutual inductance model.

The **Breaker** keeps its state during simulation. The **Scope** records the voltage on the first axis and the currents of the first and the second windings on the second axis. To fulfill simulation under open **Breaker** (Initial status—*Open*), one can see that the current in the first winding rises by exponent and reaches 0.95 of the steady state in 0.03 s after the step. To set **Breaker** in the closed state and to repeat simulation, one can see the current pulse in the second winding; the current shape in the first one differs from an exponent, and the value of 0.95 is reached in 0.052 s.

The second part of the model simulates a resonant circuit that, with the second winding open, has the resonant frequency of 50 Hz. The resistance of the windings is reduced in ratio 10, in order to improve the quality factor. **Scope1** records the current of the resonant circuit. To fulfill simulation with **Breaker** open, the slightly damped oscillations with the period of 0.02 s can be seen. To repeat simulation with **Breaker** closed, one can see that the process dies down for 0.1 s, and the frequency of oscillations increases to about 115 Hz. This frequency corresponds to the following approximate estimate. The operational resistance of the coil with two equal windings that is viewed from the terminals of one winding when another is shorted out is [4]

$$Z = R + sL_s + (R + sL_s)\frac{sL_m}{R + sL_1} \tag{2.13}$$

When R is small,

$$Z = sL_s\frac{(L_1 + L_m)}{L_1} = s\frac{\left(L_1^2 - L_m^2\right)}{L_1} \tag{2.14}$$

which in the case under consideration gives $Z = 0.19$ s, whereas under the second winding open, $Z = sL_1 = s$. Thus, the equivalent inductance of the resonant circuit decreases in ratio 5.26, and the resonance frequency increases in ratio $\sqrt{5.26} = 2.29$ correspondingly and is estimated as $2.29 \times 50 = 114.5$ Hz.

The third part of the model demonstrates the response of the coil with three windings to the sinusoidal signal of 50 Hz. The parameters of the block **Mutual Inductance** are the same as in the first model part. Fulfilling simulation with open and closed states of **Three-Phase Breaker1**, one can see that in the second case the current value in the first winding increases from 0.3 to 0.65 A.

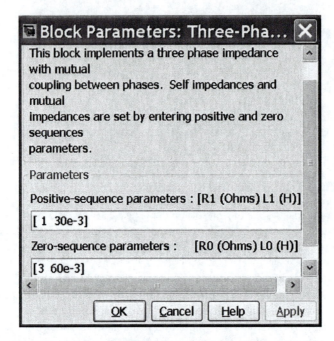

FIGURE 2.22   Dialog box of model Three-Phase Mutual Inductance Z1–Z0.

The last block considered in this section is **Three-Phase Mutual Inductance Z1–Z0**. Its picture is given in Figure 2.19d. This block models a three-phase balanced inductive and resistive impedance with mutual coupling between phases. This block realizes the same function as the three-winding **Mutual Inductance** block but is more convenient for employment under simulation of transmission lines. The parameters of the block are usually calculated by computing the three-phase line parameters (Chapter 6). The dialog box of the block is shown in Figure 2.22.

2.3   TRANSFORMERS

There are the following blocks in SimPowerSystems that simulate various types of transformers:

1. Linear Transformer
2. Saturable Transformer
3. Multiwinding Transformer
4. Three-Phase Transformer 12 Terminals
5. Three-Phase Transformer (Two Windings)
6. Three-Phase Transformer (Three Windings)
7. Zigzag Phase-Shifting Transformer
8. Grounding Transformer
9. Three-Phase Transformer Inductance Matrix Type (Two Windings)
10. Three-Phase Transformer Inductance Matrix Type (Three Windings)

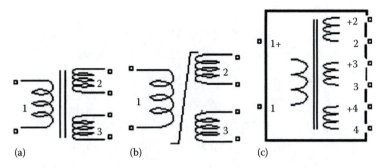

FIGURE 2.23 Pictures of the single-phase transformer models. (a) Linear Transformer; (b) Saturable Transformer; (c) Multiwinding Transformer.

The pictures of the first three transformers are shown in Figure 2.23. The first transformer can have not three but two windings, which is indicated in its dialog box. The transformer impedances: the leakage inductances and resistances of the windings, active and inductive resistances of the magnetization circuit—can be given in pu or in SI; this is fixed in the first line of the dialog box. In the first case, these parameters are defined in the ratio to the base resistance of the corresponding winding; for the magnetization circuit the base values are the resistances of the first winding. For the winding with the rated voltage U_i (V) of the transformer with power S (VA) and frequency f, the base values are

$$R_b = \frac{U_i^2}{S}, \quad L_b = \frac{R_b}{2\pi f} \tag{2.15}$$

For the model of **Saturable Transformer**, the same equivalent circuit is used, but there is an opportunity to simulate saturation both with and without taking hysteresis into account. Hysteresis modeling demands additional computation time; therefore, it has to be used only in specific cases.

In the last versions of SimPowerSystems, the dialog box of this and some other transformers have changed. Now it has three pages called *Configuration, Parameters,* and *Advanced.* The first page defines the structure of the block and some of its main characteristics, the second one its parameters, and the third page some additional characteristics. On the first page of the transformer under consideration, the existence of the third winding and necessity to simulate saturation can be marked; the name of the file with extension .mat, in which the parameters of the hysteresis curve are stored, and the measured quantities are also specified (Figure 2.24a).

There is the special line *Saturation characteristic* on the second page, Figure 2.24b, in which the magnetization curve is defined as a collection of pairs: magnetization current—main flux in the earlier accepted units (pu or SI). The first point is the pair [0 0; ...]. If the residual flux Φ_{res} is modeled, two pairs with zero abscissa are defined: [0 0; 0 Φ_{res}; ...]. A piecewise-linear approximation is carried out under simulation.

If the option *Simulate hysteresis* is marked, the main flux linkage of the transformer is computed by integration of the voltage across the magnetization circuit,

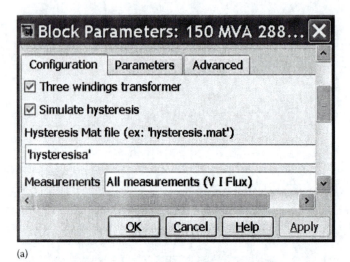

(a)

(b)

FIGURE 2.24 Dialog box of the Saturable Transformer: (a) the first page; (b) the second page.

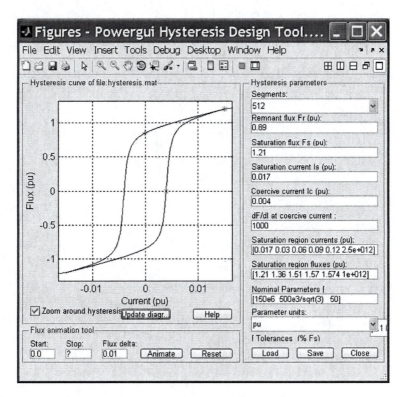

FIGURE 2.25 Powergui window under forming of the hysteresis curve.

and magnetization current is calculated from the found flux linkage. Both major and minor hysteresis loops are simulated.

In order to create the file describing the hysteresis, it is reasonable to use the item *Hysteresis Design Tool* of **Powergui**. When this option is activated, the window that is shown in Figure 2.25 appears. In the right area of the window, the parameters of the curve are specified; in the left area, the curve is displayed corresponding to these parameters. We note that the option *Zoom around hysteresis* is marked, with which the loop takes up the entire screen that makes tuning of the curve easier. Without this option, all the workspace of the magnet system appears on the screen.

The built curve has to be saved on the disc with the help of the command File/ Save the model under some name. The file receives extension .mat. This name must be put into the line *Hysteresis Mat file* on the first page of the dialog box with marked function *Simulate hysteresis*. It is worth noting that under hysteresis simulation, the active resistance of the magnetization circuit takes into account only eddy-current loss, without hysteresis loss.

There is one more interesting possibility of this item. There is an option *Flux animation tool*, the additional fields *Start*, *Stop*, *Flux delta*, and two buttons *Animate* and *Reset*. With the help of these instruments, it is possible to observe the minor hysteresis loops. If, for example, to set *Start* = −0.8, *Stop* = 0.8, and to fulfill the

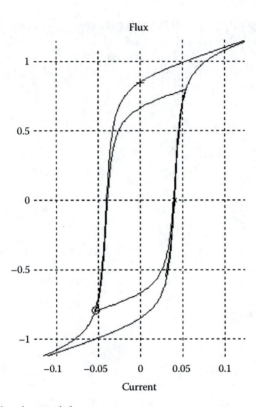

FIGURE 2.26 Minor hysteresis loop.

command *Animate*, and afterward, *Start* = 0.8, *Stop* = −0.8, the picture of the minor loop appears, which is shown in Figure 2.26.

On the third page, the option *Break Algebraic loop in discrete saturation model* can be marked. The fact is that under saturation simulation in the discrete mode, there is a direct connection between the output and input quantities at the same time. Such a relationship is called an algebraic loop, and its existence slows down simulation. The considered option gives the opportunity to break this loop, but under discrete simulation, the input quantity will define the output quantity on the next step that makes the precision worse and can result in oscillations under big sample time.

The model2_4 is intended for the study of the characteristics of the saturable transformer with hysteresis. The transformer parameters are given in Figure 2.24, and hysteresis loop in Figure 2.25. The transformer is supplied from the controlled voltage source with an external control. The source of the sinusoidal voltage with the amplitude of 1.41 288.7 kV and the frequency of 50 Hz defines the supply voltage. The **Timer** schedules the following amplitudes of voltage U: $U = 0.8\ U_n$ at $0 < t < 0.1$ s, $U = U_n$ at 0.1 s $< t < 0.2$ s, $U = 1.5\ U_n$ at 0.2 s $< t < 0.3$ s. The elements that are set between the supply and the transformer do not affect its operation, but provide a smooth start and give the opportunity to increase the sample time. The transformer operation can be observed both in the no-load condition and under load. For that, it is set 0 or 1 corresponding to the dialog box of **Breaker**, in the line *Initial state*.

In order to observe the process, the voltage measurement option is set in the **Controlled Voltage Source**, and all quantities are measured in the transformer block. The measured quantities are processed by the block **Multimeter** in such an order: the transformer flux, pu, the magnetization current without iron losses, A, the total excitation current including iron losses, A, the supply voltage, V, the primary transformer current, pu. These quantities are recorded by **Scope** and some of them by the graph plotter. Besides, there is another block **Multimeter1** that measures the voltage and the current of the primary transformer winding; these values in SI are used for the active and reactive power computation.

To fulfill simulation with the selected option *Simulate hysteresis* and with the secondary winding open, one can observe on the graph plotter drawing of the hysteresis loops. There are three loops. Process in time is observed by **Scope**. The fraction of the magnetization current curve is given in Figure 2.27. Its essential increase at $t = 0.2\,\text{s}$, when the supply voltage increases drastically, can be seen. The active power at $t = 0.15\,\text{s}$ (under rated voltage) is 935 kW. Let us calculate losses in the resistor R_m. If the base value $R_b = 288.7^2 \times 10^6/150 \times 10^6 = 556\,\Omega$, then $R_m = 556 \times 500 = 0.28 \times 10^6\,\Omega$ and the losses are $P_1 = 288.7^2 \times 10^6/0.28 \times 10^6 \approx 300\,\text{kW}$; hence, hysteresis losses are 635 kW.

If the simulation is repeated without the option *Simulate hysteresis*, the graph plotter displays the piecewise-linear curve. **Scope** shows that till $t = 0.2\,\text{s}$, the currents are sinusoidal. The power is 310 kW at $t = 0.15\,\text{s}$, which is in conformity with the calculation here. Thus, in order to appreciate the active losses in the magnetization circuit correctly, when the hysteresis is not modeled, it is necessary to set $R_m = (500/935)\,300 = 160$ pu.

These experiments can be repeated when **Breaker** in the secondary winding is closed. The magnetization currents do not change noticeably, but the influence of nonlinearity on the primary current reduces essentially.

Multiwinding Transformer simulates a transformer with the changing number of windings both on the primary (on the left) and on the secondary (on the right) sides. The equivalent diagram is the same as for the linear transformer, but saturation can be

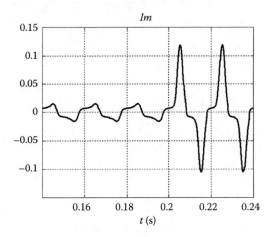

FIGURE 2.27   Changing of the transformer magnetization current in time.

taken into account additionally. The main special feature of the transformer is that to the first primary (upper on the left) or to the first secondary (upper on the right), the evenly spaced taps can be added, so the voltage between two next taps is equal to the winding total voltage divided by the number of taps plus 1. The total resistance and leakage inductance of the winding are evenly spaced along the taps. This transformer model is usually used for the simulation of the transformer that is controlled under load with the aim to keep constant the bus voltage. The requested range of the voltage changing is ±(10%–20%) of the rated value. The tap number is 6–10. Usually, the transformer has the main primary or secondary winding that is calculated for the rated voltage and the adjusting winding with taps that can be connected aiding or anti-aiding to the main winding. Under aiding connection to the primary winding, the output transformer voltage reduces, and under anti-aiding connection increases. Under connection of adjusting winding in series to secondary, the effect of its commutation is contrary.

This type of transformer unit has a drive for switching-over taps. Drive is controlled by the voltage regulator. On the whole, this unit is a rather complex device, and there is the special model *Three-Phase OLTC Regulating Transformer* (*OLTC* is an abbreviation for *On Load Tap Changer*) in SimPowerSystems.

This model is not included in the library of SimPowerSystems and is in the demonstration model *OLTC Regulating Transformer*. We name this model model2_5. Whoever wants to use it can copy it in his model. It is worth noting that in the library of SimPowerSystems, there is the model *Three-Phase OLTC Regulating Transformer* that operates in the Phasor mode only. The employment of this model can be reasonable in the complex systems with several controlled transformers, when simulation time can turn out too large.

Let us consider model2_5 in detail. If you select the option *Look Under Mask*, it would be seen that the device has the block of **Multiwinding Transformer** with taps for each phase, the tap switching block, and the control block. The diagram of the phase A main circuit is shown in Figure 2.28. There are seven taps that give eight steps. The adjusting winding can be connected as aiding or anti-aiding to the primary one that, keeping in mind the possibility to connect the latter directly to the supply, provides 17 voltage steps. The switching-over aiding–anti-aiding is realized with the help of a two-pole switch Reverse. If it is in position 1, the current runs through the upper contact, switches BK1 or BK2, one of the closed switches B0–B8, and the low contact of Reverse to the terminal H2 of the primary winding. Thus, the primary and adjusting windings are connected as aiding. If the switch Reverse is in position 2, the circuit flows: phase A—the low contact of the switch—the end terminal of the adjusting winding—one of the closed switches—switches BK1 or BK2—the upper contact of the switch Reverse—the terminal H2 of the primary winding. Thus, the primary and adjusting windings are connected as anti-aiding.

The transition from step to step is carried out in the following order. Under request to switch over, the control block sends the logic signal for turning off both switches BK1 and BK2. Afterward, the signal is formed to turn on the next step, and some time later, to turn off the previous one. After this, the switches BK1, BK2 close again.

The diagram of the voltage regulator is shown in Figure 2.29. If the difference between reference V_{ref} and measured V voltages is more than the dead band, then logic signal 1 appears at the output of one of the blocks "comparison with 0."

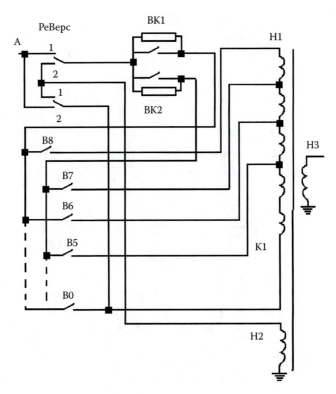

FIGURE 2.28   Diagram of the phase A main circuit of Multiwinding Transformer.

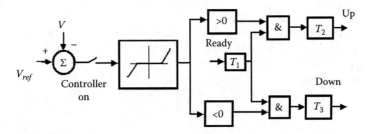

FIGURE 2.29   Voltage regulator of *OLTC Regulating Transformer.*

If $V > V_{ref}$, the low block is active, and, when the signal "Ready" with duration more than T_1 (set 60 ms) has been formed, after delay T_2 (set 1 s) the logic signal "Up" appears. This means that additional taps have to be connected, and the secondary voltage decreases. If $V < V_{ref}$, the logic signal "Down" appears under the same conditions. This means that a part of the taps has to be taken out of the operation, and the secondary voltage increases.

The main circuit switches are controlled according to the circuits shown in Figure 2.30. The signals "Up" or "Down" are differentiated, and the value 1 is added or subtracted in the block Count, depending on the logic signal "Down" (subtraction under 1). After changing the value Count, the signal "Ready" disappears for

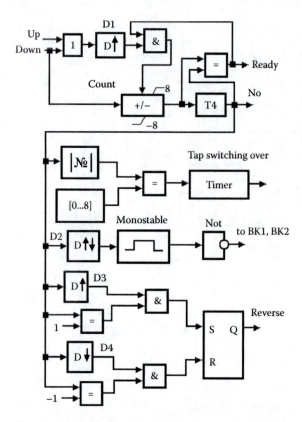

FIGURE 2.30 Control of the main circuit switches of *OLTC Regulating Transformer.*

3–10 s, and then the cycle repeats. After the same time, the signal № of the new tap appears. This signal is used for forming signals that control the states of the switches. In order to determine which tap switch has to be turned on, the signal № is compared with whole numbers 0, 1,.., 8, and the switch is closed, whose number coincides with №. The unit **Timer** forms the necessary sequence of the commands for turning off the previous switch and turning on the new one: at first, the latter turns on, and then, the former turns off. When the number № changes, the differentiator D2 forms with the help of the monostable flip-flop, the short signal of the logic 0 that turns off the auxiliary switches BK1 and BK2.

The required connection of the primary and adjusting windings is fulfilled with the help of R-S trigger. If the number № changes from 0 to 1, the differentiator D3 and the element "comparison with 1" are active, the trigger sets in 1 and fixes the switch Reverse in state 1 that corresponds to aiding winding connection (or confirms this state). Under № changing from 0 to −1, the trigger resets and fixes the switch Reverse in state 2 that corresponds to anti-aiding winding connection.

Before beginning of employment of the model, its parameters must be fixed. The dialog box has two pages: *Transformer parameters* and *OLTC and Voltage Regulator parameters,* which are selected in the menu, option *Show.* On the first page, the usual transformer parameters are set.

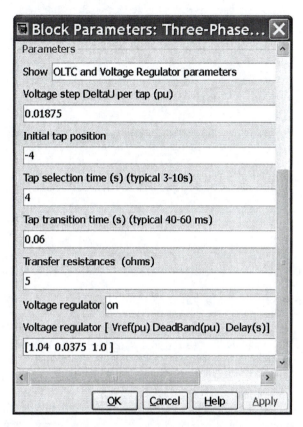

FIGURE 2.31   The second page of the dialog box of *OLTC Regulating Transformer*.

The second page is shown in Figure 2.31. In the line *Voltage step*, the voltage between next taps is given; thus, the total controlled range is ±8 0.01875 = ±0.15 = ±15%. The next line—*Initial tap position*. It is fixed −4. The time of mechanical travel between the next taps (*Tap selection time*) is equal to the delay T_4 in Figure 2.32 and is set to 4 s. *Tap transition time* is the time of the open state of the switches BK1 and BK2, and this time is equal to the time while the monostable flip-flop is active (is fixed 60 ms). *Transfer resistances* are the resistance of the resistors that are set in parallel to the contact BK1, BK2. The *Voltage regulator* indicates if the voltage regulator is on or off. In the last line, the voltage regulator parameters are defined: the voltage reference, the dead band, the delay times (T_2, T_3 in Figure 2.29).

Although the number of taps is taken as seven, and this number cannot be changed in the dialog box, the procedure for changing this parameter is given in the option *Look Under Mask/Procedure to change the number of taps*.

The model2_6_phasor and the model2_6 give the examples of this transformer employment. The former operates in the Phasor mode, and the latter in the discrete mode. They differ with the blocks that form voltage feedback. Under carrying out simulation, it is seen that the first model works much faster, than the

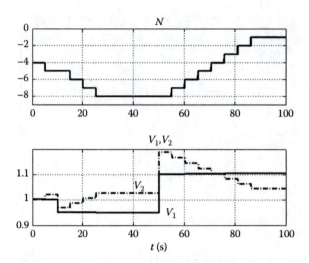

FIGURE 2.32 Voltage regulation in the network with *OLTC Regulating Transformer*.

second one, which, of course, gives better ideas about the processes in the system. The three-phase programmable voltage source with rated voltage of 345 kV changes its output from 1 to 0.95 at $t = 10$ s and from 0.95 to 1.1 at $t = 50$ s. Three-phase series *R-L* circuit models the source internal impedance. The transformer 50 MVA, 345/24.5 kV is set in the center of the three-phase transmission line of 100 km length. The line is simulated by the blocks **Three-Phase PI Section Line**. The parameters of the first line section are taken from Table 2.1, and the parameters of the second section are taken from the computation of such a line that is fulfilled in Chapter 6.

Initially, the adjusting winding is in the position -4. It is seen after simulation that at $t < 10$ s, the source voltage is 1 and output voltage is 1.02. At $t > 10$ s, the taps begin to switch over and set in the position -8 (Figure 2.32), the source voltage is 0.95, and the secondary one 1.03. The difference between this value and the reference 1.04 is lesser than half of the dead band, equal to $0.0375/2 = 0.01875$. At $t > 50$ s, when the transient ends, the taps set in the position -1, the source voltage is 1.105, and the secondary one 1.045. In such a way, the transformer output voltage is controlled with defined precision. The same results are obtained for model2_6.

TABLE 2.1
Typical Parameters of Transmission Lines

Voltage (kV)	220–230	330–345	500	750–765	1100–1150
R (Ω/km) 10^{-2}	5–6	2.8–3.7	1.8–2.8	1.17–1.2	0.005
L (mH/km)	1.27–1.29	0.97–1	0.86–0.93	0.87	0.77
C (F/km) 10^{-9}	8.91–8.94	10.6–12	12.6–13.8	13–13.4	14.6
Z_c (Ω)	377–387	285–300	250–278	255–260	230–250
NP (MW)	125–140	360–420	900–1000	2160–2280	5260–5300

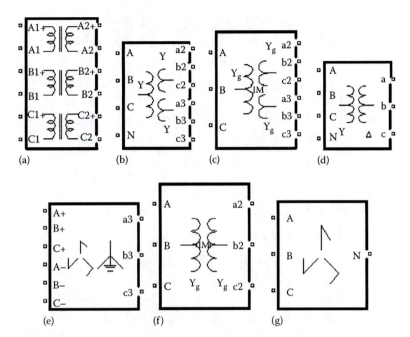

FIGURE 2.33 Pictures of the three-phase transformer models. (a) Three-Phase Transformer 12 Terminals; (b) Three-Phase Transformer (Three Windings); (c) Three-Phase Transformer Inductance Matrix Type (Three Windings); (d) Three-Phase Transformer (Two Windings); (e) Zigzag Phase-Shifting Transformer; (f) Three-Phase Transformer Inductance Matrix Type (Two Windings); (g) Grounding Transformer.

The pictures of the three-phase transformers are given in Figure 2.33. **Three-Phase Transformer 12 Terminals** consist of three single-phase two-winding transformers, and all 12 terminals are accessible. The block **Three-Phase Transformer (Two Windings)** models the three-phase transformer, using three single-phase ones. It is possible to simulate saturation and hysteresis. Each of the windings can be connected as Y (floating neutral), Yn (accessible neutral), Yg (grounded neutral), D1 (D lags Y by 30°), and D11 (D leads Y by 30°). The block picture changes for each variant. The dialog box has three pages; the first one is shown in Figure 2.34. The other pages contain data with which we had a matter earlier.

It is worth noting that not always are all necessary transformer parameters known. For example, from a catalogue (beside the power P_n and voltages V_1, V_2) the impedance voltage of the transformer e_{sc} (%), the no-load current I_{nl} (%), the no-load losses P_{nl} (W), and the short-circuit losses P_{sc} (W) are known. Then, the leakage inductances in pu are $L_1 = L_2 = 0.5 \ e_{sc}/100$, $R_1 = R_2 = 0.5 \ P_{sc}/P_n$, $R_m = P_n/P_{nl}$, $R_{me} = 100/I_{nl}$, $L_m = R_m R_{me}/\sqrt{R_m^2 - R_{me}^2}$.

For example, it is given $P_n = 2540$ kVA, $V_1 = 6$ kV, $e_{sc} = 6.3\%$, $I_{nl} = 1.4\%$, $P_{nl} = 6.9$ kW, $P_{sc} = 17.7$ kW. Calculation: $L_1 = L_2 = 0.0315$, $R_1 = R_2 = 0.5 \ 17.7/2540 = 0.0035$, $R_m = 2540/6.9 = 368$, $R_{me} = 100/1.4 = 71$, $L_m = 72.4$.

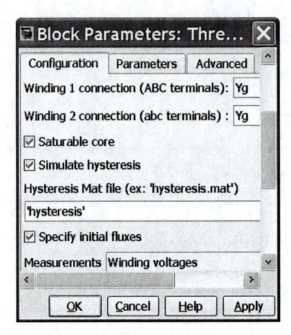

FIGURE 2.34 The first page of the dialog box of the model. Three-Phase Transformer (Two Windings).

The block of **Three-Phase Transformer (Three Windings)** is different from the previous one by the existence of the third (the second secondary) winding that affects the block picture (Figure 2.33).

The **Zigzag Phase-Shifting Transformer** model uses three single-phase three windings transformers; its primary winding is formed by connecting two windings of the single-phase transformers in the zigzag configuration. On the second page of the dialog box, in addition to usually defined parameters, the phase shift γ of the secondary voltage about the primary voltage is indicated.

The model2_7 gives an example of this transformer employment. It contains four transformers; each of them provides the shift γ of the preceding one. The first and the third transformers are modeled with the blocks **Three-Phase Transformer (Two Windings)** connected as Y/Y (or D1/D1) and Y/D11, respectively, and the second and the fourth transformers are modeled with the blocks **Zigzag Phase-Shifting Transformer,** whose secondary windings are connected as Y + 15° and D11 + 15°, respectively. The transformers have the power of 15 kW, the voltage of 380/315 V. The transformers supply the diode bridges (Chapter 4) that are loaded on the 20 Ω resistors $R1$–$R4$. The bridges are connected in series and supply the 20 Ω load R0. The **Scope** fixes the A–B transformer voltages, **Scope1** records the resistor currents, and **Scope2** fixes the currents in the supply and in phase A of the transformers. To carry out simulation, one can see that the secondary phase-to-phase voltages are distorted essentially, owing to the commutation processes in the bridges. In order to see their undistorted values, it is necessary to break connection between the voltage sensors and the bridges; in this way, one can make sure that the secondary voltage of

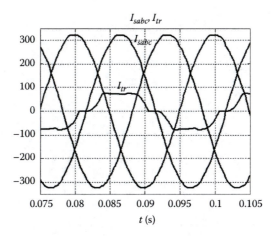

FIGURE 2.35 Primary (I_{tr}) and source (I_s) currents in the scheme with phase-shifting transformer.

each bridge leads the voltage of the antecedent bridge by 15°. **Scope1** shows that the current pulsation of each bridge is 6 A under the mean value of 18 A, that is, 33%, whereas the current pulsation in the R0 is 1.3 A under the mean value of 72 A, that is, about 2%. **Scope2** shows that the currents in the primary windings I_{tr} are distorted essentially, whereas the current in the common supply source I_s is almost sinusoidal (Figure 2.35).

If to select the option of Powergui *FFT Analysis*, and in the right fields of the opening window to choose *Structure: ScopeData2, input 2, Start time 0.1, Number of Cycles 4, Max Frequency* 1000 Hz, the curve of the current I_{tr} appears in the left area on top of the window; on execution of the command *Display,* the Fourier plot appears in the left below. THD = 17.09% for the first and for the third transformers and THD = 16.1% for the second and for the fourth ones. If to select the input 1 instead of the input 2, after command *Display* THD = 0.83% is found. The 11th and the 13th harmonics of the small amplitudes are seen on the Fourier plot, which must be absent theoretically. It happens because the currents in the primary transformer windings have some different shapes. Nevertheless, employment of the transformers in zigzag configuration give an opportunity to receive nearly perfect currents in the common load and in the supply without using of the filters and smoothing reactors.

Grounding Transformer is used in distribution networks for providing of a neutral point in a three-wire system, usually, for supplying of the single-phase loads connected to the ground (Figure 2.36). The transformer has three primary and three secondary windings connected in zigzag, and all six windings have the same number of the turns; in each phase, "belonging to it" primary and "foreign" secondary windings connected with opposite polarity (anti-aiding). The voltage of each of six windings is $V/3$, where V is the rated network voltage. The current distribution is shown in Figure 2.37.

The main characteristic of the **Grounding Transformer** is its zero-sequence impedance $Z_0 = R_0 + jX_0 = 2(R + jX)$, where R and X are the active and inductive resistances of one winding. The block dialog box contains the fields for indication

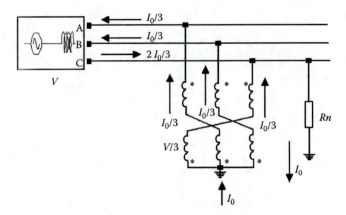

FIGURE 2.36   Diagram of Grounding Transformer connection.

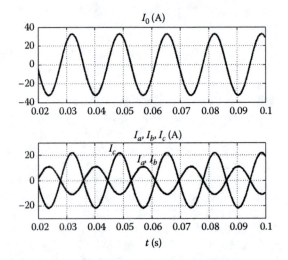

FIGURE 2.37   Currents in the single-phase load and in the supply.

of the parameter setting units (SI or pu), the rated power and frequency, the rated network voltage, and values R_0, X_0, R_m, X_m. The currents are shown in Figure 2.37 for the transformer 1 MVA, 50 Hz, each winding voltage is 8333 V, the network voltage is 25 kV, $R_0 = 0.002$ pu, $X_0 = 0.03$ pu, $R_m = X_m = 500$ pu, and the single phase load has the power of 1 MW.

The blocks of the two- and three winding transformers with inductances of the matrix type simulate these transformers taking into account the inductive coupling between windings of the different phases. The transformers can have three- or five-limb cores. Such a model can be required, for example, during investigation of the various nonsymmetric modes, specifically, of commutation processes and emergency operations in the thyristor multibridge converters. Modeling of this transformers demands rather detailed information about them that can be received either by computation with known transformer construction or by testing of the prototype. In this book, simulation of such transformers is not considered.

2.4   TRANSMISSION LINE MODELS

The transmission line is characterized by their active resistance R, the series inductance L, and the parallel capacity C. These parameters are given for a unit of the line length (m or km).

The active line resistance exceeds the DC resistance R_{dc}, owing to the skin effect and stranding. Besides, the resistance increases with the temperature rise:

$$R_t = R_{20}(1 + K\,\delta t) \tag{2.16}$$

where
   R_{20} is the resistance at 20°C
   δt is the exceeding of the temperature over 20°C
   $K = 0.0038$ for the copper conductors and $K = 0.004$ for aluminum ones

The inductance L is due to interaction of the AC current in the line with surrounding magnetic field. The phase inductance depends on the position of the phase conductors, relatively other phases and about the ground, i.e., the inductances of the phases can be different. The transposition of the different phase conductors is used for inductance balancing.

The capacity C is due to the potential difference between the line conductors, so that the conductors charge. The charge per unit of potential difference is C.

These RLC parameters are evenly distributed along a transmission line. The line voltages and currents change as the electromagnetic waves: the voltage and current values in some point depend not only on the time but on the position of this point. Let $Z = R + j\omega L$, $Y = j\omega C$, and ω is the network angular frequency. The transmission line is described by such parameters [2]:

$$\gamma = \sqrt{ZY} \tag{2.17}$$

is the propagation time,

$$\gamma = \alpha + j\beta = \frac{R}{2}\sqrt{\frac{C}{L}} + j\omega\sqrt{LC} \tag{2.18}$$

where
   α is the attenuation constant
   β is the phase constant

The quantity

$$Z_c = \sqrt{\frac{Z}{Y}} \tag{2.19}$$

is called the characteristic (wave) impedance. Under $R = 0$

$$Z_c = \sqrt{\frac{L}{C}} \qquad (2.20)$$

The general equations for currents and voltages in the transmission line are

$$I(x, s) = A_1 e^{\gamma x} + A_2 e^{-\gamma x} \qquad (2.21)$$

$$V(x, s) = -Z_c A_1 e^{\gamma x} + Z_c A_2 e^{-\gamma x} \qquad (2.22)$$

where
  s is the Laplace transformation symbol
  x is the distance from the sending line end to the considered point

then the second terms are interpreted as the incident (forward) waves I_f, V_f, and the first terms are interpreted as the reflected waves I_b, V_b. It is worth noting that the distance x is taken often from the receiving line end [2]; in this case, signs by γ in (2.21), (2.22) change. If the line is loaded by the impedance $Z_f(s)$ (under $x = d$), then, denoting $\rho_i = I_b/I_f$, $\rho_v = V_b/V_f$ where they are the reflection factors of current and voltage, it can be found:

$$\rho_i = \frac{Z_c - Z_f}{Z_c + Z_f}, \quad \rho_v = -\rho_i \qquad (2.23)$$

Under $Z_f = Z_c$, the reflected wave is absent. Such an impedance is called matched load.
  The natural or surge line load is defined as

$$NP = \frac{V_{nom}^2}{Z_c} \qquad (2.24)$$

when Z_c is determined by (2.20) ($R = 0$) and has the dimension of the pure resistance. Under transmission of such a power, the voltage amplitude does not change along the line.
  The description of the line by the wave equations is not always suitable for computation and simulation; therefore, the so-called equivalent π-model is used, which is shown in Figure 2.38, where

$$Z_\pi = Z_c \sinh(\gamma d) = \frac{Zd \sinh(\gamma d)}{\gamma d} \qquad (2.25)$$

$$\frac{Y_\pi}{2} = \frac{Yd}{2} \frac{\tanh(\gamma d/2)}{\gamma d/2} \qquad (2.26)$$

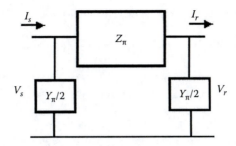

FIGURE 2.38 π-model of the transmission line.

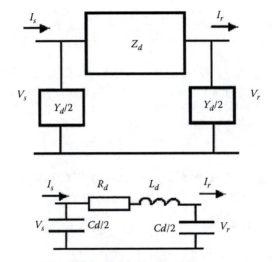

FIGURE 2.39 Nominal π-model of the transmission line.

For shorter lines, the nominal π-model can be used as shown in Figure 2.39, where

$$Z_d = Z_c \gamma d = Zd = (R + j\omega L)d \tag{2.27}$$

$$\frac{Y_d}{2} = \frac{Yd}{2} = \frac{j\omega Cd}{2}. \tag{2.28}$$

It is considered that at the frequency of 50 Hz and the line length d more than 180–200 km, the equivalent π-model must be used, under the line length more than 80 km (but less than 200 km), the nominal π-model can be used, and for shorter line, it is possible to neglect the parallel capacitance and to use the series RL model [2]. In Table 2.1, the typical parameters of overhead lines are given for the frequency of 50–60 Hz obtained from diverse sources [1,2,5,6]. Since not all sources for Z_c simultaneously give data for L and C, sometimes one can observe discrepancy between these values in the table.

SimPowerSystems has the following models of the transmission lines:

1. Model of the N-phase line with distributed parameters, **Distributed Parameter Line**
2. Model of the single-phase line with lumped parameters, **PI Section Line**
3. Model of the three-phase line with lumped parameters, **Three-Phase PI Section Line**

The block **Distributed Parameter Line** models the N-phase line with different phase parameters. If to neglect the losses, the line is characterized by three parameters: Z_c as (2.19), the velocity of wave propagation $v = 1/\sqrt{LC}$, and the length d. In the model the fact is used that in the line without losses the quantity $V + Z_c I$ in one line end reaches its another end without changing, after delay time of $\tau = d/v$. In order to take the losses into consideration, at the line ends the resistors $R/4$ and in middle of line the resistor $R/2$ are set.

The block dialog box is shown in Figure 2.40. In the first field (the term "field" is used instead of earlier used "line," in order not to confuse with considered transmission line), the number N of the line phases is selected. At that point, the model picture changes automatically. In Figure 2.41 the block pictures are given for $N = 1$ and $N = 3$. In the second field, the frequency is given.

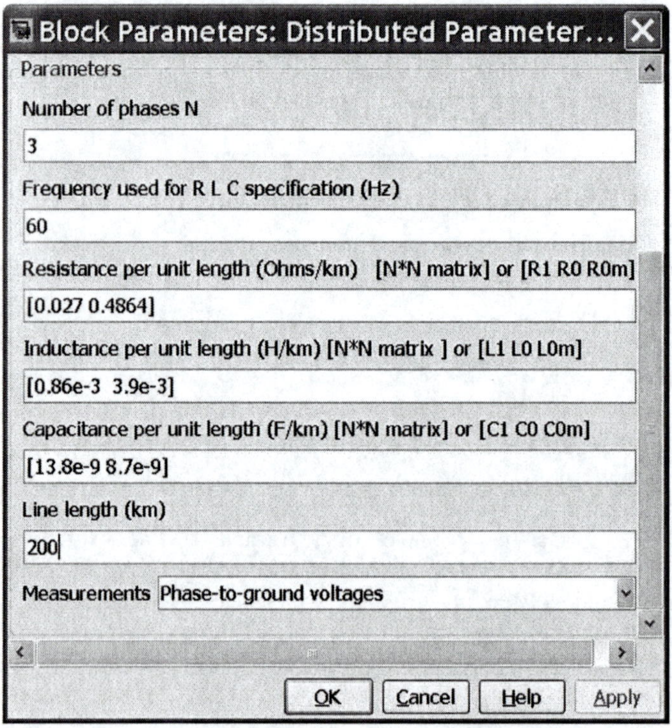

Block Parameters: Distributed Parameter... ✕

Parameters

Number of phases N

3

Frequency used for R L C specification (Hz)

60

Resistance per unit length (Ohms/km)   [N*N matrix] or [R1 R0 R0m]

[0.027 0.4864]

Inductance per unit length (H/km) [N*N matrix] or [L1 L0 L0m]

[0.86e-3 3.9e-3]

Capacitance per unit length (F/km) [N*N matrix] or [C1 C0 C0m]

[13.8e-9 8.7e-9]

Line length (km)

200

Measurements Phase-to-ground voltages

OK   Cancel   Help   Apply

FIGURE 2.40   Dialog box of the block Distributed Parameter Line.

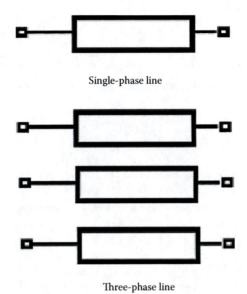

Single-phase line

Three-phase line

FIGURE 2.41 Picture variants of the block Distributed Parameter Line.

Afterwards, parameters of the transmission line are entered. In the third field, the values of the active resistance are defined, Ω/km. Under simulation of a single-phase line it is a scalar, and under simulation of the two- or three-phase line it is a two-element vector: resistances for positive- and zero sequences (as in Figure 2.40). For a nonsymmetric line, the $N \times N$ must be specified. For making the calculation simpler, there are the special command and the item in **Powergui**, with the help of which the resistance as well as the inductance and the capacitance per unit of the line length depending on its geometry and the characteristics of its conductors are calculated. These possibilities are considered in Chapter 6. In the following fields, the values of the inductance, H/km, and of the capacitance, F/km, are given in the analogous way. The last field contains indication relatively of the line phase voltage measurement with help of the block **Multimeter**. At that point, N voltages on the sending end are measured, and afterwards, N ones on the receiving end.

The block **PI Section Line** models a single-phase transmission line with lumped parameters. The model consists of N sections connected in series; each of the sections is made according to the scheme in Figure 2.39 (see Figure 2.42). The picture of the model is shown in Figure 2.43. The line parameters per unit of length, the line length, and the selected number of section N are indicated in the block dialog box. The value of N depends on the required frequency range f_{max}. The following formula is recommended [1]:

$$f_{max} = \frac{Nv}{8d} \tag{2.29}$$

where
   v is the velocity of wave propagation
   d is the line length

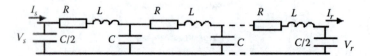

FIGURE 2.42   Model of the single-phase line with lumped parameters.

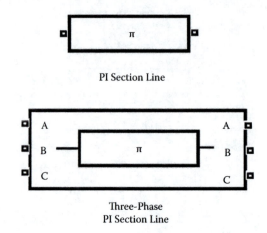

FIGURE 2.43   Pictures of the models PI Section Line and of Three-Phase PI Section Line.

For example, for the line of 150 km length, under $v = 300,000$ km/s, $f_{max} = 500$ Hz with $N = 2$. Such a frequency range can be sufficient under simulation of the machine systems but is small under investigation of the commutation processes.

The block **Three-Phase PI Section Line** models a symmetric three-phase transmission line with lumped parameters—only one section is shown in Figure 2.39 for one phase. The model picture is shown in Figure 2.43. The employment of a single section is possible under short lines and then, when only the processes at the fundamental frequency are of the utmost interest. In the other cases, several sections have to be used (see the formula 2.29).

The simple examples of using the described blocks are given in the model2_8–model2_10, bearing in mind that more complex models that contain the devices for production and consumption of the electrical energy are considered in Chapter 6.

The model2_8 simulates DC source switching on the long line. The originating wave processes are investigated, depending on the load power. The source voltage is 230 kV, the line length is 200 km, and the line parameters are taken from Table 2.1 for this voltage ($Z_c = 380\ \Omega$). The first two outputs of the block **Multimeter** measure the input and output (on the load) voltages, and the other two outputs measure the input and output line currents. The **Breaker** turns on at $t = 0.05$ s.

To fulfill simulation with the load resistance $R = Z_c = 380\ \Omega$, one can see that voltages and currents have rectangle shapes (the reflected wave is absent), and they are delayed on the receiving end by 0.675 ms, which corresponds to the theoretical value: $\tau = 200\sqrt{8.9 \times 1.3 \times 10^{-12}} = 0.68 \times 10^{-3}$.

The voltages and currents under $R = 2, Z_c = 760\ \Omega$ are shown in Figure 2.44. In the steady state, the input and output currents are equal, but the output voltage is

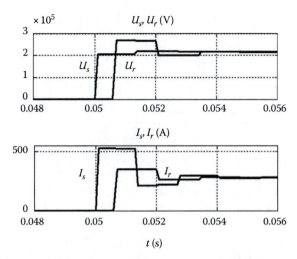

FIGURE 2.44 Processes in the long line with DC voltage under $Z \neq Z_c$.

less than the input voltage by 3 kV, because of the voltage drop in the line. In the transient, the voltages and currents change stepwise, owing to the interaction of the incident and reflected waves. The amplitude of the first voltage wave at the output is bigger than the amplitude of the input wave, and the situation is opposite for the currents. Under repeating of simulation with $R = 190\ \Omega$, it is seen that the amplitude of the first voltage wave at the output is less than amplitude of the input wave. As before, the interaction of the incident and reflected waves can be seen.

Let us carry on simulation under the load inductance 1 H, $R = 380\ \Omega$. The voltage and current waves are shown in Figure 2.45. It can be seen, specifically, that the load voltage is twice as large as the source voltage. Such a situation is typical, for example, for the induction motors (IM) supplied from the voltage source inverters (VSI) and speeds up an insulation aging.

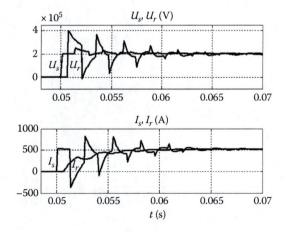

FIGURE 2.45 Processes in the long line with DC under inductive load.

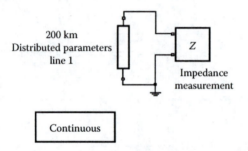

FIGURE 2.46   Diagram of the model for measurement of the frequency characteristics.

The transmission line frequency characteristics are of great interest. The reader can make the simple model shown in Figure 2.46. It is called model_use1. The line parameters are taken from the model2_8.

To open the window **Powergui**/*Impedance vs Frequency Measurement* and to select modes and range as in Figure 2.47, after clicking on the button *Display* the amplitude- and phase-frequency characteristics appear as shown in Figure 2.47. An alternation of the poles and zeroes is seen every 740 (2 × 370) Hz. The first pole is situated on the frequency $f = 1/(4\tau) = 370\,Hz$ ($\tau = 0.675\,ms$).

The model2_9 simulates the same line, but supplied by AC voltage with amplitude of 1.41 × 230,000/1.73 V. The block **Series RLC Branch** ($R = 5\,\Omega$, $L = 5/314\,H$) takes into account the source internal impedance. The **Breaker** is closed at $t = 0.24\,s$. The line is connected to the parallel L-R load with the natural load of 140/3 MW = 46.7 MW per phase. One can make sure after simulation that the wave effects are absent under switching, and the voltage amplitudes on the line input/output are 184 kV/178 kV. Increasing the active load power twice (to 2 46.7 MW),

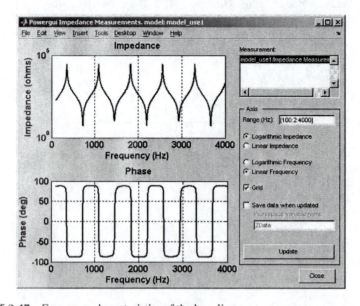

FIGURE 2.47   Frequency characteristics of the long line.

one can observe the load voltage drops to 160 kV and wave effects appear. To reduce the active power to 46.7/4 MW and the reactive one to 5/3 Mvar, the voltage amplitudes on the line input/output are 187 kV/189 kV. The dependence of the voltage along the line on the load makes it necessary to take special measures for improving the voltage profile, which is considered in Chapter 6.

If to replace the distributed line model by the π-nominal model (**PI Section Line**) as in the model2_10 and to carry on simulation, it would be seen that in the steady state the results are close enough to the obtained ones for the previous model. For example, under $P = 46.7$ MW and $Q = 10/3$ Mvar the voltage amplitudes on the line input/output are 184/174 kV, i.e., the same practically, but the transients, under **Breaker** switching on, are totally different, just under rather a large number of sections (for example, 10).

2.5   MISCELLANEOUS

The elements from the folder *SimPowerSystems/Elements* not included in the previous paragraphs are considered here, namely:

1. Breaker
2. Three-Phase Breaker
3. Three-Phase Fault
4. Surge Arrester
5. Neutral
6. Ground
7. Connection Port

and the block **Ideal Switch** from the folder *SimPowerSystems/Power Electronics*.

The pictures of the blocks are shown in Figure 2.48.

The block **Breaker** models a switch, whose commutation process is controlled either by the internal timer or by external signal Simulink. In the first case, the initial state of the switch and the vector of the times, when the switch state changes for opposite, are fixed. In the second case, the additional input appears on the block picture, as shown in Figure 2.48, the signal of the logic 0 turns off, and the signal of the logical 1 turns on the breaker. It is important to note that the actual contact separation takes place at the first current crossing of the zero level after the appearance of the switch-off signal. In the closed state, the contact resistance is R_{on}. If the breaker is closed in the initial state, SimPowerSystems initializes automatically all the model states corresponding to this state. The Rs-Cs snubber is used when the breaker connects in series with the inductance circuit.

The block **Three-Phase Breaker** consists of three single-phase ones and has the same peculiarity. It is necessary to mark in the dialog box, which phases switch over, when the given commutation times come. The unmarked phases stay in the initial state.

The block **Three-Phase Fault** uses three blocks **Breaker**, which can close and open individually, in order to simulate the short-circuits phase—phase, phase—ground or their combination. If ground short-circuit is not simulated, the resistance about ground R_g is 1 MΩ. The dialog box is shown in Figure 2.49. At first, it is marked to which phases the given commutation times are related. The initial states are opposite to the

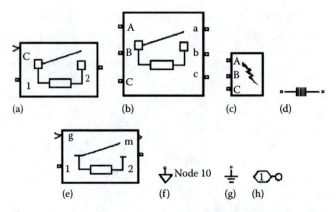

FIGURE 2.48 Pictures of the "miscellaneous" blocks. (a) Breaker; (b) Three-phase Breaker; (c) Three-Phase fault; (d) Surge Arrester; (e) Ideal Switch; (f) Neutral; (g) Ground; (h) Connection Port.

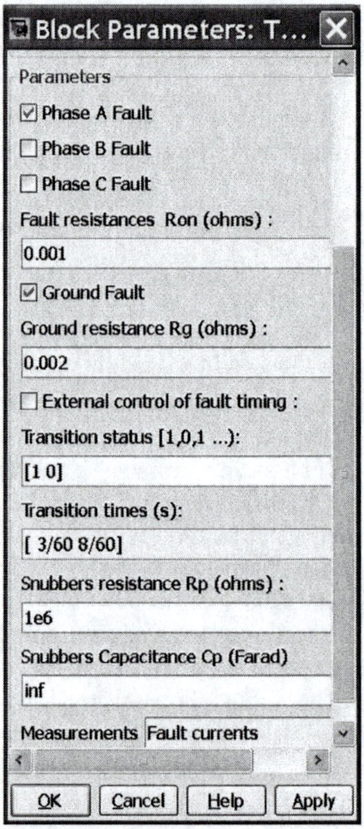

FIGURE 2.49 Dialog box of the block Three-Phase Fault.

first state, indicated in the line *Transition status*. The first state in this line is usually 1; it means that the initial states of all the contacts are open. It is possible also to set the first state 0, and, consequently, to begin simulation with the initial state 1, i.e., from the short-circuit state. If the option *Ground Fault* is marked, it is necessary to fix R_g, as in Figure 2.49. If this option is not marked, and two or three phases are marked, the line-to-line fault is simulated. The block can be controlled by the external signal.

The block **Surge Arrester** (varistor) is intended for modeling of a nonlinear resistor that protects the line elements from the overvoltage (surge). Its dialog box is shown in Figure 2.50.

The dependence of the voltage V on flowing through the varistor current I can be approximated with three nonlinear segments:

$$\frac{V}{V_{ref}} = K_i \left(\frac{I}{I_{ref}} \right)^{1/\alpha_i} \tag{2.30}$$

where $V_{ref} = V_0$ is the voltage when the varistor conducts the current; the voltage is limited approximately at this level. This voltage corresponds to the current $I_0 = I_{ref}$, usually, $I_{ref} = 500$ or 1000 A, K_i, α_i ($i = 1, 2, 3$) are the factors that are different on the

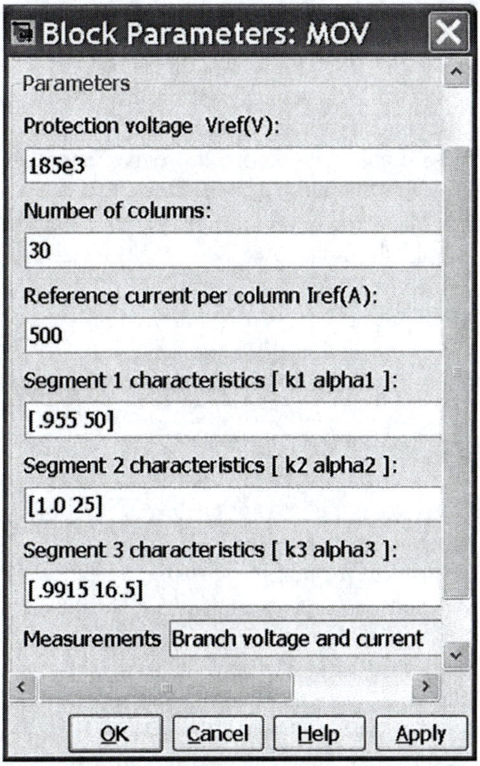

FIGURE 2.50 Dialog box of the block Surge Arrester.

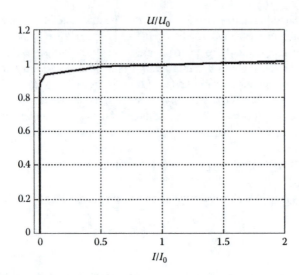

FIGURE 2.51 Model Surge Arrester curve.

different segments of the curve (2.30). The values given in Figure 2.50 are typical for the metal-oxide varistors. The points of the transition from one segment to the other are determined by the crossing of individual curves. For given parameters, they are $I/I_{ref} = 0.1$ and $I/I_{ref} = 1.5$. The corresponding dependence is shown in Figure 2.51. If it is necessary to dissipate the high power, several varistors are connected in parallel (the line *Number of columns*).

The model2_11 shows employment of the **Surge Arrester**. The transmission line consists of two sections, and the length of each is 50 km; the line is loaded on the active-inductance load. Only a single phase of the transmission line is simulated. The line is supplied with three-phase voltage 500 kV. The capacitor is set in the line to realize the series voltage drop compensation. The **Surge Arrester** whose parameters are given in Figure 2.50 is set in parallel to the capacitor. At $t = 0.05$ s the ground fault occurs in the line end that is cleared at $t = 0.15$ s.

If to fulfill simulation without **Surge Arrester**, it would be seen that the voltage on the capacitor reaches its maximum at $t = 0.073$ and is equal to 730 kV, and the maximum of the current flowing through capacitor reaches 22 kA at $t = 0.067$ s. With the **Surge Arrester**, the capacitor voltage is limited to 150 kV under the current of 10 kA. The maximal current in the varistor is 10 kA/20 = 500 A. The graph plotter draws the voltage versus the current for the **Surge Arrester**.

The block **Neutral** provides a common node in the circuit with the definite number. This block can be used for connection of several points of the scheme without drawing of the connection line. If the node number is 0, the connection with ground is realized. The block number is defined in its dialog box.

The block **Ground** models connection to the ground.

The block **Connection Port** is used as the point for connection of the SimPowerSystems blocks to the subsystem that contains the blocks of SimPowerSystems too.

The block **Ideal Switch** can be used for simplified modeling of the semiconductor devices (Chapter 4) and for modeling of a switch in the main circuits that, unlike **Breaker**, permits a commutation with current. **Ideal Switch** is modeled as the resistor R_{on} in series with a switch that is controlled by the signal g. The block does not conduct current under $g = 0$ and conducts in any direction under $g > 0$. The switching-over occurs instantly. The block contains the snubber that consists of R_s and C_s connected in series; this snubber can be connected in parallel to the switch contacts.

If the option *Show measurement port* is marked, the output m appears on the block picture; the vector signal (current, voltage drop) returns on this output.

REFERENCES

1. MathWorks, SimPowerSystems™, User's Guide, 2004–2011.
2. Kundur, P., *Power System Stability and Control*, McGraw-Hill, New York, 1994.
3. IEEE Task Force on Load Representation for Dynamic Performance, Load representation for dynamic performance analysis, *IEEE Transactions on Power Systems*, 8(2), 472–82, May 1993.
4. Kovacs, K.P. and Racz, I., *Transiente Vorgänge in Wechselstrommaschinen*, Verlag der Ungarischen Akademie der Wissenschaften, Budapest, Hungary, 1959.
5. Alexandrov, G.N., *Transmission of the Electrical Energy* (in Russian), Polytechnical University, St Peterburg's, Russia, 2009.
6. Kothari, D.P. and Nagrath, I.J., *Modern Power System Analysis,* Tata McGraw-Hill, New York, 2003.

3 Measuring and Control Blocks

3.1 MEASUREMENT OF MAIN CIRCUIT QUANTITIES

SimPowerSystems™ contains several blocks that are intended for measurement of currents, voltages, and some other features in the main circuits and for the transformation of these characteristics into signals that are available for Simulink® blocks [1]. These blocks are

1. Voltage Measurement
2. Current Measurement
3. Three-Phase *V–I* Measurement
4. Multimeter
5. Impedance Measurement

The blocks are in the folder *SimPowerSystems/Measurements*. Pictures of the blocks are given in Figure 3.1.

The block **Voltage Measurement** measures the voltage instantaneous value between two points of the model. When the block is utilized in the model in the Phasor mode, an opportunity appears to select a mode of the voltage vector representation: the complex number as is accepted in MATLAB®, the real and the imaginary parts as two-dimensional vector, the module and the angle as two-dimensional vector, and only the module.

The block **Current Measurement** measures the instantaneous value of the current that flows in the circuit. For use of this block in the Phasor mode see previous paragraph.

Three-Phase *V–I* Measurement is connected in series with three-phase devices. In such a case, phase-to-phase or phase-to-neutral voltages and the line currents can be measured, and the block can form the voltage and current values in both SI and pu units. In the last case, additional lines appear in the dialog box for indication of the base values. As the actual voltage base value, the amplitude of the phase voltage that corresponds to the rms value of the phase-to-phase voltage V is given in the dialog window, and as the current base voltage, the line current amplitude is taken that corresponds to the given value of the power P and of the voltage V:

$$V_{base} = \frac{V\sqrt{2}}{\sqrt{3}} \quad I_{base} = \frac{\sqrt{2}P}{\sqrt{3}V} \tag{3.1}$$

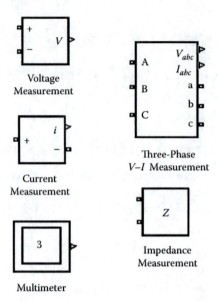

FIGURE 3.1   Pictures of the measurers of the main circuit quantities.

The dialog box is shown in Figure 3.2. There are two options for output of the measured quantities. If the option *Use a label* is not marked, then the outputs for vectors of voltages and currents appear on the block picture, as shown in Figure 3.1. If these options are marked, as in Figure 3.2, lines that indicate the names of these outputs appear, and these outputs disappear from the block picture. In order to read these signals, the blocks Simulink **From** with the same names are set in appropriate places for their use.

When the model operates in the Phasor mode, the form of presentation of the complex value has to be given, as described previously.

The block **Multimeter** is intended for the measurement of the voltages, currents, and some other quantities of the blocks; in order to measure these features, the option *Measurements* is activated. If we put this block in the model, in which such blocks are present, we click on its picture to open it, and a window appears that consists of two parts. In the left part, a list appears containing all quantities that can be measured in this model. The quantities that have to be measured must be duplicated on the right. For that purpose, they must be first selected using the mouse in the left part and the button >> clicked. Selected quantities appear as the vector at the block output in the order in which they are written. For changing the order and the sign, the buttons Up, Down, Remove, ± are used. When the model operates in the Phasor mode, the form of presentation of the complex values has to be given. If we choose the option *Plot selected measurements*, it is possible to create the plots illustrating change in these quantities without using the block **Scope**.

One of the problems with using this block is the correct definition of the polarity of the measured quantities. When the blocks for the voltage and the current measurement are used, the signal polarity is shown on the block picture, unlike the use of **Multimeter**. For three-phase transformers, the signal polarity can be found by signal labels in **Multimeter**; for example, U_{an}: voltage phase A–neutral, plus on

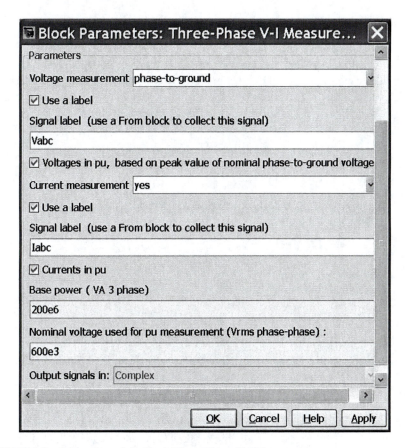

FIGURE 3.2   Dialog box of the block Three-Phase *V–I* Measurement.

A terminal; U_{ab}: phase A–phase B line voltage, plus on A terminal; I_{cn}: winding *C* current under *Y* connection, plus on *C* terminal; I_{bc}: current that flows from B to C under Delta connection, plus on B terminal; and so on. For the load blocks and some others, the polarity is determined by the block orientation. In order to find the block orientation, this block must be selected, and the command get_param(gcb, 'Orientation') has to be executed in the command window. The natural orientation (in library) is right for the horizontal blocks and down for the vertical ones. Under the right orientation, the currents flow from left to right, and for the voltages plus is on left, and under the left orientation—in the opposite direction. Under the orientation Down, currents flow top-down, and for the voltages plus is on top.

The block **Impedance Measurement** measures the impedance between two circuit points as a function of the frequency. It contains the current source of the variable frequency I_z connected to the block input/output and the meter of the voltage difference V_z between these points; the impedance is computed as V_z/I_z. The module and phase of impedance are shown using the **Powergui** item *Impedance vs. Frequency Measurement tool*. The example of the block utilization is described in Figures 2.46 and 2.47.

3.2 METERS WITH EMPLOYMENT OF SIMULINK® BLOCKS

SimPowerSystems has a number of blocks that use the main circuit quantities measured by the devices described in Section 3.1 in order to produce the signals of the states, parameters, quantities, and so on, which are used for observing and control. These blocks are built by using Simulink blocks and are in the folders *SimPowerSystems/ Extra library/Measurements* or/*Discrete Measurements*. Hereinafter, the following blocks from the first folder are considered:

1. abc_to_dq0 Transformation
2. dq0_to_abc Transformation
3. Active & Reactive Power
4. Three-Phase Instantaneous Active & Reactive Power
5. dq0-Based Active & Reactive Power
6. Three-Phase Sequence Analyzer
7. RMS
8. Fourier
9. Total Harmonic Distortion
10. Mean Value

The pictures of the first six blocks are given in Figure 3.3.

The first block fulfils Park's transformation from stationary condition to rotating condition, with the speed ω (rad/s) reference frame by formulas:

$$V_d = \frac{2}{3}\left[V_a \sin x + V_b \sin\left(x - \frac{2\pi}{3} \right) + V_c \sin\left(x + \frac{2\pi}{3} \right) \right] \qquad (3.2)$$

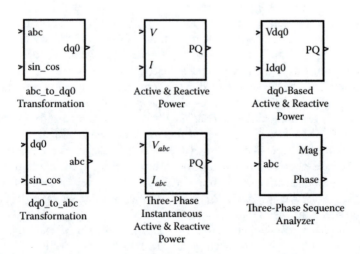

FIGURE 3.3   Pictures of the measures of points 1–6.

$$V_q = \frac{2}{3}\left[V_a \cos x + V_b \cos\left(x - \frac{2\pi}{3}\right) + V_c \cos\left(x + \frac{2\pi}{3}\right)\right] \qquad (3.3)$$

$$V_0 = \frac{1}{3}\left[V_a + V_b + V_c\right], \quad x = \omega t \qquad (3.4)$$

The inputs are the vectors $[V_a\ V_b\ V_c]$ and $[\sin \omega t \cos \omega t]$; the output is the vector $[V_d\ V_q\ V_0]$. The second block fulfils the inverse Park's transformation by the following formulas:

$$V_a = V_d \sin x + V_q \cos x + V_0 \qquad (3.5)$$

$$V_b = V_d \sin\left(x - \frac{2\pi}{3}\right) + V_q \cos\left(x - \frac{2\pi}{3}\right) + V_0 \qquad (3.6)$$

$$V_c = V_d \sin\left(x + \frac{2\pi}{3}\right) + V_q \cos\left(x + \frac{2\pi}{3}\right) + V_0 \qquad (3.7)$$

The inputs are the vectors $[V_d\ V_q\ V_0]$ and $[\sin \omega t \cos \omega t]$; the output is the vector $[V_a\ V_b\ V_c]$. The block **Active & Reactive Power** computes the active and reactive powers as

$$P = 0.5\ VI \cos \Delta\varphi \qquad (3.8)$$

$$Q = 0.5\ VI \sin \Delta\varphi \qquad (3.9)$$

where
   V and I are the amplitudes of the fundamental voltage and current harmonics that
      are computed by the **Fourier** blocks (see the following)
   $\Delta\varphi$ is the angle between harmonics

The inputs of the block are the voltage and the current of the circuit, for which the powers are calculated; the output is the vector $[P\ Q]$. In the dialog box, the fundamental frequency is defined.

There is a discrete version of the block. It has an additional output for the vector $[V\ I]$, and the sample time has to be indicated. There are two lines in which the initial values of the amplitudes and angles of the voltage and currents can be put into. The corresponding power values are held till the end of the first fundamental frequency period. Besides, there is the analogous block in the folder *Extra library/ Phasor Library*, which is used only in Phasor mode simulation.

The block **Three-Phase Instantaneous Active & Reactive Power** uses the relationship:

$$P = V_a I_a + V_b I_b + V_c I_c \qquad (3.10)$$

$$Q = 0.577 \, (V_{bc}I_a + V_{ca}I_b + V_{ab}I_c) \tag{3.11}$$

The inputs are the vectors of the phase voltages and currents; the output is the vector $[P \; Q]$.

Besides, there is the analogous block in the folder *Extra library/Phasor Library*, which is used only in Phasor mode simulation.

The block **dq0-Based Active & Reactive Power** uses the vectors $[V_d \; V_q \; V_0]$ and $[I_d \; I_q \; I_0]$ as the inputs:

$$P = 1.5 \, (V_dI_d + V_qI_q + 2V_0I_0) \tag{3.12}$$

$$Q = 1.5 \, (V_qI_d - V_dI_q) \tag{3.13}$$

The block **Three-Phase Sequence Analyzer** calculates the positive- (direct), negative- (reverse), and zero-sequence components of the three-phase signal. Remember that the voltage phasors $\mathbf{V}_a$, $\mathbf{V}_b$, $\mathbf{V}_c$ (or other quantities) in the any nonsymmetric three-phase system may be expressed as a sum of three vectors in the symmetric three-phase systems that are called "the sequence components":

$$\mathbf{V}_a = \mathbf{V}_1 + \mathbf{V}_2 + \mathbf{V}_0 \tag{3.14}$$

$$\mathbf{V}_b = a^2\mathbf{V}_1 + a\mathbf{V}_2 + \mathbf{V}_0 \tag{3.15}$$

$$\mathbf{V}_c = a\mathbf{V}_1 + a^2\mathbf{V}_2 + \mathbf{V}_0 \tag{3.16}$$

$$a = e^{j2\pi/3}$$

The sequence components are calculated by known phase phasors:

$$\mathbf{V}_0 = \frac{1}{3}(\mathbf{V}_a + \mathbf{V}_b + \mathbf{V}_c) \tag{3.17}$$

$$\mathbf{V}_1 = \frac{1}{3}(\mathbf{V}_a + a\mathbf{V}_b + a^2\mathbf{V}_c) \tag{3.18}$$

$$\mathbf{V}_2 = \frac{1}{3}(\mathbf{V}_a + a^2\mathbf{V}_b + a\mathbf{V}_c) \tag{3.19}$$

The space vector $\mathbf{V}_1$ turns in the same direction; as the space vector in the symmetric system, it defines the positive sequence; the space vector $\mathbf{V}_2$ turns in the opposite direction; it defines the negative sequence, and the vector $\mathbf{V}_0$ defines zero sequence.

The frequency, the harmonic number n ($n = 1$ for fundamental), and the type of the sequence—positive, negative, 0, or all three simultaneously—are specified in the dialog box. The vector of three-phase quantities enters the block. The block has two outputs: one for the amplitude of the selected sequence and the other for its phase angle.

It is reasonable to use the block as a voltage feedback sensor for voltage regulation tuned with $n = 1$ for positive sequence (see model2_6) as well as for the study of the method symmetric components. For this purpose, a simple model, model3_1, is made. Three single-phase sources of 100 V facilitate changes in the amplitude and angle of each phase voltage independently. The block **Mutual Inductance** is used for the load. The self-impedances of all phases are equal to $X_s = 10\ \Omega$, and the mutual impedances, $X_m = 9\ \Omega$. **Multimeter** measures three-phase supply voltage and three load currents. The first block of the **Three-Phase Sequence Analyzer** processes the measured voltages, and the second one, currents. The analyzers compute all three sequences.

For the voltage phase angles of $0°$, $-120°$, $120°$, and for the amplitudes of 100, 110, 90 V, the symmetric components according to (3.17) through (3.19) are

$$V_1 = \frac{100}{3}(1 + 1.1a^{-1}a + 0.9a^{-2}a^2) = 100$$

$$V_2 = \frac{100}{3}(1 + 1.1a^{-1}a^2 + 0.9a^{-2}a^1) = j5.77$$

$$V_0 = \frac{100}{3}(1 + 1.1a^{-1} + 0.9a^{-2}) = -j5.77$$

It is known from the symmetric component theory [2] that for the symmetric circuit with a mutual inductance, the impedances for the positive and negative sequences are $X_s - X_m = 1\ \Omega$, and for zero sequence, $X_s + 2X_m = 28\ \Omega$. With the help of **Scope**, one can see after simulation that the amplitudes of the voltage sequences correspond to the items measured as discussed previously; their phase angles are $0°$, $90°$, and $-90°$, and the current amplitudes are 100, 5.8, and 0.207 A, respectively, which correspond to computation too.

Let the voltage amplitudes of all phases be equal and phase angles of B and C phases be $-130°$ and $-230°$. Then

$$V_1 = \frac{100}{3}(1 + e^{-j130}a + e^{-j230}a^2) = 99\ V$$

$$V_2 = \frac{100}{3}(1 + e^{-j130}a^2 + e^{-j230}a) = 10.5\ V$$

$$V_0 = \frac{100}{3}(1 + e^{-j130} + e^{-j230}) = 9.52\ V$$

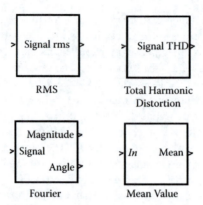

FIGURE 3.4   Pictures of the measures of points 7–10.

and, respectively, currents are $I_1 = 99\,\text{A}$, $I_2 = 10.5\,\text{A}$, $I_0 = 0.34\,\text{A}$. One can make sure after simulation that these values are realized in real-time situations. If we need to set the phase angles of $0°$, $-120°$, $-240°$, but different self-impedances of 11, 10, 9.5 Ω, the following values of current symmetric components are yielded after simulation: $I_1 = 100.5\,\text{A}$, $\varphi = -90°$, $I_2 = 37.8\,\text{A}$, $\varphi = 110°$, $I_0 = 1.36\,\text{A}$, $\varphi = 49°$.

The block has the discrete version in the folder *Discrete Measurements*; besides, in the folder *Extra library/Phasor Library,* there is the analogous block **Sequence Analyzer** that functions only in the phasor simulation mode. The example of this block utilization is provided in model2_6_Phasor.

The pictures of four remaining blocks from the list previously mentioned are shown in Figure 3.4. The block **RMS** calculates it over a running window of one cycle of the specified fundamental frequency; that is, for the last T s, $T = 1/f$, f is the fundamental frequency. In the discrete version of the block, the sample time and the initial block input are specified additionally, and the corresponding output is held during the first period T.

The block **Fourier** performs Fourier analysis of the input signal over a running window of T width. The fundamental frequency and the harmonic number, whose amplitude and phase are formed at the block outputs, are specified in the dialog box. There is the discrete version of the block.

The block **Total Harmonics Distortion** computes the total value of distortion factor (THD) of the periodic signals as

$$\text{THD} = \frac{\sqrt{I_{rms}^2 - I_1^2}}{I_1} \tag{3.20}$$

where
I_{rms} is rms value of the input signal
I_1 is rms of its first harmonic

There is also the discrete version of the block.

The block **Mean Value** calculates the mean value of the input signal over a running window, whose duration (in seconds) is defined in the dialog box. The same block for the phasor mode is in the folder *Extra library/Phasor Library*. There is also the discrete version of the block, in which a running window is equal to $T = 1/f$.

In addition to that described previously, the folder *Extra library/Discrete Measurements* contains other useful blocks. The following are considered:

1. Discrete Three-Phase Total Power
2. Discrete Variable Frequency Mean Value
3. Three-Phase Positive-Sequence Fundamental Value
4. Discrete Three-Phase Positive-Sequence Active & Reactive Power
5. FFT
6. Discrete PLL-Driven Fundamental Value
7. Discrete Three-Phase PLL-Driven Positive-Sequence Fundamental Value
8. Three-Phase PLL-Driven Positive-Sequence Active & Reactive Power

The block pictures are given in Figure 3.5. The first block measures an active power of three-phase system of voltages and currents that may contain harmonics. The power is computed by averaging the instantaneous power (that is equal to the sum of the products of voltages and currents' instantaneous values of all phases) in a running window over one period of fundamental frequency width. The block inputs are

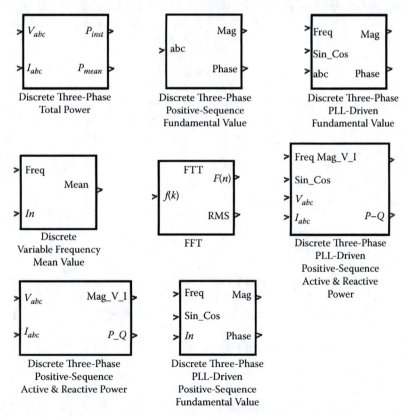

FIGURE 3.5 Pictures of some blocks for discrete measurement.

the vectors $[V_a\ V_b\ V_c]$, $[I_a\ I_b\ I_c]$, and the block outputs are the values of the instantaneous P_{inst} and mean P_{mean} powers. The initial values may be indicated (see previous discussion).

The block **Discrete Variable Frequency Mean Value** calculates the mean value of the input signal (*In*) in a running window of $1/F_{req}$ width, where F_{req} is the signal at the input Freq. Thus, the variable frequency of the input signal is possible. The initial values of *In* and F_{req} are specified in the dialog box. The possible minimal frequency is put into also, in order to determine the buffer size of the Simulink block **Variable Transport Delay** that is used for building the considered block.

The block **Three-Phase Positive-Sequence Fundamental Value** computes the amplitude (Mag) and phase (Phase) of the positive sequence fundamental harmonic by measurement of the vector $[V_a\ V_b\ V_c]$, whose components may contain harmonics and form nonsymmetric system.

The block **Discrete Three-Phase Positive-Sequence Active & Reactive Power** calculates the active and reactive powers of the positive-sequence fundamental harmonic in the nonsymmetric three-phase circuit, in which voltages and currents may have harmonics. The outputs are the vectors [Mag_V Mag_I] (these are the fundamental harmonic amplitudes of the voltage and current positive sequence) and [P Q].

The block **FFT** (fast Fourier transformation) computes Fourier transformation of an input signal that may be a vector of dimension N_{sig}. This signal enters the input $f(k)$ with the sample time of T_s. At the output $F(n)$, with the time interval that is equal to the specified value *FFT Sampling time*, the matrix of the size $[N_h, N_{sig}]$ appears. This matrix consists of harmonic amplitudes of the order from 0 to $N_h - 1$, $N_h = 1/(2f_1 T_s)$, where f_1 is the fundamental frequency. At the output RMS, with the same time interval, the matrix of the size $[5, N_{sig}]$ of the rms values appears where the first value is the total signal rms, the second value is the positive component, the third value is the fundamental harmonic rms, the fourth value is the high-harmonic rms, and the fifth value is the total rms without positive component. Simultaneously, the spectrum is displayed as a plot.

The dialog window is shown in Figure 3.6. Lines 1–4 were explained previously. In what follows, the base values for harmonics are specified as a vector of dimension N_{sig}. If they are equal to 1, the absolute values are displayed. There is the opportunity to show rms values on the plot; at this, it has to be indicated in the line *Decimals to display RMS* how many cycles of the Fourier coefficient computation pass between the successive renewals of the information on the plot. The information about rms is not displayed under setting of 0 in this line. In the line *Mask*, as the vector of dimension N_{sig}, the multiplication factors are given, by which the Fourier coefficients are multiplied under plotting, which give the opportunity to modify plot appearance. If, for example, the factor is −1, the corresponding Fourier coefficients are directed down on the plot, as in Figure 3.7. In the last line, the plot number is specified that gives the opportunity to use several **FFT** blocks in the same model.

As an example of block employment, model3_2 is considered, which is a modified version of the demonstrational model power_fft.mdl. At the input of **Gain**, the following harmonic signals sum up (amplitude/frequency): 100/50, 40/250, 25/550, 15/950. The amplitude of the fifth harmonic can be modulated with a frequency of 5 Hz,

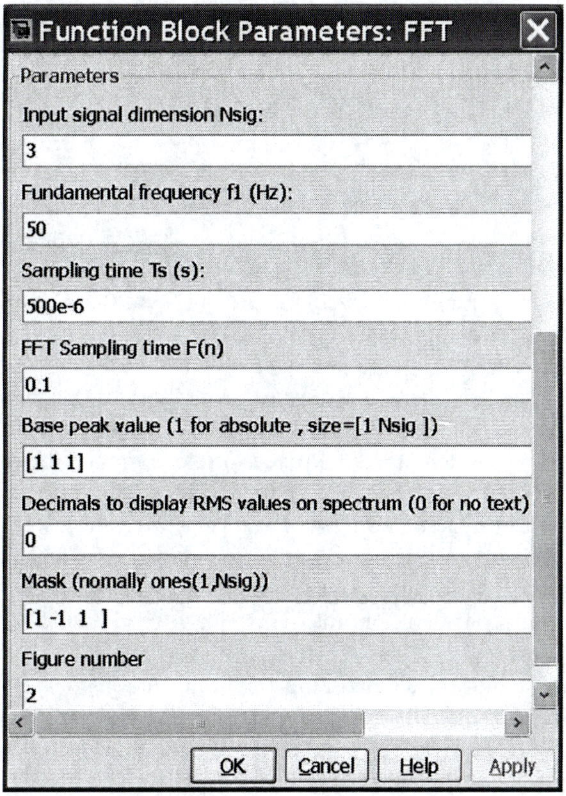

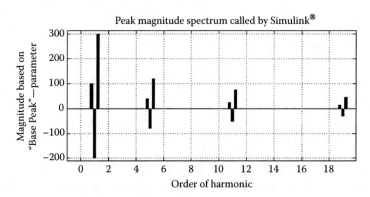

FIGURE 3.6 Parameter window of the block **FFT**.

FIGURE 3.7 Plot of the Fourier coefficients.

and the modulation amplitude is 0.5. The **Gain** is the vector amplifier with component multiplication factors 1, 2, 3, so that the input $f(k)$ is the three-dimensional vector. The dialog box of **FFT** is shown in Figure 3.6. The block **Display** from Simulink collection is set at the FFT output. Besides, **FFT** output is stored in the workspace. A purpose of the blocks **Demux** and **Scope** will be explained hereinafter.

To execute simulation when the amplitude of the source **0.5, 5 Hz** is zero, the information about amplitudes of the harmonics 0–19 for all three signals appear on the blocks **Display**.

Besides, the plot of the Fourier coefficients that is shown in Figure 3.7 appears on the screen.

Let us modify how the model is made, for model3_2_1: the amplitude of the source **0.5, 5 Hz** is 0.5; the factor of **Gain** is 1 (scalar); for **FFT**, the values of N_{sig}, of *Base peak value* and of *Mask* are equal to 1 (scalar); *FFT Sampling time* is 0.01 s; and the input of **Mux** that picks out sixth signal is connected to output $F(n)$. During simulation, it can be seen on the Fourier coefficient plot that the amplitude of the fifth harmonic changes continuously from 20 to 60. On the **Scope**, the oscillations of the fifth harmonic amplitude in the same limits with the period of 0.2 s are seen. So the block can be used "on-line," for example, for filter parameter tuning.

The last three blocks receive a part of the input signals from PLL (Phase-Locked Loop) units. These units are utilized for the computation of the frequency and the phase of a harmonic signal, whose measurements are carried out with noise or are distorted with higher harmonics. They are the devices with feedback and are considered in the next paragraph. Their outputs are the input signal frequency f and the signals $\sin(2\pi ft)$ and $\cos(2\pi ft)$ that determine the output phase. In the blocks under consideration, they enter input Freq and, as the vector [sin, cos], input Sin_Cos.

The first of these blocks calculates the amplitude and the phase of the first harmonic. It contains the said two blocks **Discrete Variable Frequency Mean Value** D1, D2. At D1, D2 inputs, values f and products $2V \sin(2\pi ft)$, $2V \cos(2\pi ft)$ enter where V is the signal at the input *In* of **Discrete PLL-Driven Fundamental Value**. The output D1 defines the real part and the output D2 defines the imaginary part of the input fundamental harmonic that converts afterward in the module and in the phase.

The next block **Discrete Three-Phase PLL-Driven Positive-Sequence Fundamental Value** contains, in addition to the previous one, the block **abc_to_dq0 Transformation**. Three-phase signal comes to the input abc as the vector $[V_a \ V_b \ V_c]$ and transforms in the V_d, V_q components. After averaging, V_{dav} defines the real part of the output signal and V_{qav} the imaginary part, which converts afterward in the module and in the phase.

The last block **Three-Phase PLL-Driven Positive-Sequence Active & Reactive Power** contains two previous blocks. The three-phase signals of the voltage $[V_a \ V_b \ V_c]$ and current $[I_a \ I_b \ I_c]$ enter their inputs. The fundamental harmonic amplitudes of the positive sequences of the voltage V and the current I come to the output Mag_V_I of the block under consideration, and the power values are computed by the formulas

$$P = 1.5 \, VI \cos \Delta\varphi \tag{3.21}$$

$$Q = 1.5 \, VI \sin \Delta\varphi \tag{3.22}$$

where $\Delta\varphi$ is the voltage and current phase difference.

3.3 CONTROL BLOCKS

A number of blocks that carry out the functions of control and regulation are in the folders *SimPowerSystems/Extra library/Control Blocks* and */Discrete Control Blocks*. These blocks may be divided in two groups: the blocks of common application and the blocks for special application. Only the first group is considered in this chapter, and the rest are considered together with the systems for which they were developed.

In the folder *Control Blocks*, the common application blocks are as follows:

1. Sample & Hold
2. On/Off Delay
3. Edge Detector
4. Monostable Flip-Flop
5. Bistable Flip-Flop
6. Timer
7. First-Order Filter
8. Second-Order Filter
9. Three-phase Programmable Source
10. One-phase PLL
11. Three-phase PLL

There are discrete variants for a number of these blocks. They are different from the continuous ones given the necessity to specify a sample time. The pictures of the first six blocks are given in Figure 3.8. The first block output follows its input while at the block input S the logic signal 1 holds ($S = 1$). When S is going to 0, the block output holds its input at this moment of time. The initial output value can be specified in the dialog box.

The next block is intended for delay of the input logical signal for specified time. In the dialog box, besides *Time Delay* and the input initial value before simulation starts *Input at t = −eps*, the block operation mode is specified: *On delay* or *Off delay*. In the first case, the input = 1 appears at the output after *Time Delay* and is held in this state while (or until) the input = 1. When the input is going to 0, the output is going to 0 without delay. For the *Off delay* mode, in this description, 1 has to be replaced by 0 and 0 by 1. The example of the block utilization is given in model2_5.

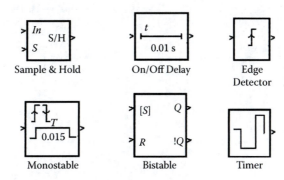

FIGURE 3.8 Pictures of the control blocks of points 1–6.

The block **Edge Detector** compares continuously the current value of the input signal with the previous one. The block can be tuned for generation of output logical signal 1 under increasing the input signal (*Edge: Rising*), when it is decreasing (*Edge: Falling*) or when there is any inequality between the current and previous values (*Edge: Either*). For the logical input signals, when the specified event takes place, the logical 1, whose duration is equal to 1 integration step, appears at the block output. For the continuous signals, the block responds to changes in their direction. If, for example, we give the sinusoidal signal at the block input and select the option *Edge: Rising*, the rectangular oscillations take place at the output, whose values are 0 and 1 and their first harmonic leads the input signal by $\pi/2$.

For the block **Monostable Flip-Flop**, as well as for the previous one, the direction of the input changing is specified, for which the blocks respond. If this event takes place, the block outputs the logical 1 during the time that is fixed in the parameter window *Pulse duration*. The example of the block utilization is given in model2_5.

The block **Bistable Flip-Flop** is an *R-S* flip-flop that sets the output $Q = 1$ under $S = 1$, $R = 0$, resets $Q = 0$ under $S = 0$, $R = 1$, and holds Q under $S = 0$, $R = 0$. The state under $S = 1$, $R = 1$ is defined by the selected option *Priority to input*: If *Set* is selected, $Q = 1$, and if *Reset*, $Q = 0$. We have the opportunity to fix the initial flip-flop state.

The block **Timer** generates certain signals at fixed times. In the first line of the dialog box, the times are given when the output changes, and in the second line, the output values at these times are given; thus, both quantities as vectors have the same dimension. The example of the block utilization is given in model2_4.

The pictures of the remaining five blocks are shown in Figure 3.9. **First-Order Filter** realizes the transfer function of the first-order element, whose form depends on the filter type selected in the dialog box: *Lowpass* or *Highpass*. In the first case,

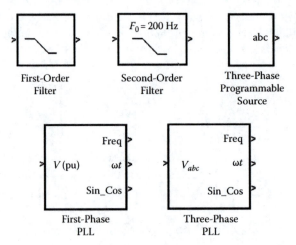

FIGURE 3.9 Pictures of the control blocks of points 7–11.

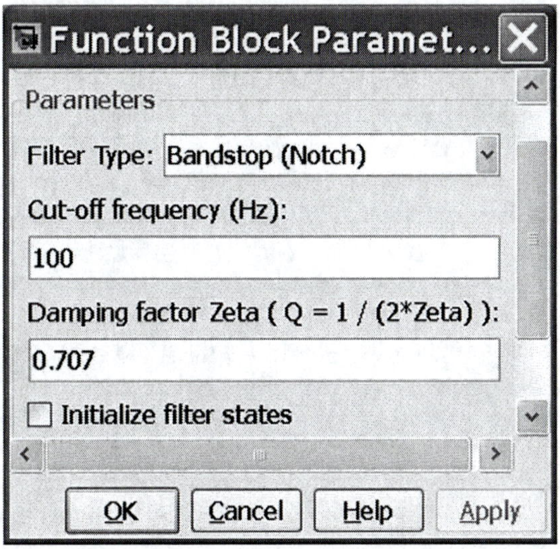

FIGURE 3.10 Dialog box of the second-order filter.

when $\omega \rightarrow \infty$, the amplitude-frequency characteristic tends to zero, the phase-frequency characteristic tends to $-\pi/2$, and the step-response tends to 1 when $t \rightarrow \infty$. In the second case, when $\omega \rightarrow \infty$, the amplitude-frequency characteristic tends to 1, the phase-frequency characteristic tends to 0, and the step-response tends to 0 when $t \rightarrow \infty$. The time constant of the filter T is set in the dialog box; the amplitude-frequency characteristic is equal to 0.707 when $f = 1/(2\pi T)$.

This dialog box has two additional possibilities. To mark the option *Plot filter response*, the Bode diagram and the step response are plotted on the screen. The frequency range and its increment have to be specified. If we mark the option *Initialize filter response*, then lines appear that indicate the initial input values.

The block **Second-Order Filter** gives more possibilities for processing the signals. The dialog box is shown in Figure 3.10. In the first line, the filter type is selected. Other options include *Lowpass, Highpass, Bandpass*, and a rejection filter *Bandstop*. The filter cut-off frequency is fixed in the next line. For *Lowpass* and *Highpass* filters, the amplitude-frequency characteristics are equal to 0.707 at this frequency; this characteristic has a maximum at this frequency for *Bandpass* and a minimum for *Bandstop* filters. The requested damping factor is fixed afterward. The filter characteristics can be plotted, and the initial input values can be given, as in the case of **First-Order Filter**.

Model3_3 is used in order to better understand the filter operation. The signal *Xin* is a sum of the four signals of frequencies 50 (fundamental harmonic), 250, 550, and 950 Hz. It enters the filters of the first and second orders. THDs are measured both in *Xin* and in filter outputs (**Display, Display1, Display2**). The amplitudes of the selected harmonics are computed also (**Display_1, Display_2, Display_3**).

Let us consider that it is necessary to pick out the fundamental frequency in *Xin*, which means the filters must transmit the signal of 50 Hz with minimal distortion and must suppress to the maximum possible extent the signals with frequencies in the range of 250 Hz or higher. The cutoff filter frequency is accepted to be 100 Hz; filter type is Lowpass, which for the first-order filter means $T = 1/(2\pi 100) = 1.6$ ms. The voltages of the sinusoidal signals are set, according to the frequency growth, as follows: 100, 40, 30, and 10 V. After simulation, it is seen that THD = 0.51 for *Xin* and 0.17 and 0.06 for the filters of the first and second order, respectively. One can see a more effective smoothing by the second-order filter, looking at the curves on the **Scope**.

Let the inverse problem be actual: an extraction of the high-frequency components. Since it is desirable that the signals with frequencies in the range of 250 Hz and higher are reproduced precisely, the cutoff frequency is taken as 150 Hz ($T = 1.1$ ms). The filter type is Highpass. After simulation, the blocks **Display_1**, **Display_2, Display_3** show that with the amplitude of the input fundamental harmonics of 99.75 V, at the output of the first filter, this amplitude decreases to 32.63 V, and at the output of the second filter, to 11.1 V. One can see a more effective suppression of the fundamental harmonic by the second-order filter, looking at the curves on the **Scope**.

The filter operation as a band filter is demonstrated with the help of model3_4. The sinusoidal signal generators have the same amplitude of 40 V. The block **FFT** is set at the filter output. The sampling time is 50 μs, and *FFT Sampling time* is 0.1 s. The filter has the cutoff frequency of 250 Hz (it means we try to pick out the fifth harmonic), type—Bandpass. After simulation, the following harmonic amplitudes are seen on **Display1** (Harmonic number/Amplitude): 1/11.3, 5/40, 11/25.2, 19/14.85. The fifth harmonic has the maximum amplitude, but the values of rest of the harmonic amplitudes are essential too. It means that in the case under consideration, the second-order filter is inadequate. If the filter type is replaced with Bandstop one, the harmonic amplitudes after simulation are as follows: 1/38.7, 5/0, 11/31.8, 19/31.14. Thus, the fifth harmonic disappears, and the values of rest of the harmonic amplitudes are close to the actual ones. So, the second-order filter, as Bandpass one, functions successfully.

The block **Three-Phase Programmable Source** fulfils the same functions as the block **Three-Phase Programmable Voltage Source** discussed in Chapter 2 and has the same dialog box. The difference is that the latter forms the voltages that are compatible with the block SimPowerSystems, and the former gives the three-phase signal as a vector [a, b, c] that is compatible with the block Simulink.

The last two blocks are the Phase-Locked Loop blocks, **PLL** that are used for computation of the frequency and phase of a harmonic signal. They are devices with feedback.

The block diagram of **One-Phase PLL** is shown in Figure 3.11. The processed harmonic signal, usually the voltage in pu $V = V_m \sin\theta_1$, enters the block. The signal

$$V_m \sin\theta_1 \cos\theta_2 = 0.5V_m[\sin(\theta_1 - \theta_2) + \sin(\theta_1 + \theta_2)]$$

is formed at the output of the multiplier M. This signal comes to the block **Discrete Variable Frequency Mean Value** that was described in the previous paragraph. The second term, having the high frequency $\omega_1 + \omega_2$, is filtered by this block, and at

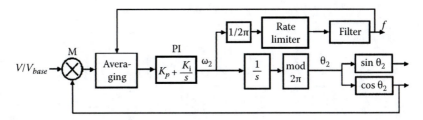

FIGURE 3.11   Block diagram of the one-phase PLL.

the input of PI-controller the signal generated, but with small error, is proportional to the difference between the input voltage phase and the computed angle. Feedback effect tends this difference to zero, $\theta_2 \to \theta_1$. Since angle θ_2 is formed as the integral of PI-controller output, this output is proportional to the frequency ω_2; that is, it is asymptotic to ω_1. Division in modulo 2π limits θ_2 by the values of $0-2\pi$. The value f (Hz) is formed after limitation of its changing rate and its filtration.

The dialog box contains the lines for fixing the initial values of the phase (deg) and frequency (Hz) and the controller gains. Block functioning is shown in Figure 3.12. The initial values of the phase and frequency are set to 0 and 60 Hz, respectively, and the input signal has the phase of $\pi/6$ and frequency of 50 Hz. As a noise, the fifth harmonic is added with an amplitude of 0.1. In the upper figure, it is seen that the computed frequency tends to the actual value; in the middle figure it is seen that for 0.1 s, the difference between the fundamental harmonic and the output $\sin\theta_2$ decreases to zero; and in the lower figure, it is seen that the signal ωt is synchronized with crossing zero by the input voltage.

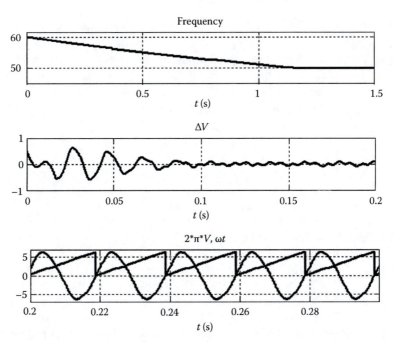

FIGURE 3.12   Transient response in one-phase PLL.

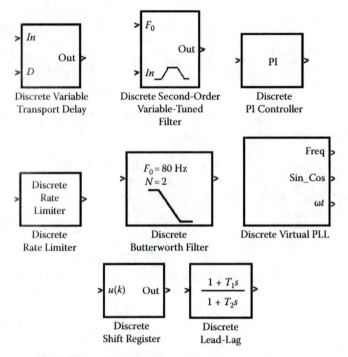

FIGURE 3.13   Pictures of the discrete control blocks.

The **Three-Phase PLL** differs from **One-Phase PLL** in that its input unit is not multiplier, as in Figure 3.11, but the block **abc_to_dq0 Transformation**. The q-component of the output is utilized. If we substitute into (3.3) $V_a = V_m \sin\theta_1$, $V_b = V_m \sin(\theta_1 - 2\pi/3)$, and $V_c = V_m \sin(\theta_1 + 2\pi/3)$, $x = \theta_2$, the output $V_q = V_m \sin(\theta_1 - \theta_2)$ and may be used for angle adjustment.

A number of additional blocks are included in the folder/*Discrete Control Blocks*, whose pictures are given in Figure 3.13:

1. Discrete Variable Transport Delay
2. Discrete Rate Limiter
3. Discrete Shift Register
4. Discrete Second-Order Variable-Tuned Filter
5. Discrete Butterworth Filter
6. Discrete Lead-Lag
7. Discrete PI Controller
8. Discrete Virtual PLL

The first block delays the input signal *In* for the time determined by the second input *D*. This value may change in time.

The second block is a discrete copy of the continuous block **Rate Limiter** from Simulink.

Discrete Shift Register models a shift register with the serial input and parallel output. If the vector signal of N_{sig} dimension comes at the input, the block stores the last $N \times N_{sig}$ inputs, and the number N is indicated in the dialog box. These signals appear at the block output as a vector of $N N_{sig}$ dimension that is arranged as follows: $[u_1(k), u_1(k-1), u_1(k-(N-1)), u_2(k), u_2(k-1), \ldots]$. The sample time and initial register content are specified in dialog box.

Discrete Second-Order Variable-Tuned Filter differs from those described previously with the additional input $F0$, at which point the signal of the filter cutoff frequency comes in. This signal may change in time.

The block **Discrete Butterworth Filter** models the Butterworth N-order filter. The available filter modifications are the same for the block **Second-Order Filter**. While modeling the band filters, the filter order doubles automatically. The filter of Butterworth has the maximum flat (has no ripples) frequency response in the transmission band and rolls off toward zero with a relatively small slope in the attenuation band. The filter order N, the cutoff frequency (Hz), the bandwidth for the band filters that corresponds to the level of 0.707, and the sample time are specified in the dialog box. As for the filers described previously, there is the opportunity to put in the initial values and plot the filter characteristics.

In order to better understand the advantages of the said filter, it may be recommended to the reader to replace in model3_4 the second-order filter with the filter of Butterworth in the mode of the bandpass filter with a cutoff frequency of 250 Hz and bandwidth of 50 Hz. When $N = 1$, the following harmonic amplitudes are seen: 1/1.6, 5/39.9, 11/4.5, 19/2.2; these are much better than the ones obtained via the second-order filter. When $N = 2$, the following values are produced after simulation: 1/0.07, 5/40, 11/0.52, 19/0.13, that is, nearly ideal filtration. However, the filter transient response has the essential fluctuation.

The block **Discrete Lead-Lag** realizes in the discrete form the unit with the transfer function

$$W(s) = \frac{1+T_1 s}{1+T_2 s}. \tag{3.23}$$

Time constants T_1 and T_2 are given in the dialog box.

The block **Discrete PI Controller** models the discrete proportional-plus-integral controller with the fixed sample time. The block output is a sum of the two signals: the proportional channel output with gain K_p and the integral channel output that has an integrator of Simulink with gain K_i. The values K_p and K_i are indicated in the dialog box. The upper and lower limits are specified here too; these limits affect both the integrator output and the total signal.

The last block **Discrete Virtual PLL** has no input signals. It simulates the operation of the actual PLL by use of parameters specified in the dialog box.

REFERENCES

1. Mathworks, SimPowerSystems™, User's Guide, 2004–2011.
2. Gross, C.A., *Power System Analysis*, John Wiley & Sons, New York, 1986.

4 Simulation of Power Electronics Devices

4.1 MODELS OF POWER SEMICONDUCTOR DEVICES

SimPowerSystems/Power Electronics includes various models of semiconductor devices [1]:

1. Diode
2. Thyristor
3. Detailed Thyristor
4. GTO
5. MOSFET
6. IGBT
7. IGBT/Diode

The pictures of the devices are shown in Figure 4.1. For all devices, the option *Show measurement port m* is marked. The dialog boxes contain a number of the repeated lines (fields), so the dialog box of the IGBT-transistor is described first, and afterwards, the distinguishing features are indicated for some devices.

The block **IGBT** models the transistor with the parallel connected snubber that consists of the connected in series R_s and C_s. In the conducting state, IGBT has the interior resistance R_{on}, the inductance L_{on}, and the source of DC voltage V_f that models the forward conduction drop. The values of R_{on} and L_{on} cannot be set to zero simultaneously. In switch-off state, IGBT has infinitely large resistance. IGBT turns on when the voltage collector-emitter is positive and as big as V_f, and the gate signal $g > 0$. It turns off when the voltage collector-emitter is positive and $g = 0$. IGBT remains in the turned-off state for negative voltage collector-emitter. IGBT process of turning-off consists of two segments: collector current decreasing from I_{max} to $0.1I_{max}$ during the fall time T_f, and subsequent current falling to zero during the tail time T_t. This IGBT model does not take into consideration the device geometry and the complex physical processes that take place in it. The block is modeled as a current source and cannot be connected in series with an inductance, a current source, or with an open circuit without snubber.

The block dialog box is shown in Figure 4.2. The parameter designations correspond to those given previously. If the initial current $I_c > 0$, simulation begins from the turn-on state, but such a mode is not recommended. If the option *Show measurement port* is marked, the output *m* appears on the block picture, with the vector signal: the device current and the voltage drop.

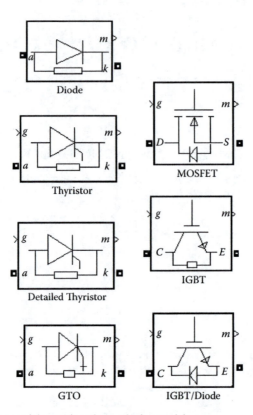

FIGURE 4.1   Pictures of the semiconductor device models.

The block **Diode** is modeled as IGBT, with the difference that it turns on as soon as the voltage V_{ak} becomes positive and turns off as soon as this voltage becomes negative. In the dialog box, the lines for T_f and T_t are absent.

MOSFET transistor is controlled by the signal $g > 0$, if the device current $I_d > 0$. Parallel to the transistor, in opposite direction, the internal diode is connected that conducts under the negative transistor voltage ($V_{ds} < 0$). The block is modeled as a series connection of the variable resistor R_t, of the inductance L_{on} and of the switch that is controlled by the gate signal ($g > 0$ or $g = 0$). The transistor turns on when $V_{ds} > 0$ and $g > 0$. If $I_d > 0$, the transistor turns off when $g = 0$. If $I_d < 0$, the current flows through the diode and the transistor is turned off. $R_t = R_{on}$ under $I_d > 0$, and $R_t = R_d$ under $I_d < 0$. The block contains R_s, C_{sc} snubber.

GTO thyristor, unlike the usual thyristor, can be turned off by the gate signal. It is modeled as a series connection of the resistance R_{on}, of the inductance L_{on}, of the forward voltage source V_f and of the switch. The thyristor turns on when the anode-cathode voltage $V_{ak} > V_f$ and $g > 0$. When g goes to zero, the process of turning-off begins for IGBT as well. After current is reduced to zero, the thyristor remains in the nonconducting state.

The block **Thyristor** is modeled as GTO thyristor. It turns on under the same conditions but cannot be turned off by the control signal. Under $g = 0$, it remains in the conducting state, while the anode-cathode current $I_{ak} > 0$.

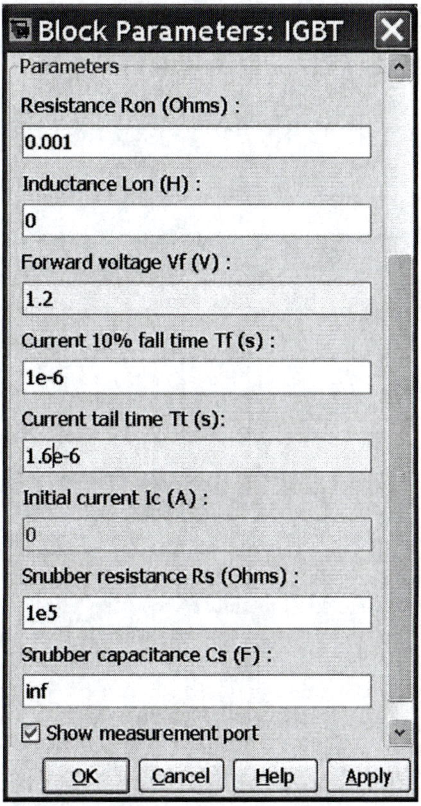

Block Parameters: IGBT ✕

Parameters

Resistance Ron (Ohms) :

0.001

Inductance Lon (H) :

0

Forward voltage Vf (V) :

1.2

Current 10% fall time Tf (s) :

1e-6

Current tail time Tt (s):

1.6e-6

Initial current Ic (A) :

0

Snubber resistance Rs (Ohms) :

1e5

Snubber capacitance Cs (F) :

inf

☑ Show measurement port

OK   Cancel   Help   Apply

FIGURE 4.2   IGBT model dialog box.

The block **Detailed Thyristor** contains in its dialog box the additional lines for the latching current I_l and for the turning-off time T_q. With the thyristor turned-on, the firing signal duration $g > 0$ has to be sufficient for fulfilling of the condition $I_{ak} > I_l$ (for the block **Thyristor** is accepted $I_l = 0$). If when turned-on this condition is not reached, the thyristor turns off after setting $g = 0$. The turn-off time T_q is the carrier recovery time: it is the time interval between the instant when I_{ak} reaches zero and the instant when the thyristor is capable of withstanding positive voltage V_{ak} (for the block **Thyristor** is accepted $T_q = 0$). It is seen that such important characteristics of the thyristor as the dependence of its behavior on the rate of change of the forward current and of the voltage are not taken into consideration.

The block **IGBT/Diode** is a simplified model of the pair: IGBT (or GTO, or MOSFET) Diode that has only one parameter: the interior resistance (the snubber parameters also).

Together with separate devices, the folder SimPowerSystems/Power Electronics contains blocks that consist of the following devices, for example:

1. Universal Bridge
2. Three-Level Bridge

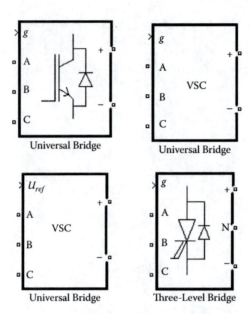

FIGURE 4.3   Pictures of the blocks **Universal Bridge** and **Three-Level Bridge**.

Universal Bridge simulates a semiconductor converter that has two to six devices connected to the bridge circuit. According to indication in the dialog box, the block can have one bridge arm (two devices connected in series with three terminals), two bridge arms (single-phase bridge circuit), and three bridge arms (three-phase bridge circuit). According to this, the block picture (Figure 4.3) has two DC terminals (+, −) and one, two, or three AC terminals. The devices shown in Figure 4.1 can be used in this block (except the block **Detailed Thyristor**) and the block **Ideal Switch** as described in Chapter 2 as well. With a choice of the specific device, its depiction appears on the bridge picture, and the content of the dialog box can change too. This block is a base for building of the rectifier models and of the two-level converters—rectifiers and inverters. For simulation acceleration, two modifications of this block are developed in addition, which are shown in Figure 4.4 too and are considered further.

Gate signals ordered as a vector come at the input g. Their order corresponds to the order of the device numbering in the bridge circuit. Numbering is different for the diodes and the thyristors (Figure 4.4a), and for the rest of the devices (Figure 4.4b).

The dialog box of the block while using of IGBT is shown in Figure 4.5. Unlike in the case of the separate transistor, the forward voltage for the antiparallel diode V_{fd} is specified too. With the use of the other devices, the parameters in the dialog box change and become clear enough when compared with the dialog boxes of the separate devices. Dwell upon two important questions.

The first question is about snubber employment and its parameters. To remove the snubber, it is necessary to take either $R_s =$ inf or $C_s = 0$, and to have the purely resistive

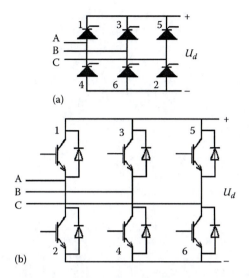

A
B
C
1
3
5
4
6
2
+
–
U_d
(a)
A
B
C
1
3
5
2
4
6
+
–
U_d
(b)

FIGURE 4.4 Device numbering in the block **Universal Bridge** (a) for diodes and thyristors and (b) for other devices.

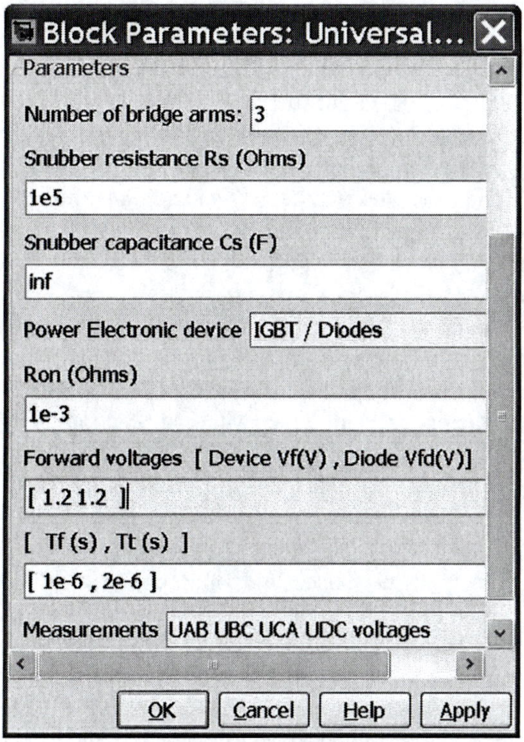

Block Parameters: Universal...
Parameters
Number of bridge arms: 3
Snubber resistance Rs (Ohms)
1e5
Snubber capacitance Cs (F)
inf
Power Electronic device IGBT / Diodes
Ron (Ohms)
1e-3
Forward voltages [Device Vf(V) , Diode Vfd(V)]
[1.2 1.2]
[Tf (s) , Tt (s)]
[1e-6 , 2e-6]
Measurements UAB UBC UCA UDC voltages
OK
Cancel
Help
Apply

FIGURE 4.5 Dialog box of the block **Universal Bridge** with of IGBT.

snubber, it is necessary to set $C_s = $ inf. The numerical instability of simulation can appear in the discrete mode when diodes and thyristors are used. To prevent it, it is recommended to use the $R_s - C_s$ snubber with the following parameters

$$C_s < \frac{0.001 P_n}{2\pi f V_n^2}, \quad R_s > \frac{2T_s}{C_s}$$

where
  P_n is the nominal converter power (VA)
  f is the fundamental frequency (Hz)
  V_n is the nominal line-to-line voltage (V, rms)
  T_s is the sample time (s)

The forced-commutated devices operate satisfactorily with the purely resistive snubber, and the gate pulses come to the devices, but when they are blocked, the bridge changes into the diode one, and $R_s - C_s$ snubber is demanded.

The second question is about the measured quantities, when the option *Measurement* is used. When the option *Device voltages* is chosen, the voltage across each device is measured. By selecting *Device currents*, the currents flowing through each the device is measured; if this current >0, it is the current through the forced-commutated device, and if this current <0, it is the current through antiparallel diode. The snubber current is not measured. By selecting U_{AB}, U_{BC}, U_{CA}, U_{DC} voltages, the AC and DC voltages at the block terminal are measured. When the option *All voltages and currents* is chosen, all the above-mentioned quantities are measured. The measurement is carried out with help of the block **Multimeter**.

In the fifth version of SimPowerSystems, two additional possibilities appear that help to select the block structure. It will be clear from Figure 5.33; the voltage source inverter (VSI) with this block forms at its output six active and two zero states of the voltage space vector. The set of the gate pulses corresponds to each state. If one neglects transients in the devices and the voltage drops across them, the values of the voltage space vector components are proportional to the inverter capacitor voltage; moreover, the proportionality factors depend on the set of the gate pulses that come in. These factors are called the switching functions. The current that the inverter draws from the inverter capacitor is defined by the inverter phase currents and the switching functions. By choosing the structure *Switching-function based VSC*, and when a certain gate pulse set comes to the bridge, at its output, with help of two blocks **Controlled Voltage Source**, the corresponding voltage space vector is formed, and with help of the block **Controlled Current Source**, the input currents are formed. When this simulation mode is used, harmonics generated by the bridge are represented correctly. When another structure *Average-model based VSC* chosen, the output inverter voltage is proportional to the input voltage in the control system U_{ref}, with the proportionality factor depending on the capacitor voltage that takes place only for the average values. In this mode, a larger sample time can be used, but the harmonics are not represented. This mode can be used for simulation of the systems with a lot of VSI.

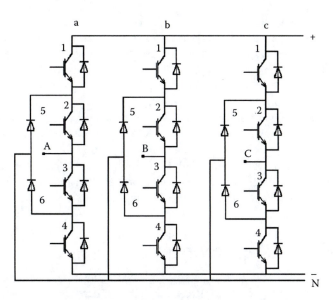

FIGURE 4.6   Diagram of the block **Three-Level Bridge**.

The diagram of the block **Three-Level Bridge** is shown in Figure 4.6. As it can be indicated in the dialog box, the bridge can have one arm (four devices connected in series with four terminals), two arms, and three arms. Accordingly, the block picture (Figure 4.3) has three DC terminals (+, −, N) and one, two, or three AC terminals. The forced-commutated devices (**GTO**, **MOSFET**, **IGBT/Diode**) and the block **Ideal Switch** can be used. By selecting a specific device, its depiction appears on the bridge picture, and the content of the dialog box can change too. This block is a base for building of the models of the three-level rectifiers and inverters that, with the same device nominal voltage, provide higher output voltage.

Two adjacent devices connected in series conduct under normal operating mode. If the devices 1 and 2 are switched, the phase terminal has the positive voltage; if the devices 3 and 4 are switched then there is negative voltage; and if the devices 2 and 3 are switched, the phase terminal is connected to the neutral N. When **Ideal Switch** is chosen, the diagram in Figure 4.7 is realized. The block dialog box, similar to the previous one, contains the lines for specifying the number of arms, device type, its parameters and indication, which block quantities have to be measured. What was said about the snubbers earlier is valid for this block too. Gate signals ordered as a

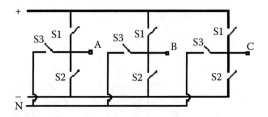

FIGURE 4.7   Diagram of the block **Three-Level Bridge** with the devices **Ideal Switch**.

vector come at the input g, in general, and in the order: 1–4 for the phase A, which remains the same for phase B and phase C (their numbers, 4, 8, and 12, depends on the number of the bridge arm). When **Ideal Switch** is used, the gate pulse 1 comes to the switch S1, the gate pulse 4 comes to the switch S2, and the logic AND of the pulses 2 and 3 comes to the switch S3.

In the line *Measurement* one can select: *None*—no measurement; *All device currents*—the currents of the all semiconductor devices are measured in the order: phase A, transistors 1–4; phase B, transistors 1–4; phase C, transistors 1–4; phase A, antiparallel diodes 1–4; neutral clamping diodes 5, 6; these apply to both phase B and phase C; *Phase-to-neutral and DC voltages* on the block terminals are measured; with the option *All voltages and currents*, all afore-mentioned currents and voltages are measured. For **Ideal Switch**, currents flowing through the switches (maximum nine) are measured.

4.2 CONTROL BLOCKS FOR POWER ELECTRONICS

The converters must have at their outputs the electric quantities (voltages, currents) with certain parameters: amplitude, frequency, phase. These parameters are provided by sending to these converters the trains of the firing pulses with definite character-istics: frequency, duration, arrangement in time. At the same time, the references for the requested electric parameters are generated by the control system in analogous or digital form. Thus, units that convert the reference signals into demanded pulse sequence are necessary. The blocks that are included in SimPowerSystems for this aim can be divided into two groups: the pulse generators for the converters with thyristors and the pulse generators for the converters with the forced-commutated devices (with pulse-width modulation—PWM).

The first group includes

1. Synchronized six-pulse Generator
2. Discrete Synchronized six-pulse Generator
3. Synchronized 12-pulse Generator
4. Discrete Synchronized 12-pulse Generator

The block pictures are shown in Figure 4.8. The first two blocks are intended for con-trol of the converters, whose main circuits are carried out according to the diagram in Figure 4.4a. The output voltage in the continuous current mode is

$$E_d = E_{d0} \cos \alpha \qquad (4.1)$$

$$E_{d0} = 1.35 U_1 \qquad (4.2)$$

U_1 is the rms of the phase-to-phase voltage, α is the firing angle that is counted off from the natural firing point, when the thyristor anode voltage becomes positive (the corresponding phase-to-phase voltage crosses zero level). Therefore, output volt-age control is carried out by a change in the instant when the firing pulses are sent

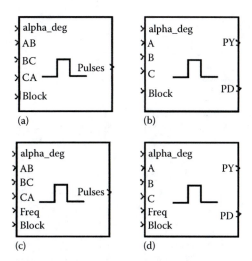

FIGURE 4.8 Pictures of pulse generators for the converters with thyristors. (a) Synchronized 6-Pulse Generator; (b) Discrete Synchronized 6-Pulse Generator; (c) Synchronized 12-Pulse Generator; (d) Discrete Synchronized 12-Pulse Generator.

to the thyristors. The block inputs are the phase-to-phase voltages and the requested angle α (°). The pulses are generated only when the input *Block* = 0. The block output is the vector of six pulses, whose numeration corresponds to the thyristor numeration in the bridge diagram. Pulse parameter is its duration. As it follows from the thyristor description given previously, this duration must be sufficient, in order to ensure that the thyristor current exceeds the latching current I_l.

The successive thyristors in the bridge are fired every 60°. Under small current and (or) small load inductance, the duration of the current flow through thyristor can be less than 60°; on current decay, the conducting thyristor is blocked. In such a case, at the instant a firing pulse is sent to the next thyristor, the load current becomes zero; it is a discontinuous current mode. In such a mode, when the firing pulse is sent to the next thyristor, it will not conduct current because, according to the bridge diagram, the current must flow through two thyristors, and there is no way current can flow in this case. For providing normal operation, it is necessary, simultaneously with firing of the new thyristor, to send the firing pulse to the previous one, which is blocked now. It can be done by two ways: to form the pulses with duration more than 60°; or, to send such a pulse to the previous one together with the short pulse to the ingoing thyristor. This way is called "double pulsing."

The frequency of the supply voltage, pulse duration, and request for double pulsing are specified in the block dialog box. The diagram of the circuit for the thyristor T1 is shown in Figure 4.9.

The natural T1 firing (firing in the diode bridge) takes place, when the phase voltages $U_a = U_c$, that is, $U_{ac} = -U_{ca} = 0$. In this moment, the element D1 forms the short pulse that resets the integrator I1, and afterwards its output rises as a linear function of time $I_1 = ft$. Simultaneously, the flip-flop Tp is reset, preparing the gate AND1 for setting. The signal I_1 is compared with the signal $\alpha°/360$ at the input of

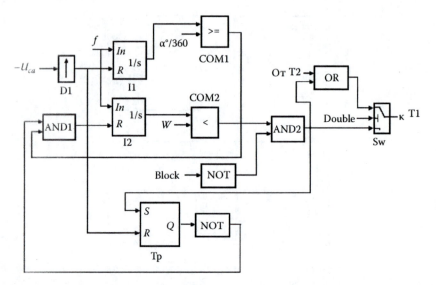

FIGURE 4.9 Firing pulse generation for the thyristor *A*.

the comparison element COM1. They will be equal in the time $t_i = \alpha$ (rad)/$2\pi f$. In this instant, the logical 1 is formed at the output of COM1 that resets the integrator I2 through the gate AND1. The logical 1 appears at the output of the comparison element COM2, because it appears on condition that the first input is less than the second one. This signal goes to the thyristor gate (control electrode) through the gate AND2 under Block = 0 and through the switch Sw. When output of I2 becomes equal to the signal *W* that defines the pulse duration, the logic 0 is set at the output of AND2, and the pulse terminates. Simultaneously, the flip-flop Tp is set, and the logic 0 is set at the output of AND1, so that the integrator output exceeds *W* and the new pulse cannot arise. If *double pulsing* is selected, the switch Sw is in the upper state, and the thyristor T1 receives the pulses not only from its circuit but also from the next one—from the circuit of the thyristor T2.

Usually, while supplying the rectifier from a three-phase transformer, the primary voltages are used for block phasing, and attention has to be paid to their correct connection. The phase-to-phase voltages that are indicated on the block picture are used only when there is no phase shift between the primary and secondary windings. For the schemes Y/D1 or D11/Y, the phase voltage U_c comes to the input CA, with the corresponding arrangement for the other inputs and for the schemes D1/Y and Y/D11—the phase voltage $-U_a$.

The discrete version of the block differs with the use of discrete integrators; in the dialog box the sample time has to be specified. The frequency is not specified because its value comes to the block input Freq as an external signal. It gives the opportunity to use this block when a rectifier is supplied from a weak source.

Synchronized 12-Pulse Generator controls two thyristor bridges that are connected to two secondary windings of the three-winding transformer; moreover, the primary and the first secondary windings have star connection, and the second secondary has delta connection, whose phase voltages lag about Y. The primary-phase voltages supply the block inputs, and the phase-to-phase voltages, which give the natural firing points when they cross the zero level, are formed in the block. The block contains two groups of the circuits similar to those shown in Figure 4.9. The firing pulses for the bridge that is connected to the star-winding are sent to the output PY, and for the second bridge—to the output PD.

For the discrete version of the block, the sample time has to be specified and the external signal F_{req} has to be set.

The pulse generators in SimPowerSystems are classified as multichannel with the linear reference signal. Their demerits are a nonlinear dependence of the output voltage on the control signal and an influence of supplying voltage frequency oscillations. The first drawback can be eliminated by applying of the function arc cosine to the input signal, and the second—by using PLL. There are many other modifications of the pulse generators. For example, as the reference signals, the corresponding sections of the supplying voltages can be used (so-called the pulse generators with cosinusoid reference voltage). They provide the linear characteristics input-output and insensitivity to the frequency oscillations but are sensible to changes in the voltage amplitude. While using microprocessors, the single-channel systems are used usually because they have fewer timer counters and low connectivity with the supply network. If the reader simulates these pulse generators, he or she must develop them on his or her own, by using Simulink® blocks.

The second group of blocks include the pulse generators for converters with PWM:

1. PWM Generator
2. Discrete PWM Generator
3. Discrete three-phase PWM Generator
4. Discrete SV PWM Generator

These blocks are used in order to receive the output voltage with the requested parameters at the VSI outputs. This voltage has to be close, to a maximum possible extent, to the sinusoid with the demanded amplitude, frequency, and phase. Since an inverter has only the limited number of states, its output voltage consists of rectangles of different duration, with several possible amplitude values, and the first (fundamental) harmonic of this voltage must have the demanded parameters. Usually, two main PWM methods are used: comparison of the reference sinusoidal signal with the triangular carrier waveform, and the space vector modulation.

The first method is illustrated in Figure 4.10. Each phase of the bridge that is shown in Figure 4.4 is controlled independently. Take into consideration phase A. The curve U_a is the reference signal in pu. When $U_a < U_t$ (U_t is the triangular carrier), the logic signal 0 is sent to the transistor 1, and it is in nonconducting state (off-state); simultaneously, the inverse signal is sent to the transistor 2, and it is in the conducting state (on-state), and the voltage is negative at the terminal A. When $U_a > U_t$, the situation is the opposite; the voltage is positive at the terminal A. Suppose that the frequency of U_t is much larger than the frequency of U_a, so that during one period of U_t that equal to $2T_0$ it is possible to accept that $U_a = m$—the constant value. Since as

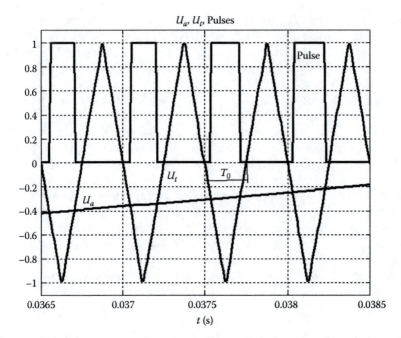

FIGURE 4.10 Comparison of the reference sinusoidal signal with the triangular carrier waveform.

shown in Figure 4.10 m is negative, the negative voltage at the terminal A is for $(1 - m)T_0$, the positive one for $(1 + m)T_0$, and the average value is:

$$U_{av} = \frac{-(1-m)T_0 + (1+m)T_0}{2T_0} = m$$

Thus, the average output voltage is proportional to the demanded signal U_a. For the three-phase inverter, three reference signals are formed, with relative shift of 120°.

The maximum output of inverter voltage takes place without modulation, when the output voltage consists of six steps; each has a duration of 60° [2]. The first harmonic amplitude of the phase voltage is

$$U_{1a} = \frac{2U_d}{\pi} \tag{4.3}$$

and the rms of the phase-to-phase first harmonic is

$$U_{1\pi,rms} = \frac{2.45U_d}{\pi} \tag{4.4}$$

When using the said modulation method, the maximal achieved voltage is 0.785 of the value given by (4.4).

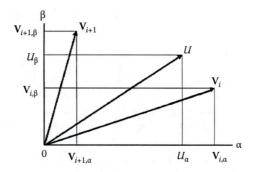

FIGURE 4.11   Forming of the resulting vector by space vector modulation.

Space vector modulation—SVM is a digital method in principle. A system operates with the time period T_c; several VSI states are selected during T_c. Since value T_c is essentially lower than the load time constants, it is assumed that the output voltage is equal to the average value of the voltages acting during T_c. When using SVM, the control system forms the requested output voltage vector **u**—in the Cartesian or in the polar coordinates. Remember that VSI can form at its output six nonzero space vectors and, by two ways, the zero vector. Two adjacent VSI states and one of zero states are used in order to receive the vector **u**. Let, for example, the vector **u** in Figure 4.11 be obtained. Let $\mathbf{V}_i$, $\mathbf{V}_{i+1}$ be adjacent to **u** vectors with the components $\mathbf{V}_{i\alpha}$, $\mathbf{V}_{i\beta}$, $\mathbf{V}_{i+1,\alpha}$, $\mathbf{V}_{i+1,\beta}$. Let vector $\mathbf{V}_i$ be used during time interval t_1 and vector $\mathbf{V}_{i+1}$—during t_2. Denote $\tau_1 = t_1/T_c$, $\tau_2 = t_2/T_c$. The value τ_1 and τ_2 are determined from the system of equations

$$\mathbf{V}_{i\alpha}\tau_1 + \mathbf{V}_{i+1,\alpha}\tau_2 = u_\alpha$$
$$\mathbf{V}_{i\beta}\tau_1 + \mathbf{V}_{i+1,\beta}\tau_2 = u_\beta$$

(4.5)

by solving which the expressions for τ_1 and τ_2 can be found [2].

During the time interval $T_c(1 - \tau_1 - \tau_2)$, one of the zero states is used. The expressions for τ_1, τ_2 in the polar coordinates are

$$\tau_1 = \frac{\sqrt{3}u}{U_d}\sin\left(\frac{\pi}{3}-\gamma\right), \quad \tau_2 = \frac{\sqrt{3}u}{U_d}\sin\gamma$$

(4.6)

The maximum value u_m can be found from the condition $\tau_1 + \tau_2 = 1$; moreover, the value u_m is minimal under $\gamma = \pi/6$. This value is

$$\frac{u_m}{U_d} = \frac{1}{\sqrt{3}}$$

(4.7)

that is, the maximal voltage in the linear region is

$$\frac{1/\sqrt{3}}{2/\pi} = 0.907$$

of the values given by (4.3), (4.4). Under a larger value u, VSI gets into the overmodulation region, in which operation is possible also but is accompanied by an increased distortion of the output current.

After the times t_1, t_2, and t_0 are computed, a problem arises to arrange in time the corresponding inverter states in the best possible way. The criteria for this are contradictory: less switching and less high harmonics in the VSI current. One of the variants is not to switch (to stay in the unchanged state) one of the VSI phase in each period T_c. It is usual to indicate VSI state by three-bit (according to the phase A B C) binary code: If the digit is 1, the upper device conducts in this phase, and if the digit is 0, the lower device conducts. Thus, there are two zero states: 000 and 111. If, for instance, the state 001 corresponds to the vector $\mathbf{V}_i$, and the state 101 corresponds to the vector $\mathbf{V}_{i+1}$, the sequence of the state changing is (in brackets the time of activity is indicated): $101(t_2/2)$, $111(t_0)$, $101(t_2/2)$, $001(t_1)$. But a low content of the high harmonics is typical under the following condition, with an increase in the switching number: $000(t_0/4)$, $001(t_1/2)$, $101(t_2/2)$, $111(t_0/2)$, $101(t_2/2)$, $001(t_1/2)$, $000(t_0/4)$.

The block pictures are given in Figure 4.12. The block PWM Generator generates gate pulses for the two-level inverter when a triangular carrier waveform is used. The block can be used for one-, two-, three-arm bridges and for 2 three-arm bridges shown in Figure 4.4b. The number of pulses generated by the block depends on the arm number. For one-arm scheme, two pulses are generated: the pulse 1 for the top device and the pulse 2 for the lower one; these pulses are complementary. For the two-arm bridge, four pulses are generated (the first and the third for the top devices, the second and the fourth for lower ones), and for three-phase bridge, six pulses are generated (the first, the third, and the fifth for the top devices; the second, the fourth, and the sixth for lower ones). When 2 bridges are controlled, 12 pulses are generated, the first 6 for the first bridge and the second 6 for the second one. The modulating

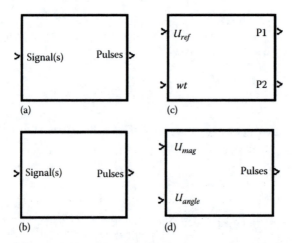

FIGURE 4.12 Pictures of the pulse generators for converters with PWM. (a) **PWM Generator**; (b) **Discrete PWM Generator**; (c) **Discrete Three-Phase PWM Generator**; (d) **Discrete SV PWM Generator**.

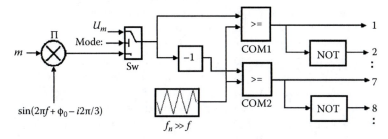

FIGURE 4.13   Block diagram of the block **PWM Generator**.

(sinusoidal) signal can be produced within the block or can be received from the external source. For one- or two-arm bridge, one signal is needed (for a single-phase two-arm bridge, the modulating signal for arm 2 has 180° phase shift relative to the modulating signal for the arm 1). For the three-phase six-arm bridge, the three modulating signals are used, with the relative shift of 120°; moreover, the modulating signal for the second bridge is inverted inside the block relative to the modulating signal for the first bridge.

The block structure is shown in Figure 4.13. If the switch Sw is in the upper position, the external modulating signal (signals) U_m is used; if it is in the lower position, the modulating signal is equal to the product of the modulation index m ($0 \leq m < 1$) that is specified in the dialog box and one or three sinusoidal signals: $\sin(2\pi f + \theta_0 - 2\pi i/3)$ where f is the frequency, θ_0 is the phase angle, and $i = 0$ for one or two arms and is equal to 0, 1, 2 for the three-phase bridge (bridges). The comparison units COM1 and COM2 compare these signals with the triangular carrier waveform with the frequency of $f_n = 1/(2T_0) \gg f$; COM2 is used only for the second bridge. The required number of the direct and inverse gate pulses is formed at the output of the comparison units. In the dialog box, the number of arms and the frequency f_n are selected. If the option *Internal generation of modulating signal* is marked, lines appeared that help to specify m, f (Hz), and θ_0 (°).

The algorithm of the block functioning does not take into account the dead-time effect under switching over of the devices of a phase, as this effect arises out of a delay during switching that is set for avoiding a shoot-through fault. This delay affects the VSI output voltage. A number of methods have been developed for eliminating this effect. Thus, it is assumed that the dead-time effect can be compensated by some way.

The discrete block differs only by computations discretely in time, and the value of the *Sample time* is specified in the dialog box.

Discrete Three-Phase PWM Generator generates gate pulses for 2 two- or three-level VSI, and the VSI output voltages are in anti-phase. Modulator uses the triangular carrier waveform. The special feature of the block is that synchronization of the modulating and carrier waveforms can be realized. The fact is that in the previous blocks the frequency of the modulating signal changes and the carrier waveform frequency is constant, so that, in the common case, a period of the former comprises nonwhole number of the periods of the latter that can result the subharmonics in the VSI current. With the use of synchronization, it is not the carrier waveform frequency that is specified but its relation to the modulating signal frequency as a whole number K_n. Such a mode demands use of PLL. The three-phase modulating signal comes to the input U_{st}, and the synchronizing signal U_t comes to the input wt. The input w disappears in the un-synchronized mode, and during the internal generation of the modulating signal all the inputs disappear. At the output P1, 6 or 12 pulses are formed for the first bridge, and so many are formed at the output P2 for the second bridge. The pulse arrangement for the two-level inverter is the same for the previous blocks, and for the three-level inverter (Figure 4.6), the first four pulses are sent to the devices 1, 2, 3, 4 of the phase A, the next four—to the such devices of the phase B, and next four—to the phase C.

While controlling the two-level inverter, the block operation does not differ practically from the said block. During synchronization, the value $\theta_n = K_n wt$, $0 < \theta_n < 2\pi$ is used, and during un-synchronization, $\theta_n = 2\pi f_n t$ is used. The triangular carrier waveform is formed as

$$U_t = \frac{2}{\pi}\arcsin(\sin\theta_n) \qquad (4.8)$$

Functioning of the block is more intricate for the three-level inverter. Its structure is shown in Figure 4.14. The signals C+ and C− are triangular that change from 0 to 1 and from −1 to 0, respectively. When $U_{st} > 0$, COM1 functions, and the logic 0 is at the COM2 output, the output of SM1 is equal to 0 or 1. When $U_{st} < 0$, COM2 functions and the logical 0 are at the COM1 output, and the output of SM1 is equal to 0 or −1. The conversion of these signals in the gate pulses is fulfilled in the decoder according to the algorithm: When SM1 = 1, the devices 1 and 2 are fired, 3 and 4 are blocked; when SM1 = 0, the devices 2 and 3 are fired, and 1 and 4 are blocked; when SM1 = −1, the devices 3 and 4 are fired, and 1 and 2 are blocked. The signals P2 are formed similarly, and the AC voltages of the bridges are in antiphase. In Figure 4.15, pulse generation for the phase A is shown.

Discrete SV PWM Generator forms the gate pulses for the two-level VSI, using the said SVM algorithm. There are three opportunities for the reference vector

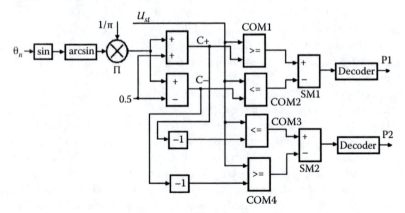

FIGURE 4.14 Block diagram of the block **Discrete Three-Phase PWM Generator** under control of the three-level inverter.

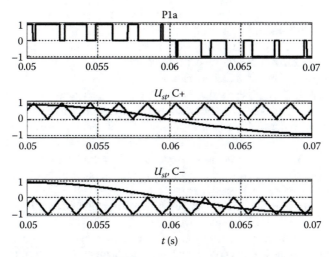

FIGURE 4.15 Gate pulse generation in the block **Discrete Three-Phase PWM Generator** under control of the three-level inverter.

specifying (option *Data type of input reference vector*): Magnitude-Angle, alpha-beta components, and Internally generated. With a choice of some way of reference, the block picture changes. When external reference is chosen, chopping frequency that is equal to $1/T_c$ and sample time are specified. When internal reference is chosen, the parameters of the reference vector are given in addition: amplitude m (pu), phase (°), and frequency (Hz).

The block gives the opportunity to choose one of the two types of state sequences during one period T_c that were described earlier (option *Switching pattern*). When *Pattern* #2, the first sequence is realized, with a decreased number of switching, and when *Pattern* #1 is chosen, the second sequence is realized, with an increased number of switching, but with less content of the high harmonics.

4.3 SIMULATION OF CONVERTER WITH THYRISTORS

A number of examples of the power electronics simulation that use the previously discussed blocks are given in this and in the following paragraphs. Simple examples for the study of electronics and more complicated models are presented. In what follows, converters with thyristors are considered.

The model4_1 demonstrates supplying of the *L-R-E* load by the 6-pulse thyristor rectifier. The latter is modeled by the block **Universal Bridge**. The firing pulses are formed by the block **Synchronized 6-Pulse Generator**. The rectifier is supplied from the network 380 V through the transformer 250 kVA, 380/410 V, E_{d0} = 1.35 × 410 = 550 V. The load parameters are 1 Ω, 0.03 H. The scope **Pulses** shows AC voltage and the firing pulses, the scope **DC**—the load current and voltage, both the instantaneous and the average values are fixed, the scope **T_Currents**—the thyristor currents and the phase A voltage, the scope **Pa,r**—the AC active and reactive power, and the scope **Is**—the transformer primary winding current.

If the pulse width is set as 10° and *Double pulsing* is not marked in the **Synchronized 6-Pulse Generator**, the switches S1 and S2 put in the lower position and α = 30° set, then, after simulation, it will be seen that the load current is practically zero. The rectifier does not start because when the firing pulse is sent to one of the thyristors, the others are blocked, and there is no way for the current to flow. For eliminating this defect, it is possible either to increase the pulse width to the value >60° (e.g., to set it equal to 70°) or to mark the option *Double pulsing*. In both cases, the load current reaches 445 A after simulation. The scope **Pulses** give the opportunity to observe how the form of the pulses changes in these cases.

If α = 95° is set and to simulation executed, the scope **DC** shows that there are the time intervals when the load current is zero; it is an intermittent current (discontinuous conducting) mode. If the simulation is repeated with α = 90°, it will be seen that the load current drops to zero but begins to grow immediately; it is a boundary between continuous and discontinuous conducting modes [2].

If the simulation is carried out with α = 30° and the thyristor currents are observed with help of the scope **T_Currents**, it can be seen that at t = 0.19 s the current begins to move from T1 to T3, but this current transfer does not occur instantly; it takes some time (in this case, 0.5 ms) for the current to flow through both thyristors, and such a phenomenon is called a thyristor commutation [2].

With help of Powergui option *FFT Analysis*, one can observe the current and voltage harmonics at the rectifier output and in the primary transformer winding under various values of α. At the rectifier output, the harmonics orders are proportional to 6, the sixth harmonic has the largest value that, for the voltage, is equal to 190 V under α = 90°. The harmonics of the orders $6n \pm 1$ (i.e., 5, 7, 11, 13, …) can be observed in the transformer input current. When α = 60°, the amplitudes of the above-mentioned four harmonics (percents of the fundamental) are, respectively, 20.5%, 13.5%, 9.1%, and 7.2%.

If the simulation is run with the switch S2 in the upper position, *emf* = −400 V will appear at t = 0.25 s; this *emf* adds to the rectifier voltage. With α = 30°, the load current in this moment increases from 445 to 820 A, and the active power increases from 201 to 356 kW. If the simulation is repeated with S1 in the upper position also, when at t > 0.25 s α changes to 120°, one can see that the current decreases to 106 A and the rectifier voltage becomes negative and is −295 V. If we look, with help of the scope **T_Currents**, at the relative arrangement on the time axes of the T1 current and the phase voltage U_a, it can be seen that at t < 0.25 s the main section of the current curve coincides with the positive section of the voltage curve, and at t > 0.25, with the negative section. It means that in the latter case the power from the DC source returns to the AC grid. The scope **Pa,r** shows that the active power changes from 201 to −30.5 kW. Such a mode of operation is called inversion mode [2]. In order to receive more complete ideas about the special features of the thyristor rectifier operation, it can be recommended to the reader to continue the model investigation using various values of α and *emf*.

The model 4_2 is complicated by adding the second thyristor bridge that is connected back to back with the first one. Such a scheme provides load current flow in both directions. In order to prevent the circulating currents and faults, the firing pulses have to be sent only to one bridge, after the conducting before bridge

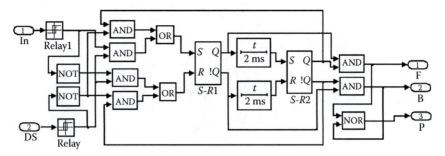

FIGURE 4.16   Diagram of the logical switching unit.

(outgoing bridge) is blocked completely. Thus, the sequence of the bridge control is as follows: current reduces to zero in the outgoing bridge by its moving in the inversion mode or by *emf* increasing up to the value that is larger than the rectified voltage; pulse removed from this bridge after current drops to zero; no-current condition that is necessary for restoring the thyristor blocking possibility; and pulse sent to the second (incoming) bridge. These actions, except of the first, are carried out by the logical switching unit LSU or by the software.

The LSU scheme that is taken from Ref. [3] and is realized in the subsystem **LSU** is shown in Figure 4.16.

Either the signal of the current direction or the search signal that is trying to find the bridge, in which the condition for current flow exists, comes to the input *In*. In model4_2, as such a signal, the grid voltage having the frequency of 50 Hz is used. Either the absolute value of the load current or the rectified current of the transformer secondary winding come to the input DS. If the current value is more than 1%–2% of the rated value (holding current), the logical 0 at the Relay output does not permit for the signal *In* to change the LSU output state. When the current value is small, the logical 1 appears at the output of one of the OR gates (depending on the *In* polarity), and the flip-flop S_R1 switches over. The logical 0 appears immediately at the output of the final gate AND that has just been active, the signal P—a pause—appears, and its count begins by the unit of the delay for turning-on. After the pause ends, the flip-flop $S\text{-}R2$ set in the new state, the logical 1 appears at the output of the other final gate AND, and the signal P is removed. The switching-over comes to the end.

In model4_2, the load current that is measured by the block **Multimeter** comes to the input DS LSU, after the filter with the time constant of 2 ms. Depending on the switch **S** position, either the signal of the requested current polarity Iz or the search signal "Hunt" comes to the input *In*.

Let us set **S** in the lower position Iz; the time vector [0 0.2 0.25] and the output vector [60 150 60] are specified in block **Timer1**, while [0 0.2] and [100 100] (*emf* constant) are specified in block **Timer2**. The block **Step** changes its output from 1 to −1 at $t = 0.2$ s. Thus, the process is simulated; when at $t = 0.2$ s, the current reference sign changes and the rectifier moves in the inversion mode in the same time. Such a situation can arise when Iz is the reference signal for the current regulator and the angle α is defined by this regulator output. After the simulation is carried out, it is seen by the scope **DC** that the current begins to decrease at $t = 0.2$ s and reaches

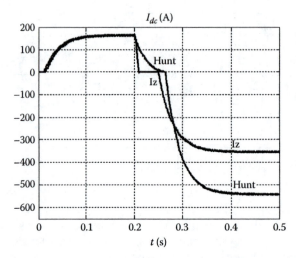

FIGURE 4.17 Current reverse. Iz—by current reference signal; Hunt—by search signal.

zero at $t = 0.242$ s. By the scope **Bridge_Contr**, one can see that the signal F disappears and the signal P appears. The signal P disappears 2 ms later, and the signal B appears. The load current increases in the negative direction (Figure 4.17).

Let us set the switch **S** in the upper position Hunt, the output vector is specified as [60 60 60] in the block **Timer1** (α constant), and the output vector is [100 300] in the block **Timer2**. Thus, at $t = 0.2$ s, the load *emf* increases and reaches a value higher than the rectified voltage that is equal to $550 \cos 60° = 275$ V; by this increase in the *emf*, the current decreases, the firing pulses are blocked at $t = 0.262$ s and the signal P appears, the firing pulses are sent to the another bridge at $t = 0.264$ s, and at $t = 0.265$ s the negative current reaches the holding current value. The switching-over process ends, and the current increases further up to the steady state (Figure 4.17). Signal Hunt < 0 at 0.262–0.272 s.

The simulation of the 12-pulse converters is considered as follows; the converter consists of two 6-pulse bridges, whose supplying voltages have the phase displacement of 30°. The bridges can be connected in series or in parallel. At the first case, a higher output voltage can be received with the same thyristors, and in the second case—a higher output current. In both cases, the load current pulsation frequency increases, which makes its filtration easier. Model4_3 simulates the connection of the bridges in series. The bridge control angles α_1 and α_2 can be the same or different. In the first case, the block **Synchronized 12-Pulse Generator** can be used. Such a variant is considered in the next paragraph. It is supposed in model4_3 that, in general, $\alpha_1 \neq \alpha_2$, so two 6-pulse generators are utilized. The bridges are supplied from the transformer with two secondary windings connected as Y and D1, respectively. The primary winding has Y connection. The winding voltages are the same, as in the previous model, and the power is 10 times as much. It has already been said in the previous paragraph that under such a connection scheme the first generator is synchronized with the phase-to-phase voltages and second generator with the phase ones. The scope **DC** fixes the load current and the bridge rectifier voltages, the scope

Udc—the output converter voltage, the scope **Source**—the transformer primary current and voltage, and the scope **Power**—the active and reactive grid powers.

By using thyristor converters, especially ones of large power, it is desirable to limit the grid reactive power. This power is proportional to $\sin \alpha$. Therefore, it is reasonable to use the angles α that are close to the boundary values $\alpha_{min} \approx 10°–15°$ and $\alpha_{max} = 180° − \alpha_{min}$. It can be achieved by using of the bridge control in turns, whose gist is the following. In order to receive a zero voltage at the output of the 12-pulse rectifier E_d, instead of $\alpha_1 = \alpha_2 = 90°$, $\alpha_1 = \alpha_{min}$, $\alpha_2 = \alpha_{max}$ are set. The bridge rectifier voltages U_{r1}, U_{r2} have the same magnitudes but opposite signs. When it is necessary to increase E_d, α_1 remains equal to α_{min} and α_2 decreases. When $\alpha_2 = 90°$, E_d is equal to a half of the maximum value, and when $\alpha_2 = \alpha_{min}$, it is equal to the maximal value. When it is necessary to decrease E_d (and to move to the inversion mode), α_2 remains equal to α_{max} and α_1 increases. When $\alpha_1 = \alpha_{max}$, E_d has the maximum magnitude and is negative.

The described control principle is realized by the subsystem **Alfa_contr**. In order to have a linear dependence of the rectifier input-output, the arc cosine transformations of the reference signal is used, which can change from −1 to 1; it is limited by the value of ±0.96 for using the inversion mode. The subsystem diagram is shown in Figure 4.18. The input signal U_{ref} that is proportional to E_d defines the setting angles α_1 and α_2. Let, for instance, $U_{ref} = 0$. Both switches Switch1 and Switch2 are in the upper position. Then $\alpha_1 = \arccos (0.96) = 16.3°$, and $\alpha_2 = \arccos (−0.96) = 163.7°$. Let $U_{ref} = 0.5$. The angle α_1 is equal as before to 16.3°, $\alpha_2 = \arccos (2 \times 0.5 − 0.96) = 87.7°$, and

$$E_d = E_{d0}(\cos(16.3°) + \cos(87.7°)) = E_{d0}$$

that is, a half of the maximal voltage under the bridge connection in series. If $U_{ref} = −0.5$, the switches Switch1 and Switch2 are in the lower position, $\alpha_2 = 163.7°$, $\alpha_1 = \arccos (0.96 − 2 \times 0.5) = 92.3°$, and

$$E_d = E_{d0}(\cos(92.3°) + \cos(163.7°)) = − E_{d0}$$

The block characteristics are shown in Figure 4.19.

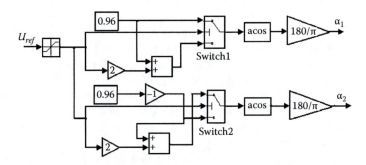

FIGURE 4.18 Firing angle reference in 12-pulse converter.

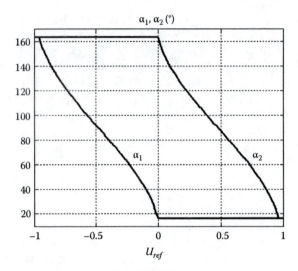

FIGURE 4.19 Characteristic of the subsystem **Alfa_contr**.

First, simulation is executed with **Sw1** and **Sw2** in the upper positions, $\alpha_1 = \alpha_2 = 90°$, and $E_{ref} = -500$ V. Pay attention that the pulse width increases to 35°. It is because the current has to flow through four thyristors; the thyristors of the second bridge are fired at 30° after thyristors of the first bridge, and by firing the second bridge thyristors, the firing pulses must be present at the thyristors of the first bridge too. By the scope **Power**, the active power $P \approx 0$ and the reactive power $S_1 = 5$ Mvar. If simulation is executed with **Sw1** and **Sw2** in the lower positions and under $U_{ref} = 0$, it is seen that $S_2 = 1.4$ Mvar, that is, with the use of control in turns, the reactive power consumption reduces by 3.5 times. If the same is carried out under other firing angles, one can receive: under $\alpha_1 = \alpha_2 = 60°$, $U_{ref} = 0.5$, $E = 0$, it is $P = 2.66$ MW, $S_1 = 4.9$ Mvar, $S_2 = 3.75$ Mvar; under $\alpha_1 = \alpha_2 = 30°$, $U_{ref} = 0.866$, $E = 400$ V, it is $P = 4.75$ MW, $S_1 = 3.02$ Mvar, $S_2 = 2.84$ Mvar; under $\alpha_1 = \alpha_2 = 120°$, $U_{ref} = -0.5$, $E = -600$ V, it is $P = -2.87$ MW, $S_1 = 4.71$ Mvar, $S_2 = 3.37$ Mvar. Thus, this scheme is effective in the cases when the rectifier operates for a long time with small voltages and heavy loads.

With help of Powergui option *FFT Analysis*, one can observe the current harmonics in the transformer primary winding under various angles α. Under symmetric control ($\alpha_1 = \alpha_2$), the harmonics orders are $12n \pm 1$, that is, the fifth and the seventh harmonics that have the lower order and the large amplitude disappears.

The parallel connection of the bridges is investigated in model4_4. The transformer power is 5 MW, and voltages and winding connection are the same, as in the previous model. The rated bridge current is 4.75 kA. The load is 5 mΩ, 0.15 mH. The firing angles $\alpha_1 = \alpha_2$, so only one block **Synchronized 12-Pulse Generator** is used. The block **Multimeter** measures the load current I_{dc} and the bridge currents I_{T1}, I_{T2}. The reactors are set in series to bridges in order to reduce the circulating current. For that purpose, it is reasonable to use one reactor with two windings; because the winding currents flow in the opposite directions, the total magnetizing current is small and the reactor can have a small air gap that decreases the reactor size (in the model, the block **Mutual Inductance** is used). We note that under accepted connection scheme, the current in the second winding flows from right to left; it gives negative values, as has already been said in the description of the block **Multimeter**, so the sign minus is set by Iw2 in the window of the measured parameters of this block. The winding self-inductance is 1 mH, and the mutual inductance is 0.9 mH.

If the simulation is executed with the switch **Sw** in the upper position and under $\alpha = 110°$, $E = 0$, it can be seen by the scope **DC** that the currents both in the bridges and in the load have the time intervals when they are zero; that is, the currents are discontinuous both in the bridges and in the load. If we repeat simulation with $\alpha = 95°$, it will be seen that the bridge currents are discontinuous and the current I_{dc} is continuous. When $\alpha = 80°$, the currents I_{dc}, I_{T1}, I_{T2} are continuous. At that point, the maximum load current is twice as large as the bridge currents; that is, the bridges operate in parallel (Figure 4.20), and the current difference is not more than 1.5%. Thus, the converter has three different operation modes, in each of them the equations that describe its work are different [3]. By the scope **Source**, one can observe the transformer primary current. The reader is recommended to execute simulation under various *emf E* values and firing angles and also to investigate an influence of the reactor winding mutual impedance on current forms, for example, to take it equal to zero (Figure 4.21).

Model4_4 demonstrates the PI current controller operation, $K_p = 0.2 \times 10^{-3}$, $K_i = 7.4 \times 10^{-3}$ (**SW** is in the lower position). For linearization of the converter input–output characteristic, the block Simulink **Trigonometry** with a function acos is set at the controller output. The initial (500 A) and the final (5000 A) current reference values in block I_{ref} correspond to the continuous mode. The reader is recommended to consider the current reference step in the discontinuous modes, under various *emf* values and when there is a decrease in the reactor winding mutual inductance down

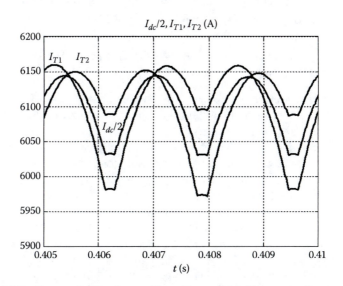

FIGURE 4.20 Currents in 12-pulse rectifier with parallel bridge connection.

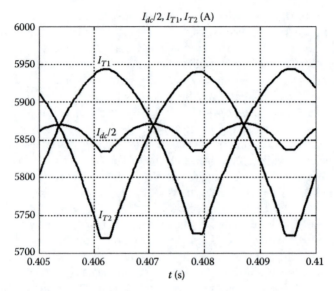

FIGURE 4.21 Currents in 12-pulse rectifier under zero mutual inductance of the reactor windings.

to zero that is equivalent to utilization of two separate reactors. In the last case, one can note that the transients of the brides are different, especially when the controller speed response is large. If, for instance, to increase the controller gains by three times, under step time $t = 0.1$ s the current I_{T1} reaches the maximum 2.5 kA and I_{T2} 3.7 kA, and then the currents converge slowly to their equal steady values. But under step time $t = 0.102$ s, the peak value of I_{T1} is larger. This phenomenon is explained by the fact that the bridges are fired not simultaneously, and a larger current has the bridge, which was fired first after the reference changing [3].

4.4 SIMULATION OF A HIGH-VOLTAGE DIRECT CURRENT ELECTRIC POWER TRANSMISSION SYSTEM

Although electric power transmission is considered in Chapter 6, high-voltage direct current (HVDC) has many peculiarities concerning power electronics; therefore, its simulation is described in this chapter.

Power transmission by DC has often advantages of power transmission by AC, especially in these cases [4]: the underwater cables, connection of two AC systems of different frequencies or when stability problems exist, and transmission of the bulk power over long distances by overhead lines. HVDC can have monopolar or bipolar configuration. In the first case, the return conductor is the earth or the metallic grounded conductor (Figure 4.22); in the second case, there are two conductors of different polarity (Figure 4.23). In this case, each terminal has two converters with equal nominal voltages, whose middle point is grounded. When one of the converters fails, the transmission line remains in operation as a monopolar. Every converter shown in Figures 4.22 and 4.23 can consist of several equal converters connected in

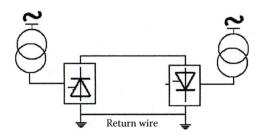

FIGURE 4.22   Monopolar HVDC electric power transmission system.

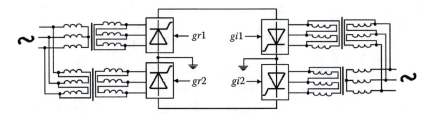

FIGURE 4.23   Bipolar HVDC electric power transmission system.

series with an aim to increase the voltage transmission level in order to diminish line current under the same transferred power and, consequently, to reduce the conductor size. Since the line current pulsation causes additional loses, the smoothing filters are set in the DC circuits. Since the use of thyristor converters generates the high harmonics in AC networks, the high harmonics filters are set in AC circuits.

The transformers powering the converters often have the taps (see **Multi-Winding Transformer** in Chapter 2) in order to control the supplying voltage. The thyristor converters consume from the network a reactive power, whose value depends on the firing angle, and for its compensation the capacitor banks are utilized.

Power-flow control in HVDC is carried out by controlling both the converter firing angles and the selected transformer taps. The control system's main problem is to prevent the large current oscillation in the line. Thus, the current controller is the main one. For the line-loss minimization, it is reasonable to have as high a voltage as possible, keeping in mind the necessity to provide the margins in the emergency modes; it means that rectifier firing angle α (lag angle) and the inverter lead angle β (the retard limit angle [2]) must be at a minimum. The operation under $\alpha < 5°$–$6°$ is not permissible usually; thus, in the nominal condition $\alpha_{min} = 15°$–$20°$ it is possible to control the power flow. The inverter angle β is set at about 15°. The following designations are accepted: $\beta = 180° - \alpha$, μ is the commutation (overlap) angle, and $\gamma = \beta - \mu$ is the turn-off angle.

Under normal operation, the constant current is kept by control of the rectifier firing angle α, and the inverter angle γ is kept constant. However, if α reaches minimum, further current control becomes impossible. In order to prevent such a situation, the inverter has the current regulator too, whose reference is by 8%–10% less than the rectifier reference current. Under normal operation, the rectifier defines the

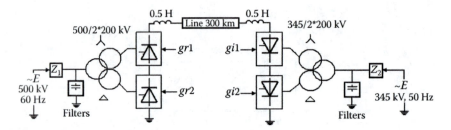

FIGURE 4.24 Main circuits of the modeled HVDC electric power transmission system.

current and the inverter defines the voltage. When α_{min} is reached, the functions of the rectifier and the inverter change mutually: the rectifier defines the voltage and the inverter the current. Instead of the angle γ control, the inverter can have the input voltage control.

The current regulator includes additional features: the limitation of line maximal current, the limitation of the line minimal current, and decreasing of the maximal current while there is a decrease in the AC voltage or the rectifier voltage. In order to prevent moving the inverter to the rectifier mode by reversing the power transmission direction during any system malfunction, the firing inverter angle α cannot be less than 90°, usually, 95°–110°. Since the HVDC line has to transfer the definite power P, the current reference is $I_{ref} = P/V_d$, and V_d is the measured DC voltage.

In SimPowerSystems the demonstrational model power_hvdc12pulse.mdl is included that shows all these special features of HVDC.

The main circuit structure is shown in Figure 4.24. The purpose of the line is to connect the networks with the frequencies of 60 and of 50 Hz; the distance between networks is 300 km. The maximal transfer power is 1000 MW (500 kV, 2 kA). For converting AC voltage into DC voltage and for reverse converting, 2 three-phase bridge converters connected in series are used in each line end. The bridge powering is carried out from the transformers with two secondary windings, whose voltages have the phase shift of 30°, so that 12-pulse rectifier scheme is realized. The primary transformer currents have the harmonics of the following orders: 11th, 13th, 23rd, 25th, and so on. The parallel capacitance and capacitance-inductance filters are set at the primary (step-up) side of the transformers for harmonic mitigation and for reactive power compensation. The transformers are supplied from AC sources 500 kV, 60 Hz and 345 kV, 50 Hz. The impedances Z_1 and Z_2 model AC systems. The smoothing reactors with inductance of 0.5 H are set at the beginning and at the end of the line. The converter control is fulfilled by sending of the firing pulses $gr1$, $gr2$ to the rectifier and $gi1$, $gi2$ to the inverter.

Each of the subsystems **Rectifier** and **Inverter** comprises one three-phase three-winding transformer and two thyristor bridges connected in series. The rectifier transformer has the power of 1200 MVA, the voltages of 500/200/200 kV, and the inverter transformer has a power of 1200 MVA, the voltages of 345/200/200 kV. The scheme does not model the transformer taps, but they are supposed: the rated primary voltages decrease (and, respectively, the secondary voltages increase under the same primary voltages) by factors 0.9 for the rectifier and 0.96 for the inverter.

The measurement of the rectified current and its reduction to pu under the base 2000 A is fulfilled in these subsystems also.

The transmission line is modeled by the block **Distributed Parameters Line** with parameters, under frequency of 60 Hz, $R = 0.015 \, \Omega/\text{km}$, $L = 0.792 \times 10^{-3} \, \text{H/km}$, $C = 14.4 \times 10^{-9} \, \text{F/km}$. Each subsystem of the filters consists of four filters: two filters of a large quality ($Q = 100$) tuned for the 11th and the 13th harmonics, the high-pass filter of small quality with large bandwidth, and the capacitor bank; the reactive power of the each filter is 150 Mvar, so that the subsystem reactive power is 600 Mvar. The units for fault simulation in the AC and DC circuits are provided.

For HVDC control, the special subsystems are developed in SimPowerSystems:

- HVDC Discrete 12-Pulse Firing control
- Discrete HVDC Controller
- Discrete 12-Pulse HVDC Control
- Discrete Gamma (γ) Measurement

The first subsystem is intended for generation of the firing pulses of two thyristor bridges, whose supplying voltages have the phase shift of 30°. It consists of the block **Discrete Synchronized 12-Pulse Generator** and of the block **PLL** that produces the synchronizing signals from the phase voltages, fulfilling their smoothing. These synchronizing signals come to the inputs A, B, C of the Generator, instead of the phase voltages. The block **PLL** contains the block **three-phase discrete PLL**; this block generates the signal ωt that is converted afterwards in three signals $\sin(\omega t)$, $\sin(\omega t - 2\pi/3)$, $\sin(\omega t + 2\pi/3)$. The subsystem input signals are the reference angle α, signal of blocking, and three-phase voltages.

The subsystem **Discrete HVDC Controller** contains the controllers and units for fabrication of the rectifier or inverter firing angles α in HVDC. It can be tuned for the operation with the rectifier, and in this case, it has a current control, or for the operation with the inverter, and in this case, besides current controller, it contains the angle γ controller and the voltage controller or one of them. The subsystem input signals are as follows: the line voltage V_{dL} and current I_d in pu, the reference values I_{d_ref}, V_{d_ref} for the controllers (in pu too), the logical signal of blocking (1—blocking, 0—work), the logical signal of the reference α setting (is used for protection), the value γ (gamma_mess) measured by the subsystem **Discrete Gamma Measurement**, and the reference of the angle changing D_alpha for inversion margin increasing during transients. The subsystem has three outputs: α, the actual current reference taking into account limitations, the code of the operation mode, namely: 0—blocking, 1—current control, 2—voltage control, 3—limitation α_{min}, 4—limitation α_{max}, 5—forcing or constant α, and 6—γ control. The subsystem dialog box depends on which mode of the operation is selected. Besides, the subsystem contains the unit for current reference limitation as a function of DC voltage. This subsystem is not described here in details because it is rather intricate and specific and its description takes a lot of place (see the section Thyristor-Based HVDC Link in Ref. [1]).

The subsystem **Discrete 12-Pulse HVDC Control** is a simplified version that unites two previous subsystems. It comprises **Discrete Synchronized 12-pulse Generator** (without PLL), the current and voltage regulators, the scheme for current reference limiting as a function of the voltage, and some others. The subsystem inputs: *Vabc* is the transformer primary phase voltages in pu, V_{dL} is the DC voltage in pu, I_d is the line current in pu, I_{d_ref}, V_{d_ref} are the current and voltage reference values, *Block* is the logical signal of blocking, and *ForcedDelay* is the logical signal of setting of the specified value α (usually, for protection). The subsystem outputs: *PulsesY*—six firing pulses for the bridge that is connected to the secondary winding with scheme Y, *PulsesD*—six firing pulses for the bridge that is connected to the secondary winding with scheme Delta, α—firing angle, $I_{d\_ref\_lim}$—actual current reference, taking into account limitations, *Mode*—the mode of operation. This block is not described too.

The subsystem **Discrete Gamma Measurement** measures for each of six thyristors of the inverter bridge the time from the moment, when current decays to zero, to the following moment of the natural firing. The frequency signal F_{req} is used for converting this time to the angle γ. The current threshold for thyristor zero-current detection is specified in the dialog box (0.001 pu is recommended).

The block diagram of the subsystem is shown in Figure 4.25. Six null-detectors Imp1 fix the instants, when the current of each thyristor crosses the threshold level. At that point, a corresponding *R-S* flip-flop is set. The flip-flop is reset, when the voltages that correspond to the natural firing points cross zero that are fixed by null-detectors Imp2. The scheme for transformation of the primary-phase voltages in the synchronizing signals depends on the secondary winding scheme. During changes in the flip-flop states, the differentiators D generate the short pulses that fix the values of the signals at inputs *In* at these moments in the discrete sample and hold units S/H. Thus, the unit S/H1 fixes the time, when the current decays, and the unit S/H2 fixes the following natural firing time for the same thyristor. At the output of the summer Sum1, the value γ expressed in seconds is formed that is stored in S/H3. In the unit "Min," the minimum of six values is determined; that value after multiplication by $360F_{req}$ gives value γ (degree). At the same time, the summer Sum2 adds all six time values; after multiplication by $60F_{req}$, this sum gives an average value γ (degree).

The subsystem **Master Control** generates the current references for the rectifier and for the inverter and controls start and stop of power transmission. The start begins with a command *Start* (at $t = 0.02$ s). At that point, the flip-flop *R-S* is set, and **Switch** in **Ramping Unit** switches over and applies the signal *Min_ref1* (accepted of 0.1) to the unit **Start-Stop Ramp**. Current reference rises to this value at the rate

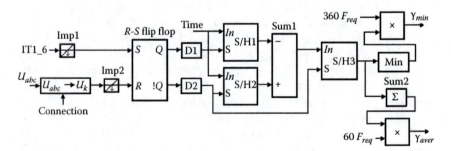

FIGURE 4.25 Circuit for angle γ measurement.

of 0.33 pu/s. At $t = 0.4$ s the additional signal appears that increases at the rate of 5 (pu/s), so that the current reference acquires the value of 1 at $t = 0.58$ s and stays on this level for 1 s. At $t = 1.4$ s the output of **Ramping Unit** begins to decrease with the rate of 5 pu/s, at $t = 1.6$ s the signal *Stop* appears that resets the flip-flop, giving zero reference. At the same time, the signals *Stop_R* and *Stop_I* are generated for blocking of the rectifier and of the inverter.

The possibility of reference step changing with help of a timer is provided in the subsystem. The reference decreases by 0.2 pu in the interval [0.7 s, 0.8 s]. In order to activate this signal, the toggle switch **Ref_Current_Step** has to be put in the upper position.

In the subsystem **Data Acquisition** the scopes are collected, which give the opportunity to observe system operation.

Experiments with this model include observation of the normal mode of the line energization, energy transmission, changing of the transmission power, the line de-energization, and investigation of the faults in DC and AC circuits. Some of the results are given in above-mentioned section of Ref. [1].

The model4_5 is the simplified version of the HVDC. It does not model the faults. Its diagram is shown in Figure 4.26. The HVDC is intended for transferring of power of 500 MVA (500 kV, 1000 A) at a distance of 450 km, from the source with the voltage of 220 kV and with the frequency of 60 Hz to the network 345 kV, 50 Hz. The units **Filters** contain three filters each one: the capacitor bank, the double-tuned for the 11th and the 13th harmonic filter, and the high-pass harmonic filter that is tuned for the 24th harmonic. **R_Control** and **I_Control** subsystems contain the subsystems **12-Pulse Firing Control**; the former has the controller of the rectifier current I_{dR}, and the latter has the controller of the voltage V_{dLI} in the end of the line (Inverter input). Besides, the subsystem **Gamma Measurement** is set in the latter for observation.

The scopes in the subsystem Data Acquisition show the voltages and currents of the sources and of the transformers (before and after Filters), voltages and currents in the DC circuit, the angles α of the rectifier and of the inverter, and the mean value of the angle γ.

The following process is simulated: the I_{dR} and V_{dLI} references are given at $t = 0.1$ s; simultaneously, signals *Block* are removed from the subsystems of firing control. The transmission line begins to transfer electrical energy. At $t = 0.8$ s the I_{dR} reference is set zero, the current in the line decreases; at $t = 1$ s the rectifier firing

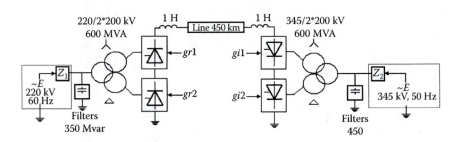

FIGURE 4.26 Diagram of the simplified HVDC.

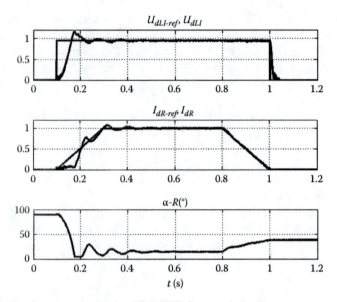

FIGURE 4.27 Transients in the simplified HVDC transmission line.

pulses are blocked and, simultaneously, the V_{dLI} reference goes to zero; at $t = 1.2$ s the inverter firing pulses are blocked.

The transient process is shown in Figure 4.27. The angle α is 12° for the rectifier, 140° for the inverter, and the angle γ is 30°; therefore, the inverter has some margin. The Figures 4.28 and 4.29 show the voltages and currents in the sources, at the input of the rectifier and at the output of the inverter (i.e., before and after filters). It is seen that the sources have sinusoidal currents and voltages.

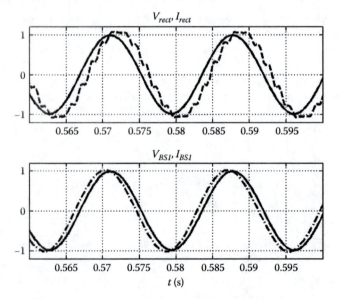

FIGURE 4.28 Voltages and currents at the input of the HVDC transmission line.

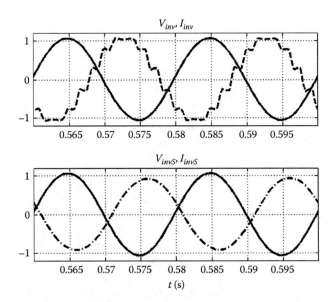

FIGURE 4.29 Voltages and currents at the output of the HVDC transmission line.

4.5 SIMULATION OF CONVERTERS WITH FORCED-COMMUTATED DEVICES

In this paragraph, the model examples of the electronics sets and the use of forced-commutated devices are considered. Model4_6 is a model of the two-level VSI, whose diagram is shown in Figure 4.4b. VSI receives the input DC voltage from the diode rectifier. Both the inverter and the rectifier are modeled by the blocks **Universal Bridge**. For smoothing of the voltage pulsation in DC link, the capacitor $C = 5\,\text{mF}$ is used. The rectifier is powered from the source 380 V, 50 Hz through the transformer 50 kVA, 380/600 V.

The inverter supplies the active load 50 kW, 380 V, 25 Hz. The filtration of the load current high harmonics is fulfilled by L-C filter with the parameters per phase 2 mH, 132 µF. The main circuits are shown in Figure 4.30. Two possibilities for gate pulse generation are provided: by the block **Discrete PWM Generator** and by the

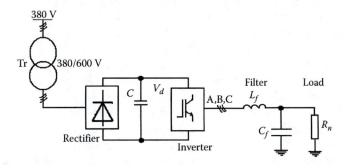

FIGURE 4.30 Two-level VSI.

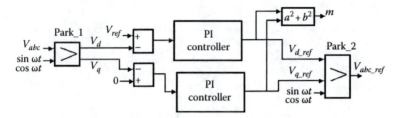

FIGURE 4.31   Controller of VSI output voltage.

block **Discrete SV PWM Generator**. The running block is selected by the switch
Sw in the subsystem **Control**. The switching frequency is 2000 Hz.

Two operation modes are provided: In an open system and in a closed one, the
choice is carried out by the switch **Sw1**. In the first case, the reference signals for gate
pulses are generated by the block **Three-Phase Programmable Source** (Chapter 3),
and in the second—with help of the load voltage controller, whose diagram is given
in Figure 4.31. The three-phase load voltage is transformed in DC voltages V_d, V_q
in the rotating reference frame, with help of the block **Park_1**. The needed signals
sinωt and cosωt, $\omega = 2\pi 25$ are produced by the block **Discrete Virtual PLL**. The
reference V_d is equal to 1 pu, and the reference V_q is 0. The reference signals for the
inverter V_{d_ref}, V_{q_ref} are formed at the outputs of PI controllers, these signals are
transformed into the three-phase signal frequency of 25 Hz with help of the block of
the inverse Park's transformation. For the block **Discrete SV PWM Generator**, this
three-phase signal is transformed in the components V_α and V_β. Simultaneously, the
signal m—the module of the voltage space vector—is formed.

A number of units are provided for process observation and its characteristic estimation. The four-axes Scope **Voltages** records DC inverter voltage, the output inverter
voltage, the load voltage, and the value m. The first trace of the scope **Current** shows
the inverter input current, and the second one—the currents of the transistors 1 and 2
of the phase A. The block **Counter** from the library *Signal Processing Blockset* is
used for count of the inverter switching number. Counter sums the pulses turning on
both transistors 1 and 2. The block **State** is designated for the inverter state observation. It has been mentioned already that the inverter states are indicated by the three-
bit code according to the phase A B C; at that, the bit is 1, when the top transistor in
this phase conducts. In the block, this binary code converts into decimal numbers
from 0 to 7. Each inverter state corresponds to the definite space vector, then, in the
counter-clockwise direction, the sequence of the states is: 4, 6, 2, 3, 1, 5 [2]; besides,
there are two zero states, 0 and 7. In addition, the pulse signal frequency of 2000 Hz
is produced to make pulse observation easier. Besides, the signal that determines
the number of the sector of 60° width, in which the wanted inverter state vector is
placed, is formed; it is needed when the space vector modulation is used.

Let us begin with simulation. The switch **Sw** is set in the lower and **Sw1** in the upper
positions that correspond to the use of PWM with the triangular carrier waveform in
the open system. In the block **Three-Phase Programmable Source**, the amplitude
of 0.8, zero phase, the frequency of 25 Hz are specified, and harmonic generation is
absent. Once simulation is complete, the first trace of the scope **Voltages** shows that

when power is turned on, the capacitor voltage rises to 870 V and afterwards reaches the steady state of 745 V. The special measures that are not considered here are used to eliminate voltage overshoot. The second trace shows that the inverter output voltage is the rectangular oscillations with frequency of 25 Hz filled with high-frequency oscillations having the amplitude of 745 V and varying durations. The third trace shows that, owing to the filter, the output voltage is near to the sinusoid with an amplitude of 515 V. From (4.4) it follows that the maximum voltage in six-step mode is 2.45 × 745/π = 581 V; in our case, the output voltage amplitude has to be of 0.785 × 581 × 1.41 × 0.8 = 514 V, which takes place actually. The first trace of the scope **Currents** shows that the inverter input current has a shape of pulses, and the second trace gives the opportunity to observe changing of the currents through transistors and diodes during generation of the output signal (the negative currents flow through diodes). The number of switching measured by **Count** is 1200 for 0.3 s, that is, 4000 s⁻¹. The scope **State** shows the inverter state changes depending on the phase reference signal. The option *Powergui/FFT Analysis* gives the opportunity to determine THD of the inverter voltage and the load current. Select Structure: *Voltages*, Input: *Vab_inverter*, Fundamental frequency: 25 Hz. After executing the command *Display*, one can find that the inverter voltage contains a lot of high harmonics (up to 90%). However, it is not only THD value that is important but also the harmonic frequency allocation because it is easier to filter the higher frequencies (by the filter or by the load inductance). To select Input: V_{ab_load}, one can see that $THD = 1.93\%$ only.

If we set the amplitude equal to 1 in the block **Three-Phase Programmable Source**, the scope **Voltages** shows $V_{dc} = 714$ V; the amplitude of $V_{ab\_load} = 615$ V corresponds to the said calculation: 515 × 714/(0.8 × 745) = 616 V.

It is interesting to observe system behavior under the modulation factor $m > 1$. For example, if we set the amplitude = 1.1 in **Three-Phase Programmable Source**, it can be found by the scope **Voltages** that the amplitude of $V_{ab\_load} \approx 672$ V. It is seen by the scope **Currents** that the time intervals exist, when one of the transistors is turned on permanently and conducts the current, and another is blocked; that is, the number of switching decreases. By the scope **Count**, the number of switching is 2900 s⁻¹. With the help of *FFT Analysis* option, one can find that the amplitude of the load voltage first harmonic is 649 V and THD increases to 2.72%.

In order to expand the upper level of the output voltage, it is possible to add the sinusoidal signal of the zero sequence to the amplitude of 0.25 and triple frequency to the reference three-phase signal [2]. For that purpose, in the dialog box of the block **Three-Phase Programmable Source**, the option *Harmonic generation* has to be marked; the additional window that appears shows the vector of the parameters for harmonic *A* is set as [3 0.25 0 0] and the amplitude of the harmonic *B* is 0. As a result of the simulation, one can find that the first harmonic amplitude of the load voltage increases to 665 V and THD decreases to 1.62% with the number of switching of 3500 s⁻¹.

For investigating the block **Discrete SV PWM Generator**, the switch Sw is set in the upper position, the amplitude is 0.4 in the block **Three-Phase Programmable Source**, and the option *Harmonic generation* is eliminated. The option *Pattern #1* is chosen in the block **SV PWM**. One can see by the scope **Voltages** after the simulation is executed that $V_{dc} = 783$ V; with the help of the *FFT Analysis* option, the amplitude of the fundamental harmonic $V_{ab\_load} = 319.3$ V, *THD* 2.61%, and the switching number is $4046\,s^{-1}$. By the scope **State**, one can watch the inverter states, when the requested voltage vector lies in various sectors. In sector 0, for example, the sequence of the states is: 0–1–3–7–3–1–0, and in sector 1: 0–2–3–7–3–2–0.

If the simulation is repeated with the option *Pattern #2*, it can be found that the amplitude of the fundamental harmonic is $V_{ab\_load} = 308\,V$, *THD* = 3.6%, and the switching number is $2750\,s^{-1}$. Hence, the decrease in the switching number is attained at the expense of deterioration of the voltage shape. With an increase in the output voltage, the difference in these two modulation methods reduces. So under $m = 0.8$, THD in V_{ab_load} is 1.34% and 1.7% and the switching number is 4000 and $3673\,s^{-1}$ for *Pattern #1* and *Pattern #2*, respectively, and under $m = 1$, 1.87%, $4000\,s^{-1}$ and 1.37%, $3390\,s^{-1}$, respectively.

For simulation of the closed system, the switch **Sw1** is set in the lower position. In the block **Vref**, the initial value is set as 0.7, the final value as 1.1, and the time of changing as $0.2\,s$. With **Sw** in the low position, the resulting processes are shown in Figure 4.32.

With help of this model, new possibilities of the block **Universal Bridge** modeling for VSI can be studied. If, for example, having copied the model under other name, to use the structure *Switching-function based VSC* with the option *Measurement "Voltages"* in this block, to select two voltages U_{ab}, U_{bc} in the dialog box of the block **Multimeter** on the left, to move them on the right and to place above (in order not to change the de-multiplexer scheme), one can see after simulation that processes are alike but saving in the simulation time is small in this case.

In model4_6a, the structure *Average-model based VSC* is used. One can see that the subsystem **Control** becomes simpler essentially. If the switch **Sw1** is set in the upper position and the simulation carried out, it is seen that the inverter output voltage is sinusoidal with the amplitude of $515\,V$, whereas the first harmonic amplitude in model4_6 is $509.4\,V$ under the same condition. The simulation time decreases drastically. Hence, this structure can be used successfully in cases where the processes within inverter are not of interest.

Model4_7 is a modification of the previous one used for study of the three-level VSI. Instead of the block **Universal Bridge**, the block **Three-Level Bridge** with

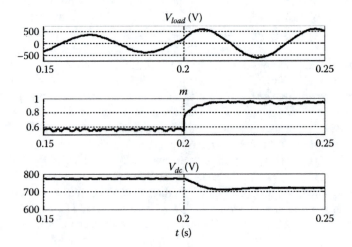

FIGURE 4.32 Processes in VSI under voltage control.

IGBT/Diode is used, two capacitors are connected in series in the DC link, and their common point is connected to the neutral block point. Two capacitor voltages come to the fourth input of the scope **Voltages**, the scope **Currents** records the currents in the transistors and diodes of the phase A and the VSI phase A current and the modulating signal for the phase A. The gate pulses are generated by the block **Discrete Three-Phase PWM Generator** with the external modulating signal.

If the simulation is carried out with amplitude of the three-phase source equal to 1, it can be seen by the scope **Voltages** that DC voltage is 715 V and the capacitor voltages oscillate with the triple frequency and with swing of 30 V. The VSI output voltage is nearer to sinusoid as for two-level VSI. With help of the option *Powergui/ FFT Analysis*, one can find that the fundamental harmonic amplitude of the VSI output voltage V_{ab_inv} = 618.1 V, *THD* = 35.4%, and for V_{ab_load}, *THD* = 1.11%.

Observing the currents in different model elements with the help of the scope **Currents**, one can get an in-depth familiarity with the processes in the three-level inverter. Thus, it is reasonable to use the inductive load. For that purpose, the capacitive reactive power Q_c has to be set as zero in the block **LC Filter** (the small active power, for example, P = 10 W has to be set), and the inductance of series circuit has to be increased to 20 mH. In Figure 4.33, the section of the resulting processes is shown for the amplitude of three-phase source of 0.5. One can distinguish

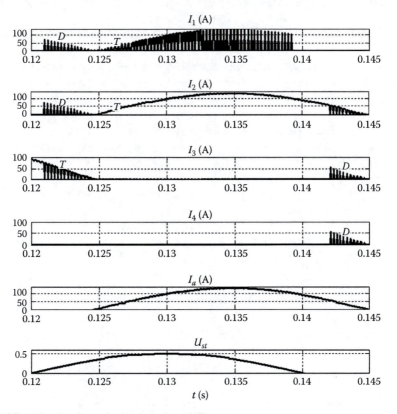

FIGURE 4.33 Processes in the three-level VSI.

three segments: 0.12—0.125 s, 0.125—0.14 s, and 0.14—0.145 s. On the first one, the phase A current $I_a < 0$, and the modulating signal (MS), crossing 0, becomes >0. The current flows from the load to the rectifier terminal "+" either through antiparallel diodes D1, D2 or through T3 (and the neutral clamping diode) to the neutral N. On the second segment, both I_A and MS > 0. The current flows either from the source terminal "+" through T1, T2, or from the point N through the neutral clamping diode and T2. On the third segment, I_A > 0 and MS < 0. The current flows either from the source terminal "−" through diodes D3, D4, or from the point N through the neutral clamping diode and T2.

Using several three-level bridges, the AC voltage with specified amplitude and frequency and with very small THD can be produced. The model4_8 comprises two equal subsystems, which are supplied from the bipolar source ±9650 V with the common point. Each subsystem consists of 2 three-level bridges having the common supply, with GTO-thyristors. The bridges feed **Three-Phase Transformer 12 Terminals**. The bridges are controlled by the block **Discrete Three-Phase PWM Generator** with the external modulating signal. The generator outputs are in antiphase. Each primary transformer winding is connected to the terminals of the same name of the both inverters, the secondary windings of the both subsystems are connected in series, and, on the whole, as wye with an insulated neutral. The load is resistive, 2.4 MW under the voltage of 750 V. The scope **Voltage_inv** fixes the inverter phase-to-phase voltages; the scope **Voltage_load** records the phase-to-phase voltage of the first transformer (on the load) and at the output of the second one.

Let us set the amplitudes of three-phase sources of the modulating voltages $m = 0.9$, the phase 0 in the first subsystem, and change the phase of the modulating voltage in the second subsystem, measuring, with help of the option *Powergui/FFT Analysis*, the amplitude of the voltage fundamental harmonic on the load and THD of this voltage. The dependence received is given in Table 4.1.

Hence, it is reasonable to accept the modulating voltage phase in the second subsystem equal to 30°. To compensate for the voltage decrease, take $m = 0.9 \times 1.07$ in the second subsystem. The simulated processes are shown in Figures 4.34 and 4.35. It is seen that, although the voltage shapes differ from a sinusoid in various parts of the scheme, the load voltage is in fact sinusoidal: under the fundamental harmonic amplitude 841 V, *THD* = 1.35%; moreover, most significant harmonic is the 79th with the amplitude of 1.25%. From the harmonics of the low order, the fifth one is most significant with the amplitude of 0.13%.

TABLE 4.1

Dependence of the Load Voltage Fundamental Harmonic and THD

Phase (°)	0	7.5	15	22.5	30	37.5
Voltage (V)	840	838	833	824	812	796
THD (%)	2.19	2.11	1.93	1.69	1.52	1.51

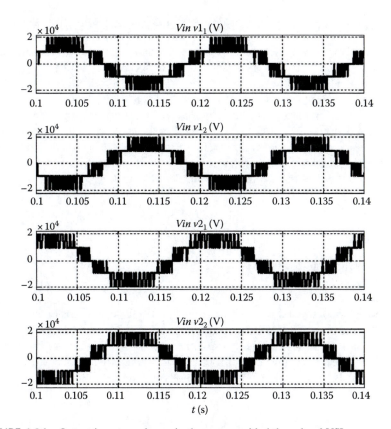

FIGURE 4.34 Output inverter voltages in the system with 4 three-level VSI.

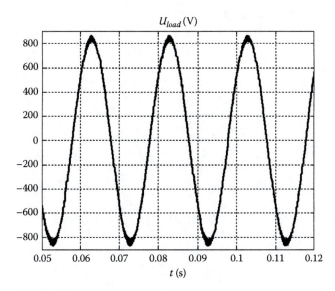

FIGURE 4.35 Load voltage in the system with 4 three-level VSI.

4.6 CASCADED H-BRIDGE MULTILEVEL INVERTER SIMULATION

The cascaded multilevel inverters consisting of the several single-phase bridges connected in series (Figure 4.36) are developed and are investigated intensively at present. They are called multicell too.

If the transistors (GTO, IGCT thyristors) 1 and 4 of the bridge **a1** are fired, "contribution" of this bridge to the load voltage is $+E$; if the transistors 2 and 3 are fired, "contribution" is $-E$; and if 1 and 3 or 2 and 4 are fired, "contribution" is 0. Hence, the load voltage changes from $-nE$ to nE, where n is the number of the bridges in a phase, in steps of E; that is, $2n + 1$ levels can be fabricated, including the zero level. RMS of the output phase-to-phase voltage is

$$U_l = 2kmnE \tag{4.9}$$

where
   m is the modulation factor
   $k = 0.612$

The merits of these inverters are as follows: the possibility to have the high-level output voltage with the limited voltage rating of one bridge, reducing THD of the output voltage, and a modularity of the construction that makes production and service lighter. The three-phase inverters of such a construction are produced by Siemens ("ROBICON Perfect Harmony") for the phase-to-phase voltages 2.3–13.8 kV,

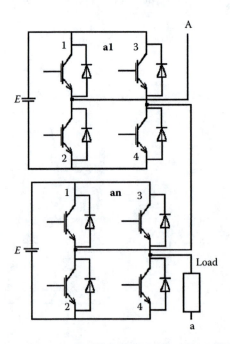

FIGURE 4.36 Diagram of the cascaded H-bridge multilevel inverter phase.

power to 31 MVA; under the voltage 2.3 kV $n = 3$, 4.16 kV $n = 4$, 7.2 kV $n = 6$. The disadvantage of this scheme is the need to have many isolated DC sources.

In the said inverters of Siemens, the sources are three-phase diode rectifiers that are supplied from the secondary windings of the multiwinding transformer; in order to reduce the harmonic content in the primary current, the voltages of its secondary windings are shifted by $60°/n$ (Figure 4.37). It is seen that the transformer turns out to be rather intricate. Besides, an inversion mode is not possible. There are elaborations of the schemes with the single-phase rectifiers and with the active three- and single-phase rectifiers permitting an inversion. In Figure 4.38a, the diagram of one of the nine blocks shown in Figure 4.37 is given, and in Figure 4.38b, the diagram of the block with three-phase active rectifier is shown.

The number of the secondary windings can be reduced by using an induction motor (IM) with six terminals, whose model is considered in Chapter 5. In such a case, each three-phase winding can supply one block in each phase (Figure 4.39).

The so-called asymmetric inverters form a special class of the converters; they have unequal DC sources in a phase. Let, for example, the DC voltage be E in block **a1**, $3E$ in block **a2** and $9E$ in block **a3** (in general, $3^k E$, $k = 0, 1, ..., n - 1$). Then the phase voltage can change from $-13E$ to $13E$ in step of E; that is, there are 27 (in general, 3^n) voltage levels with the maximum value of $(3^n - 1)/2E$. The block **a3** is the main one, and the others—subsidiary, or auxiliary. A lot of output voltage levels give the possibility to have this voltage with very small THD; the drawback of the scheme is that the modularity of the converter is lost, since the block of each level has to be designed separately. Besides, even in the absence of inversion mode in

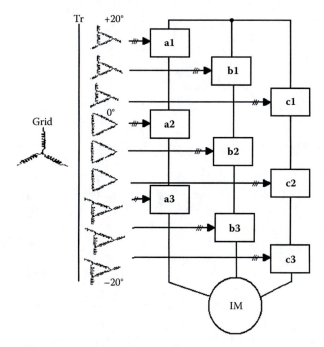

FIGURE 4.37 Cascaded multilevel inverter without inversion.

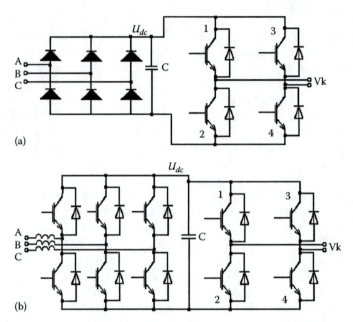

FIGURE 4.38 Diagram of a block of the cascaded multilevel inverter. (a) without inversion, (b) with inversion.

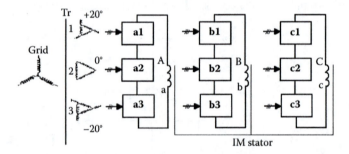

FIGURE 4.39 Cascaded H-bridge multilevel inverter and IM without common point.

general, the subsidiary blocks must permit the operation in the inversion mode. Let us assume, for example, it is necessary to have the output voltage equal to $7E$; then the block **a3** has to provide $+9E$, the block **a2** $-3E$, and the block **a1** $+E$; that is, the block **a2** has to operate in the inversion mode.

If the output frequency changes in the small bound, the output transformers can be used, whose primary windings have equal number of turns and the turns of the secondary windings are in the relationship w^3 (Figure 4.40). All the blocks have the same transistor voltage rating. This scheme provides 3^n voltage level too; furthermore, the inverter maximum voltage can be selected at will, at the expense of the turn number choice. Moreover, one DC source for all three phases can be utilized.

The gate pulses for the inverters according to Figures 4.37 and 4.38 are generated by the comparison of the input reference signal (sinusoidal or with addition of the

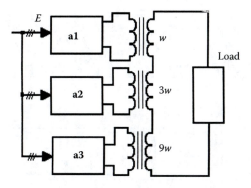

FIGURE 4.40 Diagram of a phase of the cascaded inverter with transformer.

third harmonic) with the triangular carrier waveform having frequency F_c; at that, both the single-polar and bipolar carrier waveform can be used. Each block is controlled independently, and triangular carrier waveforms of the blocks are shifted one after the other by $2\pi/n$, in order to provide the minimal THD value for the inverter phase-to-phase voltage. The pulse generation for these cases is shown in Figures 4.41 and 4.42, respectively. In the first case, the phase voltage harmonics are concentrated around the frequency nF_c; in the second case—around the frequency $2nF_c$ which makes its filtration easier, but the average switching frequency is twice as much. The phase reference signals are shifted one after the other by $120°$.

In the SimPowerSystems version 5.2 and in the following, there is a demonstration model of the cascaded inverter under $n = 5$ *Five-Cell Multi-Level Converter*.

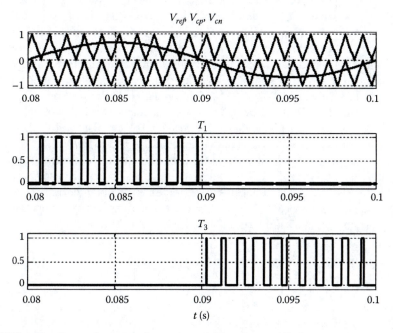

FIGURE 4.41 Gate pulse generating under a single-polar modulation.

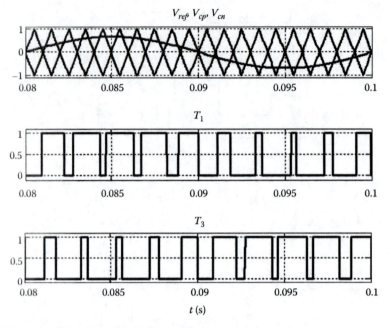

FIGURE 4.42   Gate pulse generation by a bipolar modulation.

The inverter is supplied from the network of 6.6 kV and is loaded with the passive load $R = 10\ \Omega$, $L = 1$ mH per a phase. The inverter model consists of three similar subsystems—phases, and each of them contains five blocks according to the diagram in Figure 4.38a, which are powered from the phase-shifting transformers with the relative phase displacement 12°. The transformer secondary voltage is 1320 V, so that $E \approx 1.35 \times 1320 = 1780$ V (simulation shows that, taking into account voltage drops on the scheme elements, $E \approx 1700$ V). The source of the reference modulating voltage with the frequency of 60 Hz has the modulation factor $m = 0.8$, so that one can expect $U_l = 2 \times 0.612 \times 0.8 \times 5 \times 1700 = 8323$ V. The modulation of the each bridge is carried out by the subsystem **PWM** realizing the bipolar modulation with $F_c = 600$ Hz. After completing simulation, applying the option Powergut/FFT to the curve of the output voltage (the third trace of the scope), it can be found that rms is $11{,}730/1.41 = 8{,}320$ V that is very close to the calculation and the harmonics are concentrated around the frequency $2n \times 600 = 6$ kHz. It is worth mentioning that simulation runs rather slow even for the passive load: when using *Accelerator*, the relation of the simulation time to the model time is about 500. More complicated models are considered in Chapters 5 and 6 for simulation of the electrical drives and STATCOM sets.

4.7   FOUR-LEVEL INVERTER WITH "FLYING" CAPACITOR SIMULATION

Flying Capacitor Inverters receive such a name because the terminals of the capacitors that are connected to the inverter switches do not connect directly to the source terminals and the capacitor voltages are determined by the processes in the inverter

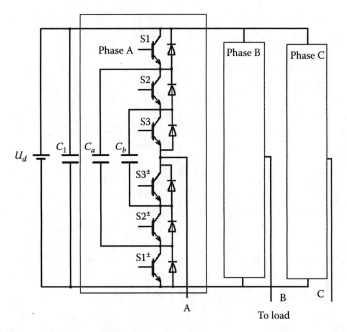

FIGURE 4.43 Inverter with "flying" capacitor.

during the switching-over of its switches. Such an inverter with four levels is produced by one manufacturer. The diagram of this inverter is shown in Figure 4.43.

The diagram of one phase shown in Figure 4.44 gives the opportunity to understand the operation of such an inverter better. Adjacent switches (S1 and S1*, S2 and S2*, S3 and S3*) do not conduct at the same time and are not blocked at the same time; that is, one of the two switches conducts always. The voltages applied to the blocked switches are

$$(U_1)_{off} = U_d - U_{ca}, \quad (U_2)_{off} = U_{ca} - U_{cb}, \quad (U_3)_{off} = U_{cb} \tag{4.10}$$

Since half of all switches are blocked, having summed up these equalities, we find that the voltage applied to the blocked switch is $U_d/3$, and the capacitor voltages are

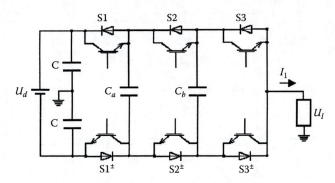

FIGURE 4.44 Diagram of a phase of the inverter with "flying" capacitor.

$U_{ca} = 2U_d/3$, $U_{cb} = U_d/3$. For receiving these voltages, the relative times of turning-on of each switch have to be equal, and the phase shift between turning-on of the next switches has to be 120°.

This scheme (Figure 4.43) provides the output voltages of $\pm U_d/2$ and $\pm U_d/6$. For example, during switching of S1, S2, S3, $U_l = U_d/2$, and switching of S1*, S2*, S3*, $U_l = -U_d/2$. The voltage $U_l = U_d/6$ can be acquired by three ways: S1 = 1, S2 = 1, S3 = 0 when $U_l = U_d/2 - U_d/3 = U_d/6$; S1 = 1, S2 = 0, S3 = 1 when $U_l = U_d/2 - 2U_d/3 + U_d/3 = U_d/6$; S1 = 0, S2 = 1, S3 = 1 when $U_l = -U_d/2 + 2U_d/3 = U_d/6$. The redundant states can be used to balance the voltages across the capacitors.

Generation of the gate pulses for one phase is shown in Figure 4.45. The system has a feature of self-balancing, when the afore-mentioned voltage values, being set beforehand, are restored after their deviations caused by changing of the supply voltage and the load parameters. The self-balancing process is defined mainly by the current high harmonics that appear during inverter switch operation. Therefore, the result and the speed of this process depend on these harmonics, which can be filtered by the inductive load. Besides, these harmonics can be small, when the modulation factor is small too. The auxiliary R-L-C-circuit, whose resonant frequency is equal

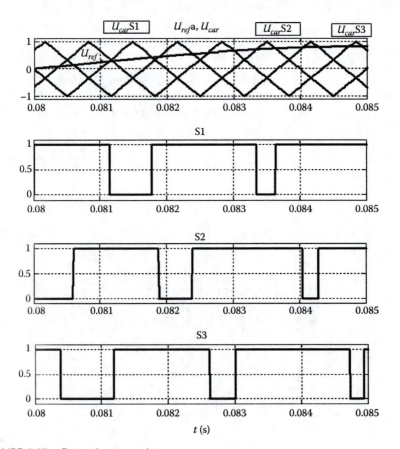

FIGURE 4.45 Gate pulse generation.

to the switching frequency, can be set at the inverter output. There are a number of papers in which the self-balancing process is considered analytically, but the expressions received are intricate and are derived with a number of assumptions, so a fit simulation is a necessary step during inverter design [5].

The voltage controllers can be used in order to keep up the voltages across the "flying" capacitors and to accelerate the self-balancing process. The equations for the voltages are [6]

$$\frac{dV_{c1}}{dt} = \frac{I_1}{C_1}(u_1 - u_2), \quad \frac{dV_{c2}}{dt} = \frac{I_1}{C_2}(u_2 - u_3) \tag{4.11}$$

where u_1, u_2, u_3 are equal to 1 or 0; they are the states of the switches S1, S2, S3. If the average values are considered for the switching period, the average values of the control signals that define these states, for instance, the voltage controller outputs, can be interpreted as u_1, u_2, u_3. It follows from these equations that control of dV_{c1} and dV_{c2} is reasonably easy to achieve by using u_1 and u_3, respectively, because in this case, the controllers are decoupled. From these equations, it follows also that the action direction and its intensity have to be dependent on the load phase current value and its sign; moreover, under a small current value, the control gets ineffective and has to be turned off. Hence, the control circuit shown in Figure 4.46 is used. The devices Cp1, Cp2, Cp3 compare the reference signal U_{ref} with the triangular carrier waveforms that are shifted by 120°, as shown in Figure 4.45. For the switches

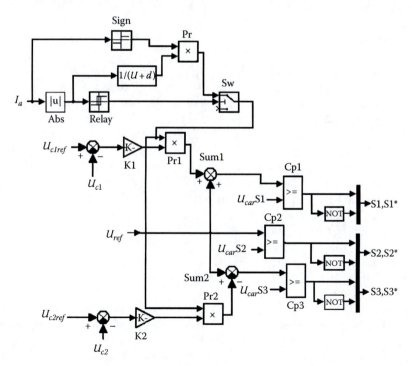

FIGURE 4.46   Inverter voltage control diagram.

S1 and S3, the correcting signals are added to U_{ref}. These signals are formed as the outputs of the capacitor voltage proportional controllers with gains K_1 and K_2 multiplied by the output of the device, whose output signal is inversely proportional to the load current, taking into account its sign (a small signal d is added in order to avoid division by 0). When the load phase current is small, the controller is turned off by the switch Sw.

The model4_9 is intended for study of the main system features. Each of three subsystems—phases is realized according to the diagram in Figure 4.44. The intermediate single-phase transformers with the diode rectifiers are set in the subsystems for the preliminary charge of the "flying" capacitors to the voltages of 1400 and 700 V. Under choice of the transformers that are supplied from the secondary winding of the main transformer, it is taken into account that the main transformer operates in the no-load conditions during the charge mode. After elapsing of the charge time (1.2 s), the transformers disconnect from the capacitors. The scope **V_C** records the capacitor voltages.

The network of 6 kV, through three-winding transformer with secondary voltages of 750 V and two diode rectifiers connected in series, powers the inverter. The resistor 1 Ω that is shorted-out during the operation by the breaker **B1** 0.1 s later after turning on AC voltage serves for the limitation of the capacitor starting currents. Pay attention to the fact that the midpoint of the input capacitors can be connected to ground with the breaker **B2**.

The three-phase load has the parameters 3 Ω, 0.3 H. The auxiliary R-L-C-circuit with the resonant frequency of 2 kHz can be connected by the breaker **B3**. The scope **Load** records the phase and the phase-to-phase load voltages and the load current, the rectifier capacitor voltages, the scopes **IGBT_V** and **IGBT_I** fix the voltages, and currents of the phase A switches S1, S2, S3.

The subsystem **Control** contains three-phase source of the PWM modulation signal with the possibility to add the third harmonic, the blocks for generation of the triangular carrier waveforms, and the devices for comparison of these signals that form the gate pulses for each phase according to Figure 4.45. The carrier waveform shift is realized by the delay blocks with the delay time of $0.33/F_c$ where F_c is the switching frequency. It is accepted $F_c = 2$ kH (see option *File/Model Properties/ Callbacks*). During the capacitor charging, all the gate pulses are blocked with help of **Switch**. With help of the switch **Mode_Switch**, the various modes of the reference signal can be chosen. In the upper position, it can be the constant amplitude and frequency, the constant frequency and changing frequency, the constant amplitude and changing frequency, and in the lower position, the simultaneous changing of the amplitude and frequency can be specified. By switches **Mswitch** and **Mswitch1**, the outputs of the inverter capacitor voltage controllers are connected. The reference values are $2/3U_d$ and $1/3U_d$ where U_d is the sum of the voltages across the rectifier capacitors.

If the simulation is carried out without voltage controllers (**Mswitch** and **Mswitch1** are in the right positions), with **Mode_Switch** in the upper position, with the amplitude of 1 and the frequency of 50 Hz for **Three-Phase Programmable Source**, without third harmonic (Gain2 = 0), when the breaker **B2** is closed, the plots of the phase and phase-to-phase load voltages and load current are obtained as

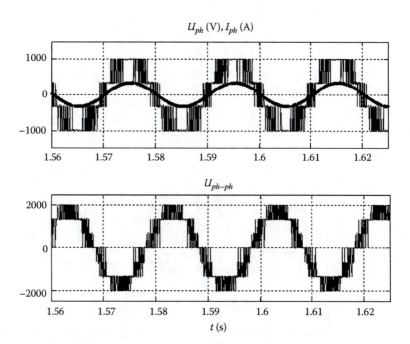

U_{ph} (V), I_{ph} (A)

U_{ph-ph}

t (s)

FIGURE 4.47 Phase and phase-to-phase inverter voltages.

shown in Figure 4.47. It is seen that the phase voltage has four levels and the phase-to-phase has seven levels. Applying the option *Powergui/FFT*, one can find that rms of the phase-to-phase fundamental harmonic (under $U_d = 2000$ V) is 1226 V, that is, the same; as for the two-level inverter, $U_{ph-ph} = 0.612\ U_d$. The following experiments are carried out with **B2** open.

Let us make the next setting: **Mode_Switch** is put in the lower position, and the amplitude of 1.155 and frequency of 5 Hz are set in the **Three-Phase Programmable Source1**; in this block, variation of the frequencies at the rate of 45 Hz/s, beginning from $t = 1.2$ s, are selected (i.e., at $t = 2.2$ s the frequency will be equal to 50 Hz). The initial condition for **Integrator** is set to 0.1, and the step value of the block **Step** is 0.9 at $t = 1.2$ s; that is, the reference amplitude reaches 1.155 with frequency of 50 Hz at $t = 2.2$ s. With **Gain2** = 0.5, the third harmonic is added. The decrease in the amplitude by step of 10% at $t = 3$ s is specified in the source of 6 kV. Figure 4.48 shows the load phase-to-phase voltage and the load current resulting from simulation with **B3** open. The voltages across the inverter capacitors are shown in Figure 4.49a. If the simulation is repeated with **B3** closed, the voltages across the inverter capacitors change as it is shown in Figure 4.49b. One can see that the auxiliary circuit improves essentially the voltage control across the "flying" capacitors. Voltage changing across the inverter switches is shown in Figure 4.50.

Let us put **Mode_Switch** in the upper position, set the amplitude of 0.1 and the frequency of 5 Hz in the block **Three-Phase Programmable Source**, and carry out simulation. The plots of the voltages across the "flying" capacitors are shown in Figure 4.51a. It is seen that under small modulation factor, the voltages recover slowly. If the simulation is repeated with active voltage controllers (the switches

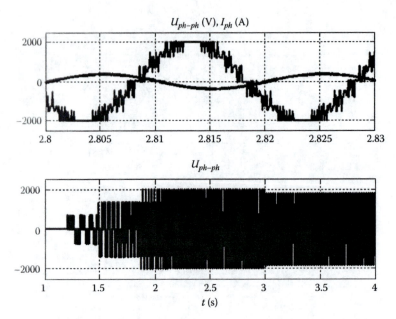

FIGURE 4.48   Processes at the inverter output when there is a change in supplying voltage amplitude and output voltage frequency.

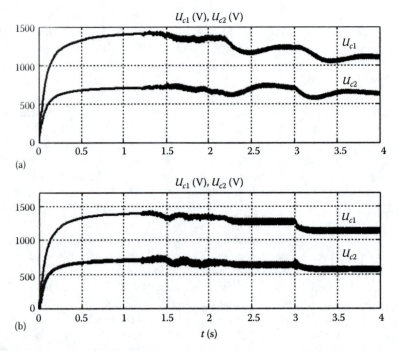

FIGURE 4.49   Voltage variations across the inverter capacitors. (a) without auxiliary circuit, (b) with auxiliary R-L-C-circuit.

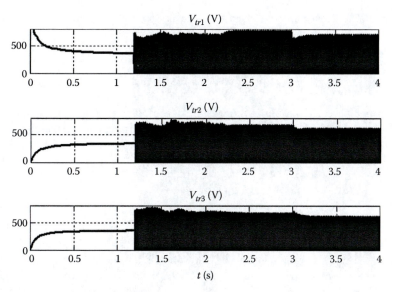

FIGURE 4.50 Voltages across the inverter switches.

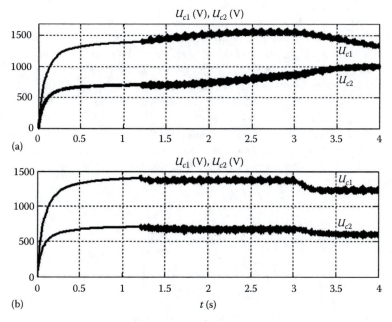

FIGURE 4.51 Capacitor voltage variations for the small modulation factor. (a) without voltage control, (b) with voltage control.

Mswitch and **Mswitch1** are moved to the left), the plots that are shown in Figure 4.51b appear. It is seen that the speed of response increases perceptibly.

In Chapter 5 use of this inverter for IM control is demonstrated.

4.8 SIMULATION OF Z-SOURCE CONVERTERS

In the usual inverters, the output voltage cannot exceed, with some factor, the voltage in DC link U_d. For example, rms of the fundamental harmonic of the phase-to-phase voltage is

$$U_{lleff} = \frac{k\sqrt{6}U_d}{\pi} \tag{4.12}$$

which is achieved by the use of PWM with the triangular carrier waveform, $k = 0.785$, on the linear section of the system characteristic (i.e., $U_{lleff} = 0.612U_d$), and addition of the third harmonic or the use of space vector modulation, $k = 0.907$.

In a number of utilizations, especially in the plants with the renewable energy sources (wind generators, fuel cells, photovoltaic systems, and others), the output voltage of these sources can change over a wide range; therefore, in order to provide at the output of the inverter (that is set at the system output and converts DC voltage into AC voltage) the demanded voltage value, the boost DC–DC converter is set in DC inverter link. In a recently developed inverter, which contains in its DC link the complex circuit of the inductive and capacitive reactances, the so-called Z-source converter, they make it possible to have the output voltage that exceeds the input voltage essentially and to further exclude the boost DC–DC converter.

The main circuit diagram of Z-source converter is shown in Figure 4.52. The circuit of the inverter usually consists of six switches. Under inverter control, the same eight states are used (six active and two zero) as in the usual two-level inverter. The difference lies in the use of the additional zero state, when both switches in a phase (or in all phases) conduct simultaneously. In the usual inverter, in such a condition, the short-circuit takes place and the switches fail because the capacitor in DC link maintains the constant voltage. In the considered scheme, the current rate of increase in the short circuit is limited by the inductances, and this current can be turned off. When turning-on switches, the currents in inductances rise because of the capacitor voltages, the voltages across the inductances increase, the diode D cuts off, and

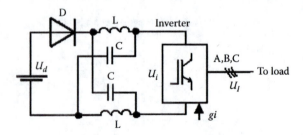

FIGURE 4.52 Z-source converter.

$U_L = U_c$, $U_i = 0$. When the switches are turned off (disconnecting of the short circuit), $U_L = U_d - U_c$, $U_i = 2U_c - U_d$. Let the turning-on time be T_0 and turning-off be T_1. The average value of the voltage across the inductance for one period is zero, then

$$U_c = \frac{T_1}{(T_1 - T_0)} U_d \tag{4.13}$$

The average value of the inverter input voltage is

$$U_{icp} = \left(\frac{2T_1}{T_1 - T_0} - 1 \right) \frac{U_d T_1}{T_1 + T_0} = \frac{U_d T_1}{T_1 - T_0} = U_c \tag{4.14}$$

Since the short-circuit mode does not necessarily affect inverter operation during the use of active states, it is realized, when the usual zero states are to be used, that $U_i = 0$ on the interval T_0 does not affect the resulting voltage applied to the phases of the load. Then [7], if $T_1 = T - T_0$ where T is PWM period,

$$U_{iekv} = 2U_c - U_d = \left[\frac{2T_1}{T_1 - T_0} - 1 \right] U_d = \frac{U_d}{1 - 2t_0} = BU_d$$

$$t_0 = \frac{T_0}{T}, \quad B = \frac{1}{1 - 2t_0} > 1 \tag{4.15}$$

Hence, when using the PWM with the triangular carrier waveform, one can write

$$U_{1eff} = 0.612 BMU_d \tag{4.16}$$

where $0 < M < 1$ is the modulation factor, that is, an amplitude of PWM input sinusoidal signal. Remember that triangular carrier waveform U_{car} changes in the limits ± 1.

There are different ways to realize the considered scheme. In the simplest case, it is fulfilled the following way. Let the maximum value M be accepted equal to $M_{max} < 1$. Then, in the intervals, when $|U_{car}| > M_{max}$, the zero inverter states are realized. Let us choose the constant value $D \geq M_{max}$. When $|U_{car}| > D$, the gate pulses are sent to all inverter switches, carrying out the short-circuit mode. Then

$$B = \frac{1}{1 - 2(1 - D)} = \frac{1}{2D - 1} \tag{4.17}$$

The maximum achieved output inverter voltage is

$$U_{max} = \frac{0.612 M_{max} U_d}{2M_{max} - 1} \tag{4.18}$$

Let, for example, be $U_d = 150\,V$ and the wanted output voltage be $U_{1eff} = 208\,V$. Solving (4.18), one can find $M_{max} = 0.64$. Note that in the usual inverter, when $M = 1$,

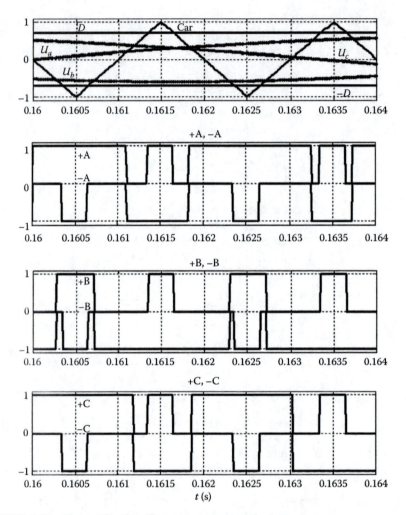

FIGURE 4.53   Generation of the Z-source converter control signals.

it would be only $U_{lleff} = 91.8\,\text{V}$. Figure 4.53 explains this method of the gate pulse generation (for clarity the pulses are directed downwards for IGBT connected to the bus "minus").

When using this method for short-circuit-mode forming, the number of times one switch is switched over (on-off) is $N = 4f\,\text{s}^{-1}$ where f is the frequency of the carrier waveform. The short-circuit mode can be achieved by switching of the switches of only one inverter phase, for example, one whose input control signal is maximum in this time and $N = (2 + 2/3)f = 2.67f$ in this case.

The model4_10 helps to study the operation of such an inverter. The reactor inductance is $160\,\mu\text{H}$, and the capacitor capacitance is $1\,\text{mF}$. It is seen in the subsystem **Control** that the gate pulses are sent to the inverter either when the input modulating signals cross the carrier waveform ($f = 10\,\text{kHz}$) or when the absolute value of carrier waveform crosses the constant level D. The subsystem **Control1** is more complicated. It contains the subsystem **Max** that, with help of the comparison units and the logical gates, forms the logical 1 at the input of the **AND** gate that corresponds to the maximum absolute value of three modulating signals Y_{ref}. The upper part of the scheme picks out the maximal positive signal, and the lower part—the maximum by absolute value of the negative signal—depending on the sign of the carrier waveform in this moment, the signal that takes part in gate pulse generation goes at the scheme output through **Switch**. Thus, the short-circuit pulses are generated with help of one inverter leg. Figure 4.54 demonstrates this method of pulse generation.

The block **Scope** fixes the inverter phase-to-phase output voltages, the inverter input voltage, and the load current. The scope **Ulf** shows the phase-to-phase voltage after the filter with a large bandwidth. The scope **Pulses** records the gate pulses of the phase A. Two first traces of **Scope1** show the reactor and capacitor currents, the next

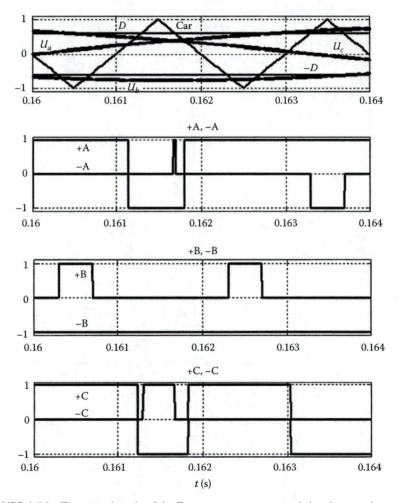

FIGURE 4.54 The second mode of the Z-source converter control signal generation.

two—the voltages across these elements and the fifth—shows the current through diode. The block **Counter** and the scope **Count** compute the number of the state changing of the inverter switch **+A** for the time determined by the block **Timer1**. The control system contains the circuits for adding of the third harmonic.

Let us carry out the simulation under the following conditions: The amplitude is equal to $M_{max} = D = 0.64$ in **Three-Phase Programmable Source**, and the third harmonic is not used (Gain = 0 in subsystem **3_Harm**), $U_d = 150$ V. **Switch** can be in any position. Processing with the help of the option *Powergui/FFT Analysis* the signals recorded on the first trace of the **Scope** or by the scope **Ulf**, one find that rms of the phase-to-phase voltage is 208.4 V, which corresponds to computation fully. The second trace of **Scope** shows that the inverter input voltage is the rectangular waveform in the range of 0–535... 541 V. The reactor, capacitor, and diode currents are shown in Figure 4.55. At the moment A, the inverter switches turn on, the reactor current increases, the voltage across it rises quickly, and the diode cuts off. At the moment B, the voltage across the reactor begins to decrease, and the conductance of the diode recovers. The capacitor voltage slight changes; it is 340.5–342 V.

Utilization of the third harmonic leads to a decrease of the PWM input signal by 1.155 times for the same inverter voltage, giving the possibility to lessen the value D and, consequently, to increase the scheme gain factor B. Then, for achievement of the wanted voltage, the modulation factor M is found from Equation 4.19

$$U_{lleff} = \frac{0.612 M U_d}{1.732 M - 1} \tag{4.19}$$

and $D = M/1.155$. In order to have $U_{lleff} = 208$ V under $U_d = 150$ V, it has to be $M = 0.775$, $D = 0.67$.

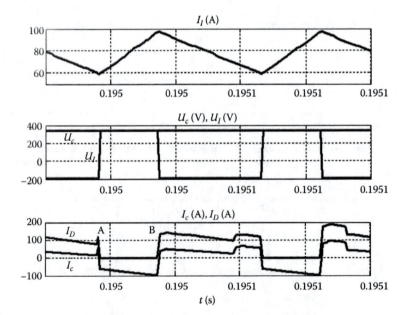

FIGURE 4.55   Currents and voltages across the inductor, capacitor, diode.

If simulation is carried out when the amplitude of **Three-Phase Programmable Source** is equal to 0.755, Gain = 0.5 in the subsystem **3_Harm**, and $D = 0.67$, it can be found by using the option *Powergui/FFT Analysis* that rms value of the phase-to-phase voltage is 208 V. By using the block **Scope 1**, it can be found that the capacitor voltage is 293–294 V, and by using the block **Scope**, the maximum inverter input voltage is 440 V that is much less compared to the voltage achieved without the use of the third harmonic and makes it possible to lower the allowable voltage for scheme devices.

As for the switching number N, one can find, fulfilling the said actions in the both positions of **Switch**, that in the lower position, when the short circuit mode is made by all inverter legs, with the interval of record equal to 0.1 s, $N = 3820$, and in the upper position, and when only one leg forms the short circuit mode, $N = 2500$. According to said relations, it has to be 4000 and 2670, respectively, which is sufficiently close to the actual values.

The output voltage control of Z-source converter is not a trivial problem because its gain factor G that is equal to $M/(1.732M - 1)$ under the short-circuit mode and equal to M out of this mode, changes as shown in Figure 4.56. If, for example, it is necessary to keep up the output voltage value when there is an increase in U_d, beginning from the small values, at first, the value M has to increase by nonlinear law (AB section) until the value $M = 1.155$ is reached, and afterwards to decrease M (BO section). Besides, there are two channels that affect the output voltage: M and D; the condition $D > M/1.155$ has to take place in the section AB, and $D > 1$ in the section BO.

One of the possible ways to control the output voltage is to combine the feed-forward and feedback control methods. If U_d can be measured or estimated, then the values M and D can be calculated by the said formulas; since these calculations are not precise, the output voltage controller is set in addition, which

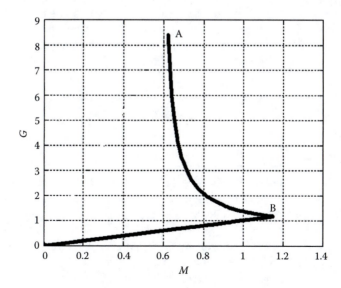

FIGURE 4.56   Dependence of the Z-source converter gain on the modulation factor.

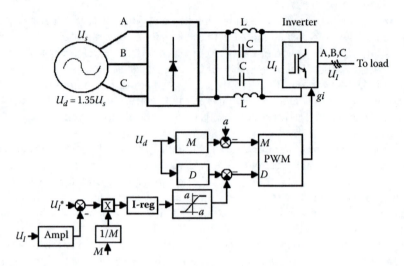

FIGURE 4.57 Z-source converter voltage control (the first variant).

impacts the factor D and operates in the bound $\pm a$. In order not to break the condition $D > M/1.155$, the value M computed according to the value U_d is reduced by the value a.

The proper block diagram is shown in Figure 4.57. The three-phase source of the variable voltage (and, perhaps, of the frequency) models, for instance, a synchronous generator with permanent magnets that is driven either by the wind turbine, or by the rotated wheels of the railway car for supplying its auxiliaries, or with help of the power take-off system from the main propulsion drive of the railway transport or vessels. It is supposed that the source output voltage can be measured or calculated. As an alternative, the diode rectifier output voltage can be measured in the time intervals, when the short-circuit mode is not used. If U_s is rms of the source output voltage, then $U_d = 1.35U_s$. The values M and D are computed with using of this quantity. Remember that while using the third harmonic

$$M = \frac{x}{1.732x - 1}, \quad x = \frac{1.634U_{leff}}{U_d}, \quad D = M/1.55$$

where U_{leff} is the demanded rms of the inverter output phase-to-phase voltage.

The computed value D is corrected by the integral controller of the output voltage **I-reg**, whose output is limited by the value a, and the calculated value M is decreased by a. For linearization of the D loop transfer function, the loop gain varies as a function of M.

When U_d, rising, exceeds the value of $1.41U_{leff}$, then the control system structure changes, $D = 1$, and the PWM input is determined by the output of the integral-proportional voltage controller (not shown in the figure).

The model4_11 simulates the described system. The power source is **Three-Phase Programmable Voltage Source**; the phase-to phase voltages of the diode bridge and its rectified voltage are measured by the block **Multimeter2**. The block **Z_in** models the source internal impedance; the capacitors C_k compensate for overvoltages that can arise during the commutation of diodes and during the short-circuit mode [8].

The signal that is equal to the maximum, by absolute value, of the instantaneous values of the phase-to-phase voltages that come to the diode bridge is used as U_d; this measuring scheme imitates three-phase diode bridge, for which $U_d = 1.35U_s$; this voltage is smoothed by the second-order filter with bandwidth of 50 Hz. The block **Multimeter** measures the output and input inverter voltages and the load current. The block **Multimeter3** measures the load-phase voltages that are filtered by the first-order filter with the time constant T_f (its value is specified in the option *Callbacks* and is accepted equal to 250 μs). This signal is used in the control system.

PWM control signals are formed in the subsystem **Control**. The block **Fcn** calculates the parameter M that, after division by 1.155 in the block **Gain1**, gives the computed value D. In the block **Sum1**, this value D is adjusted by the output of the integral controller **PI** having the limitation ±0.05. The computed value M is decreased by 0.06 in the **Sum4**. In the block **Product**, this new value M is multiplied by the three-phase signal having the amplitude of 1 and the frequency of 50 Hz that is sent to the block **PWM** through the circuits for the third harmonic addition.

The input signal of the D controller is computed as a difference of the wanted load-phase voltage $380(\sqrt{2}/\sqrt{3}) = 310\,\text{V}$ and its actual value that is measured with help of the block **Three-Phase Positive-Sequence Fundamental Value**. The linearization of the loop transfer function is realized by the block **Product1**. Furthermore, simulation showed an essential sensitivity to the errors of the numerical integration and to the controller output oscillation. In order to prevent these effects, the small step of the numerical integration is chosen, and the block **Dead Zone** with the small width (±2 V, that is, less than 1% of the controlled value) is set in the controller error circuit.

When signal U_d reaches the value of 537 V, **Switch** is put into the upper position; at this point, the output of the integral-proportional controller of the load voltage is connected to the input of PWM, whereas the value D is not less than 1 that eliminates the inverter short-circuit mode.

In Figure 4.58 the results of simulation are shown; when, in **Three-Phase Programmable Voltage Source**, the amplitude is set to 100 V, the rate of the amplitude changing is 1 pu/s (i.e., 100 V/s) and the time of variation is from 1 to 6 s. It is seen that when U_d changes by six times, the load voltage is nearly constant. The value D reaches 1 at $t = 4$ s; at this point, the controller structure changes and any further action affects only the modulation factor. The slight oscillations of the output voltage that one can observe in the figure are caused by the higher speed of the supply voltage change accepted during simulation for lessening its time; they disappear in steady states.

Together with the said combined control method, the error-closing control with two controllers can be used; at that, the first one controls the load voltage, affecting M. As for the second controller that affects D, it can be either the controller of the inverter input peak voltage U_{iekv} [9] or the controller of the Z-circuit capacitor

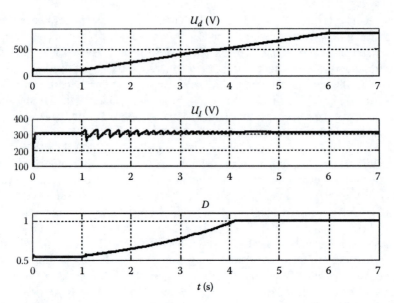

FIGURE 4.58 Input and output Z-source converter voltages with the first variant of voltage control.

voltage U_c [10]. In the first case, the controlled quantity is directly proportional to the output inverter voltage for a given modulation factor and is equivalent to the DC-link voltage of the usual inverter. However, its measurement is rather problematic because the inverter input voltage pulsates from 0 up to $U_{iekv} = U_{imax}$; besides, the pulsation of the rectified supplying voltage is affected. The voltage U_c changes much less, but under its constant value, the voltage U_{iekv} varies inversely with D. Only the second case is considered here.

The block diagram of the control system is shown in Figure 4.59. The control of the short-circuit time (short-through duty cycle) is carried out by the integral-proportional

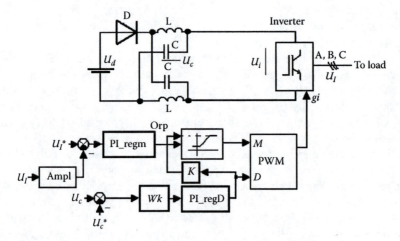

FIGURE 4.59 Z-source converter voltage control (the second variant).

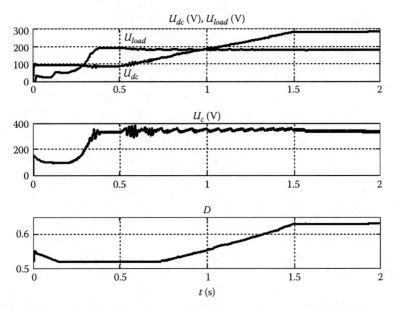

FIGURE 4.60 Input and output Z-source converter voltages with the second variant of voltage control.

controller PI_regD, which, at its input, has the phase lead compensator, according to the recommendation in Ref. [10].

The output voltage control is fulfilled by the integral-proportional controller PI_regm; furthermore, its output signal (modulation factor) is limited by the present value D with the factor $K = 1$, if PWM does not use an addition of the third harmonic, and $K = 1.155$, if this addition is used.

In order to realize the system under consideration, **Switch** is set into the upper position. The controller circuits in the subsystem **Control2** are closed some time after the model starts, when the process in the Z-circuit reaches the steady state. The scope **U-Scope** (in the subsystem) fixes the capacitor voltage and the amplitude of the load-phase voltage.

The capacitor reference voltage is 330 V. The rms of the supply phase-to-phase voltage rises from 70 V up to 210 V for 1 s. The process is shown in Figure 4.60. It is seen that during triple U_d change, the output voltage V_{load} and the capacitor voltage U_c are kept constant.

4.9 SIMULATION OF RESONANT INVERTERS

All the discussed inverters use the so-called hard switching, when the inverter switches change their states, when the currents flow through them, and when the voltage is applied across them. At that point, although either the switch current (if the switch is blocked) or the switching voltage (if the switch conducts) are equal to zero after switching-over, during switching the current and the voltage are not equal to zero, and an integral of their product defines an energy evolved during switching. The average value of this energy for 1 s (switching losses) increases proportional to the switching frequency and limits its value. At the same time, an increase in switching frequency is often desirable, because the size of the system-reactive elements reduces and the load current form improves. Besides, with the switching frequency increasing, the frequency of the acoustic noises that are caused by the force of the high harmonics on the system magnet components increases too, and these noises are made inaudible for a human ear (more than 18 kHz). During the "hard switching," because of the large values dv/dt and di/dt, a lot of electromagnetic interference are generated that can influence the operation of the electrical equipment connected to the same sources or situated nearby.

An increase in the switching frequency can be achieved by using the so-called swift switching, when the voltage or the current (or both) of the switching device are equal to zero. In order to attain such a switching, the resonant inverters can be used, in which a resonant cycle is built; during this cycle, the voltage or the current of the switching device crosses zero, and just at this moment, turning-over of the device takes place. Among the possible schemes, only VSI with a resonance in the DC link, in which the DC voltage at the inverter input, under the constant supplying voltage, fluctuates from the maximum down to zero values and switching takes place when DC voltage reaches zero level, is considered here.

Such inverters had been considered in many works in the late 1980s—early 1990s of the past century, but afterwards, certain skepticism prevailed regarding them, which was caused by the absence of the concrete sets and of the difficulties with its design. However, it appeared later that such features of these inverters, including the small value of THD in the load current and decreased electromagnet interference and acoustic noises perceived by a human ear, are very important in some special situations that promoted further development of these inverters [11,12].

One of the possible schemes of the resonant inverter model is shown in Figure 4.61. The inductor L_r and the capacitor C_r having a small inductance and a

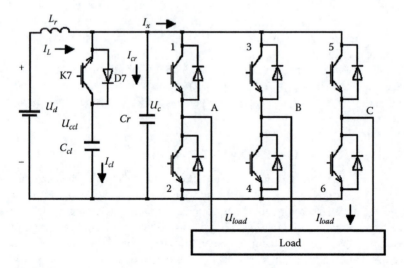

FIGURE 4.61 Resonant inverter.

small capacitance, respectively, form the resonant circuit. The transistor K7 with the antiparallel diode D7 and the capacitor C_{cl} form the clamping circuit.

Suppose first that the clamping circuit is absent. Under resonant oscillations, the voltage U_c decreases from some peak value down to zero and tries to attain a negative value, but it is impossible because of the antiparallel diodes of the inverter bridge. At that moment, it is possible to change the state of the switches "soft," with zero voltage across them, that is, without losses. However, before that it is necessary to store some amount of energy in L_r that has to provide the following resonant cycle. For this, all six inverter switches are turned on for a short time, the inductor is connected directly to DC source, and the inductor current rises rapidly. When this current reaches the requested value I_{L0}, one switch in each inverter arm is blocked according to demanded modulation mode, and the new resonant cycle begins. If $I_a = I_L - I_x$ is designated where the load current I_x is supposed to be constant, the current I_a varies as

$$L_r \frac{dI_a}{dt} + \frac{1}{C_r}\int I_a dt = U_d$$

(4.20)

$$I_a(0) = I_{a0} = I_L(0) - I_x, \quad U_c(0) = 0$$

received by integration

$$I_a = \frac{U_d}{Z_0}\sin \omega t + I_{a0}\cos \omega t, \quad Z_0 = \sqrt{\frac{L_r}{C_r}}, \quad \omega = \frac{1}{\sqrt{L_r C_r}}$$

(4.21)

$$U_c = U_d(1 - \cos \omega t) + I_{a0}Z_0 \sin \omega t$$

(4.22)

This voltage reaches its peak value

$$U_{cm} = U_d\left(1 + \frac{U_d}{D}\right) + \frac{I_{a0}^2 Z_0^2}{D}$$

(4.23)

$$D = \sqrt{U_d^2 + I_{a0}^2 Z_0^2}$$

under

$$\omega t_m = \pi - \text{arctg}\left(\frac{I_{a0}Z_0}{U_d}\right)$$

(4.24)

If, for instance, $L_r = 1.1\,\mu\text{H}$, $C_r = 0.72\,\mu\text{F}$, $U_d = 400\,\text{V}$, $I_{a0} = 160\,\text{A}$, then $Z_0 = 1.236\,\Omega$, $D = 447\,\text{V}$, $\omega = 1.12 \times 10^6\,\text{s}^{-1}$, $U_{cm} = 847\,\text{V}$, i.e. $U_{cm}/U_d = 2.11$, and $t_m = 2.39\,\mu\text{s}$.

Duration of the voltage pulsation can be taken as $2t_m$; that is, the pulsation frequency is $f = 1/(2t_m + \Delta)$ where Δ is the duration of the short-through state. Accepting arbitrarily $\Delta = 0.2t_m$, we find $f \approx 190\,\text{kHz}$. By integration of (4.22) over $[0\ 2t_m]$ and multiplying by f, the average value of the inverter input voltage can be found to be equal to 431 V in the case under consideration. Hence, the inverter switches have to withstand the voltage that is 2.11 times as large as the DC supplying voltage, but the inverter gain increases only by 8% (431/400 = 1.08). This fact prompts to limit the inverter input voltage that can be attained by the clamping circuit. The scheme operates in the following way.

When the voltage U_c, rising, reaches the value U_{ccl} that has been set across the capacitor C_{cl} during the previous cycle, the diode D7 begins to conduct, and energy transfers to C_{cl}. Since its capacitance is much more than the capacitance of C_r, the voltage across C_{cl} increases slightly. At this time, the firing gate signal is sent to K7 that does not conduct at that moment because it is bypassing D7. As the energy passes into C_{cl}, the current in its circuit I_{cl} decreases, then changes its sign and begins to flow through the transistor K7. When the energy that was acquired by C_{cl} is returned, K7 is turned off under the zero voltage across it; the voltage U_{ccl} is fixed, and the voltage U_c decreases down to zero, just as without clamping circuit. Various methods can be used for K7 turning-off: a control of a charge that comes to C_{cl}, that is, K7 is turned off, when the integral of I_{cl} returns to its value that existed before I_{cl} began to flow; a control of U_{ccl}, that is, K7 is turned off, when this voltage returns to its value that existed before I_{cl} began to flow; K7 is turned off, when the current I_{cl}, under C_{cl} discharge reaches the definite negative value, combination of some these methods.

The usual modulation methods cannot be employed in this inverter because the inverter states can be changed only at the discrete instants, when the inverter input voltage is zero. The delta modulation can be used for the output current control, whose scheme for one phase is given in Figure 4.62a. The scheme reminds the current hysteresis controller shown in Figure 5.42, but a hysteresis in the relays is absent, and the D flip-flop is set at the relay output, whose output state repeats its input state only in the instants when the synchronizing pulse C comes and during other times the flip-flop output kept up. The pulse C is generated when $U_c \approx 0$, which enables to execute switching-over of the inverter devices without losses. When it is necessary

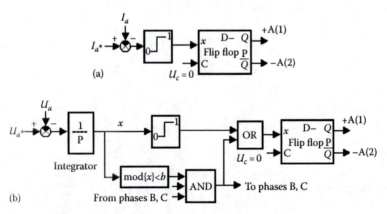

FIGURE 4.62 Current regulators for the resonant inverter. (a) delta modulation, (b) sigma-delta modulation.

to control the inverter output voltage, the PI voltage controller with the innermost delta-modulated current controller is used.

For voltage control, a sigma-delta modulation ($\Sigma\Delta M$) can be used too. It differs from delta modulation by addition of an integrator, Figure 4.62b. When only the relays are used, the inverter does not have zero states, and the phase voltages change from $+U_c$ to $-U_c$ and in the opposite direction (preferably from $+U_c$ to 0 and from $-U_c$ to 0 or in the opposite directions). In order to obtain zero states, the comparison units are used that generate the logical 1, when the output of the corresponding integrator is less than b; the AND gate unites these signals of the three phases; when the outputs of all integrators are less than b, through OR gates the inverter switches are turned on and are connected to the pole +; that is, the output voltage is zero.

The model4_12 simulates the described system. The L_r, C_r, U_d values are the same, as in the aforesaid example, $C_{cl} = 0.5\,\mathrm{mF}$, where the load consists of 2 Ω resistance and 1 mH inductance (per phase). The initial charge of the capacitor C_{cl} is carried out from the low-power source 500 V through the breaker **Br**, which is turned off after 1 ms, and at the same time, the main DC source is connected by the breaker **Br1**.

The circuits for clamping control are in the subsystem **Clamp_Contr**. With the help of **SwitchCl**, one of the two methods of K7 switching is selected. In the lower position, the integral of I_{cl} is computed; when this integral more than 0, K7 is turned on, and when the integral returns to 0, that is, the acquired charge of C_{cl} is compensated, K7 is turned off. In the upper position, K7 is turned on, when the voltage across C_{cl} exceeds the supply voltage by 80 V and the current I_{cl} reaches the value of 80 A, it is turned off, when either the voltage across C_{cl} decreases by 8 V regarding the switching on voltage or the current I_{cl} reaches the value of -660 A. By that, the additional security of the system correct operation under inaccurate tuning of the voltage control is provided.

The circuits for inverter control are in the subsystem **Inv_Contr**. Both delta modulation (the subsystem **I/U_Reg**) and $\Sigma\Delta M$ (the subsystem **U/sdm**) are provided. The choice is carried out by **SwitchD/S**. The gate pulses come to inverter through **OR1** gate either from the circuit for the initial inductor current control or from the controller of the inverter output current/voltage. The logical 1 appears at the output of **AND1** gate (**OR2** gate is used only under start), when the voltage U_c approaches zero, and the logical 0, when the inductor current I_L reaches the value that is defined by **RELAY** with an inverse output. The logical 1 comes to the control inputs of all the switches. When the logical 0 appears, the block **Diff** generates the brief signal 1 that triggers the univibrator **Monostab**; it, in turn, generates the synchronizing pulse of D flip-flops, and they switch over in the states that depend on the signs of the differences between the reference and actual load-phase currents or on the signs of integrals of the voltage difference; these states are preserved till the appearance of the next synchronizing pulse in the next resonant cycle. The D flip-flop outputs are the gate signals for the inverter switches.

The subsystem **I/U_Reg** can operate in the both modes: current control and voltage control. The first mode is realized in the upper position of **SwitchI/V**, and the amplitude and the frequency references are designated by the programmable three-phase source **Current_Ref**. In the second mode, the similar source **F_ref** defines the frequency of the output voltage; its three-phase output with the amplitude of 1 is modulated in amplitude by the output of the load voltage PI controller; this modulated signal is the reference for the load current inner controller in the lower position of **SwitchI/V**.

The subsystem **U/sdm** contains the set-point device for the amplitude and for the frequency of the three-phase voltage; the feedback signals—the phase voltages—are smoothed by the filters with the cutoff frequency of 40 kHz.

Several scopes can be used to observe system operations. The scope **Inverter** fixes the voltage U_c, the inductor current I_L, the voltage across the capacitor C_{cl}, and the gate pulses during the turned-on state of all the inverter switches. The scope **Clamp** records the current and the voltage across the pair transistor-diode of the clamping circuit and the gate pulses for K7; the scope **Load** shows the load current and voltage. The scope **Count** that shows the number of the oscillations of the voltage U_c for the certain interval, usually for 10 ms, is set in the subsystem **Clamp_Contr**.

Let us carry out some experiments with the model. Make the following setting: **Br2** is open, **SwitchI/V** is in the subsystem **I/U_Reg** in the upper position and **SwitchD/S** is in the lower position, the amplitude is 100 A, the frequency is 50 Hz in the block **Current_Ref**, the switch-on point is 120 A for the block **Relay** in the subsystem **Inv_Contr**, and the simulation time is 0.05 s. In Figure 4.63, the DC voltage U_c, the inductor current I_L, and the load current are shown that are received by simulation. The amplitude of U_c is 800–850 V, the frequency (by the scope **Count**) is $1770/0.01 = 177$ kHz that is rather near to the aforesaid calculated data (847 V, 190 kHz), and the load current is nearly sinusoidal, $THD = (0.5\%\text{–}0.6\%)$. Thus, for the low-power low-voltage VSI, when the cost of the transistors does not depend much on the rated voltage, it is reasonable to use the resonant inverters without clamping circuit.

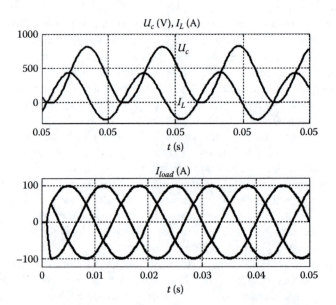

FIGURE 4.63 DC voltage, inductor current, load current without clamping circuit.

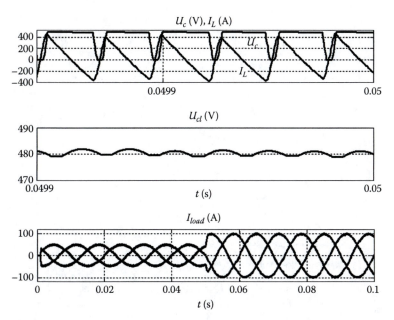

FIGURE 4.64 Processes in the resonant inverter with the first clamping variant.

Now make the following setting: **Br2** is closed; **SwitchI/V** in the subsystem **I/U_Reg** and **Switch_Cl** in the subsystem **Clamp_Contr** are in the upper positions; the amplitude is 50 A with the step to 100 A at $t = 0.05$ s in the block **Current_Ref**, the switch-on point is 160 A for the block **Relay** in the subsystem **Inv_Contr**, and the simulation time is 0.05 s. As a result of simulation, the process shown in Figure 4.64 is formed. It is seen that DC voltage is not more than 490 V, the voltage across the clamping capacitor U_{cl} is nearly constant, and the load current is nearly sinusoidal: $THD \approx 1\%$. By the scope **Count**, the oscillation frequency is 72 kHz. One can see by the scope **Clamp** that the gate pulses for the transistor K7 are given and are removed under zero voltage across it.

Let us repeat simulation with **Switch_Cl** in the lower position. At that point, for providing stable operation, it turned out that it is necessary to increase the switch-on point to 200 A for the block **Relay** in the subsystem **Inv_Contr**. The recorded process is shown in Figure 4.65. The DC voltage rises to 500–505 V, and the frequency is 80 kHz.

For simulation with the voltage controller, **SwitchI/V** is put in the lower position. In the block **U_ref** the initial-phase voltage amplitude is set to 100 V, which increased to 200 V at $t = 0.09$ s. (Note that the feedback is realized with the use of the average value of the three-phase 6-pulse rectification and when the average value is equal to 0.95 of the maximal value). During the simulation time of 0.16 s, the response curves of the voltage U_{load} (after the filter with the cutoff frequency of 500 Hz) and of the load current I_{load} that result are shown in Figure 4.66. In the same figure, the feedback signal U_{load_rect} is also shown.

When **SwitchD/S** is in the upper position, the voltage $\Sigma\Delta M$ controller is active. The response curves of the voltage U_{load} (after the filter) and of the load current I_{load}

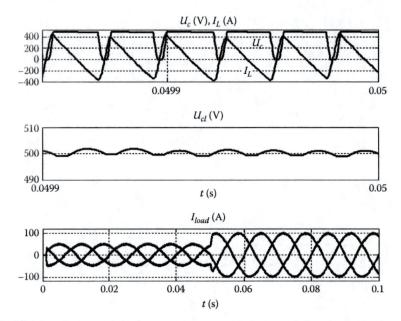

FIGURE 4.65 Processes in the resonant inverter with the second clamping variant.

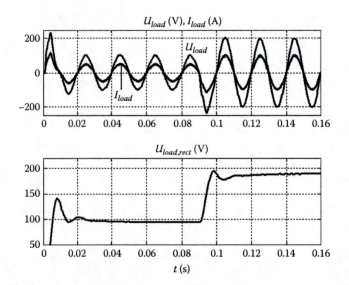

FIGURE 4.66 Voltage control with delta modulation.

are shown in Figure 4.67. At that point, $b = 0.002$, $THD = 1.24\%$ for I_{load} under its peak value of 100 A. If zero states are not used, $THD = 1.6\%$. The phase inverter voltages with and without use of zero state are shown in Figure 4.68a and b, respectively. It is seen that the phase voltage reverse is practically excluded in the latter case.

In conclusion, it is worth mentioning that simulation of the resonant inverters, because of the big oscillation frequency and of the abrupt signal change, has

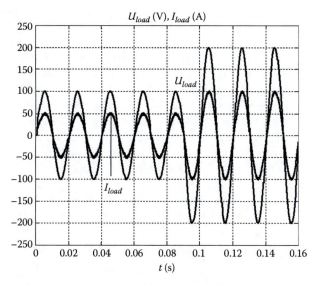

FIGURE 4.67   Voltage control with delta-sigma modulation.

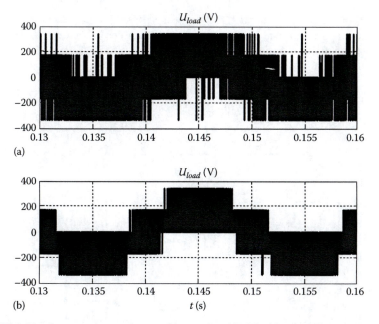

FIGURE 4.68   Inverter phase voltages without using (a) and with using (b) zero states under delta-sigma modulation.

to be fulfilled with the small integration step that leads to very large simulation time (the ratio of the simulation time to the process duration in the simulated system is of order 1000). Simulation time for the electromechanical processes with motors may range from just few minutes to couple of hours; therefore, such models are not presented here.

4.10   SIMULATION OF MODULAR MULTILEVEL CONVERTERS

The modular multilevel converters (MMC or M2C) are a relatively new class of power electronics converters. The diagram of such a converter is shown in Figure 4.69a. It consists of three similar arms (phases) connected in parallel and receives supply from the DC source of voltage V_d. Each phase contains $2n$ identical blocks (submodules); the arm midpoints are the inverter outputs. The submodule diagram is shown in Figure 4.69b. It consists of two IGBT with the antiparallel diodes and the capacitor that is charged to the voltage V_c. The IGBT are controlled in opposite phases. When T1 is switched on and T2 is switched off, the voltage between terminals X, Y is V_c; when T2 is switched on and T1 is switched off, the voltage between terminals X, Y is 0. Thus, the voltage of the semiphase V_{a1} or V_{a2} changes from 0 to nV_c. The value V_c is accepted equal to V_d/n, and control of the submodules is carried out in such a way that n submodules in one arm are in active condition at any moment. At that, the semiphase voltage changes from 0 to V_d. When the sinusoidal modulation is used, the reference voltages for the upper and the lower semi-phases of the phase A are

$$V_{ref1,2} = \frac{V_d}{2} \pm \frac{mV_d}{2}\sin \omega t \tag{4.25}$$

where $0 \le m \le 1$ is the modulation factor, for the phases B and C the reference voltages are shifted by 120° and 240°, respectively. If the third harmonic is added to the references signals, the value m can be increased to 1.155. The phase-to-phase voltage is

$$V_{ab} = \frac{V_d}{2} + \frac{mV_d}{2}\sin \omega t - \left(\frac{V_d}{2} + \frac{mV_d}{2}\sin\left(\omega t - \frac{2\pi}{3}\right)\right) = 0.866 m V_d \sin\left(\omega t + \frac{\pi}{6}\right)$$

$$\tag{4.26}$$

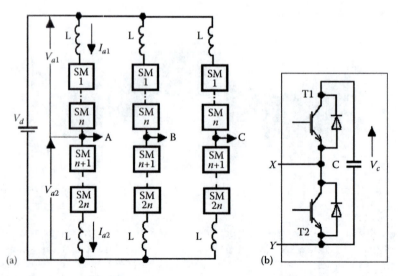

(a)   (b)

FIGURE 4.69   Modular multilevel converter. (a) general diagram, (b) diagram of one submodule.

The advantages of the scheme under consideration is [13] its modular realization enables building of converters for high voltages and is often useful for dispensing with the transformers for connection to the high-voltage network, a simple realization of the redundancy that allows for improving reliability and survivability of the system. Such converters are intended for HVDC with power up to 400MVA and voltage up to 200kV and contain up to 200 submodules in a phase and are produced by the industry.

During the development of the control system it is necessary to take into account the large number of the submodules, which makes it impossible in practice to have PWM for every submodule. The control system described as follows is a simpler version of the one suggested in Ref. [14].

With the periodicity T_m (so-called PWM period), the value of the reference arm voltage V_{ref} is fixed, and the integral and fractional parts of the ratio V_{ref}/V_c are calculated: $k = $ floor (V_{ref}/V_c), $k_{pwm} = $ rem (V_{ref}/V_c). The value k determines how many submodules have to be activated in the next period T_m, and the value k_{pwm} defines during which fraction of the next period T_m the $k + 1$ submodule has to be activated so that the average value of the voltage during this period would be equal to the reference one; that is, this submodule operates in PWM mode.

Since during most part of the modulating signal period $k + 1 < n$, the existence of the redundancy allows for its use for the purpose of balancing the capacitor voltages. This is carried out the following way.

Preliminarily, all the capacitors have to be charged to the voltage V_c. It can be done with help of a low-power source with such an output voltage that is connected instead of the source V_d. At first, the transistors T2 are switched on in all the submodules. Then, by turns, T1 is switched on in the each submodule (respectively, T2 is switched off) for the time that is sufficient for capacitor charging; afterwards, T2 is switched on again, putting the capacitor of the submodule in the storage mode. All the capacitors turn out to be charged after $2n$ cycles. All the phases can be charged at the same time. The model4_13 demonstrates this process. It takes 2s to charge the capacitor of the each submodule to the voltage of 1000 V. When this time is up, the gate pulse is taken away from this submodule and is sent to the next one. One can observe the charging process by **Scope**.

Furthermore, the system acts the following way. If the current of the semiphase >0, the submodule capacitors that are in the active state (having T1 switched on and, hence, giving a "contribution" in the phase voltage) increase their voltages, and if the current of the semi-phase <0, the voltages of the active submodule capacitors decrease. Thus, if this current > 0, k submodules with the least voltages are set in the active states, and the submodule that has the next voltage value operates in PWM mode. If this current < 0, k submodules with the largest voltages are set in the active states, and the next submodule that has a lower voltage value operates in PWM mode. Under this mode of operation, balancing of the loads on different inverter submodules takes place too.

The described operation principle is realized in model4_14. A purpose of the set is to achieve voltage control across the three-phase load 4 Ω, 20mH; $V_d = 4$kV. The set is built according to the diagram shown in Figure 4.69, under $n = 4$. Each phase of the subsystem involves use of two further subsystems that correspond to the positive (A1) and negative (A2) semiphases. Each of the semiphase contains four submodules according to the diagram in Figure 4.69b and the control circuits. The initial capacitor charging is not modeled (see model4_13); it is supposed that all the capacitors have been charged to the voltage of 1000 V already. The quantities k and k_{pwm} are formed in the semiphases A1, B1, C1, as was described earlier. The system acts with the periodicity $1/F_c$ and $F_c = 2000\,\text{Hz}$, and the signal of such a frequency is designated as T_{me} and is formed by the pulse generator in phase B. By this signal, clamping of the measured and calculated values is fulfilled with help of the zero-order hold elements.

PWM for the submodule $k + 1$ is carried out the following way. During the appearance of the successive pulse T_{me}, the integrator is reset and afterwards begins to integrate a signal that is equal to F_c numerically. While the integrator output is less than k_{pwm}, there is the logical 0 at the output of the comparison block **Comp**, the logical 1 at the output of the **NOT**-gate, and the submodule $k + 1$ in an active state. When the integrator output is getting more than k_{pwm}, the logical 0 appears at the output of the **NOT**-gate, and the submodule is deactivated. Note that for the negative semiphase, the value k is complement of 3 of the value K in the positive semiphase, and PWM is carried out after completion of PWM in the positive semiphase. Thereby, the condition that the number of the active submodules in a phase must be equal to n in any moment is realized strictly.

The subsystem **SORT** selects the active submodules in a semiphase. Its functioning can be considered with help of Figure 4.70a and b. The quantity k, the sign of the semiphase current, the PWM output, and the vector of the voltages across the semiphase submodule capacitors $\mathbf{V}_c = [V_{c1}, V_{c2}, V_{c3}, V_{c4}]$ come to the inputs of **SORT**. The MATLAB® function $[\mathbf{B}, \mathbf{IX}] = \text{sort}(\mathbf{V}_c)$ is used for ordering of the vector components. The vectors $\mathbf{B}$ and $\mathbf{IX}$ are formed when this function is carried out; the former contains the components of the vector $\mathbf{V}_c$ arranged in ascending order and the latter contains the component indices in the order of these component positions in the vector $\mathbf{B}$. If, for example, $\mathbf{V}_c = [1040, 970, 1050, 980]$, then $\mathbf{IX} = [2, 4, 1, 3]$. The commutator *Comm* changes the index order to the opposite one $\mathbf{IX}i$; the vectors $\mathbf{IX}$ or $\mathbf{IX}i$ are used depending on the current sign in the semiphase.

It is reasonable to explain the following process by an example. Let the wanted voltage be 800 V. Then $V_{ref1} = 2000 + 800 = 2800$, $k = \text{floor}(2800/1000) = 2$, and $k_{pwm} = \text{rem}(2800/1000) = 0.8$. When the successive pulse T_{me} appears, the signal is formed for switching on $k + 1$ submodule; it goes on $0.8/F_c$ (in the lower semiphase, this signal is formed after this time elapses). If the current sign >0 and the capacitor voltage distribution is the same, as it is given above, then in the positive semiphase the submodules 2 and 4 have to be switched on during the interval T_{me} constantly, and the submodule 1 has to operated in PWM mode.

The core of the following schemes is the selector *Select* that, from the input vector components, selects one corresponding to the whole number at the selector control input *CI*. The null-based indexing is used, when the first component is selected under $CI = 0$, the second one under $CI = 1$, and so on. $CI = 1$ in block 1, so that number 4 is at the *Select* output; $CI = 0$ in block 2, so that number 2 is at the *Select* output; $CI = -1$ in block 3; in order to prevent an error, the summer Sum2, under the negative signal, sends to the comparison block some number that is ineffective for comparison. Therefore, for block 1, the logic 1 is at output O1,4, and for block 2—at

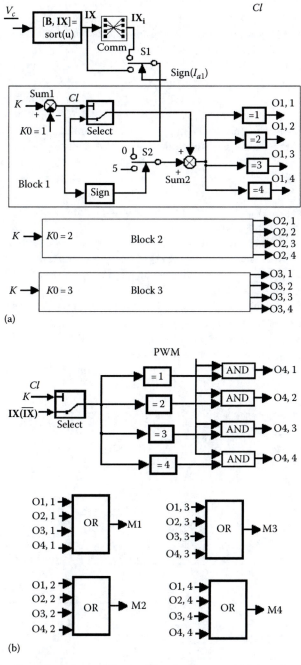

(a)

(b)

FIGURE 4.70 Control of the modular multilevel converter. (a) input circuits, (b) control pulse generation.

the output O2,2; the logical 0 are at all outputs of block 3, and there is the logical 1 at the output of the selector for PWM (Figure 4.70b), so that the signal PWM goes to the output O4.1. With the help of the OR gates, the resulting signals for control of the submodules M1–M4 are formed. From these discussion, it follows that M2 = 1, M4 = 1, M3 = 0, and M1 = 1 while PWM is active.

In addition to the foregoing elements, there is the scope **Uc_Phase** in the semiphase A1 that fixes the capacitor voltages; the scope **State**1 is set additionally in the semiphase A2 that records the gate pulses of all eight-phase submodules.

The load voltage control is carried out the following way. The subsystem **Control** generates three-phase sinusoidal signal with the amplitude of 1.15, the frequency of 50 Hz, with the addition of the third harmonic according to the scheme that has been used earlier. This signal is modulated in amplitude by the output of the voltage controller, and the modulated signal is multiplied by the phase voltage base value that is equal to $V_d/2 = 2000$ V. For a feedback, the block **Three-Phase Sequence Analyzer** that measures the first harmonic amplitude of the phase voltage positive sequence is used.

Several scopes are set in the model. The scope **Phase_V,I** records the load-phase voltages and currents, and the scope **Mag_Phase** shows the fundamental harmonic amplitude of the phase voltage positive sequence. The scope **Phase_Phase** fixes the phase-to-phase voltage V_{ab}, and the scope **Phase_A1,A2**—the voltages of the semiphases A1 and A2. The scope **Phase_I** records the currents of the semiphases A1 and A2 and their difference.

The block **Step** defines the phase voltage amplitude 1000 V, which increases up to 2300 V at $t = 0.5$ s. The amplitude of the phase voltage, the curves of this and of the phase-to-phase voltages that are obtained by simulation are shown in Figure 4.71.

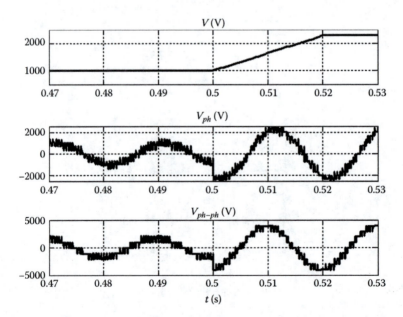

FIGURE 4.71 Converter phase and phase-to-phase voltages under reference changing.

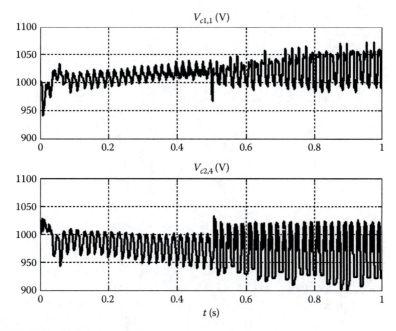

FIGURE 4.72   Voltages across the converter capacitors.

The rms value of the first harmonic of the phase-to-phase voltages that are computed with the help of the option *Powergui/FFT Analysis* is equal to 2815 V, as it has to be ($1.15 \times 0.612 \times 4000 \, V = 2815 \, V$). This voltage (under maximum modulation factor) has nine steps, each of 1000 V. Load current is sinusoidal practically, *THD* = 1%. The semiphase currents have irregular forms that are caused by the continuous commutation of the submodules, but their difference, that is, the load current, is sinusoidal, as mentioned earlier. The voltages across the first submodule of the semiphase A1 and across the fourth submodule of the semiphase A2 are shown in Figure 4.72. One can see that these voltages do not deviate much from the initial values of 1000 V.

4.11   SIMULATION OF MATRIX CONVERTERS

Conversion of AC voltage of the constant amplitude and frequency into AC voltage of the changing amplitude and frequency is carried out in VSI with an intermediate conversion in DC voltage. A disadvantage of such a system is the need for a bulky capacitor of large capacitance that has a limited service life. Therefore, the systems of a direct, without DC link, frequency conversion are under development. The direct frequency converters that use thyristors (cycloconverters) are used for many years [2]; the output frequency of these converters is not more than one third of the network frequency. Simulation of these converters is considered in the next chapter. Over the years, in connection with development of IGBT, interest toward the use of direct frequency converters is growing, as they do not place limitations (theoretically) on the output frequency—the so-called matrix converters (MC).

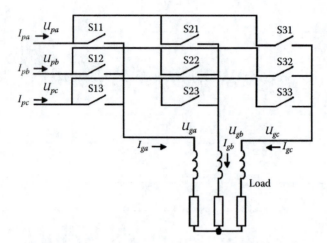

FIGURE 4.73 Matrix converter.

Such a converter consists of nine switches that have the capability to both block the bidirectional voltage and conduct the current in both directions. Each input phase can be connected to any load phase (Figure 4.73). Every switch is made up from two IGBTs and two antiparallel diodes connected back to back with the common emitter or with the common collector. When selecting admissible switch states, two factors have to be taken into account: The same load phase cannot be connected to two different input phases because these phases will be short-circuit; at least one switch in each load phase should be turned on at any instant, in order to avoid overvoltage by creating a path for the inductive load current. Thus, MC has 27 states. Six of them take place, when every load phase is connected to a different input phase, for instance, when the switches S11, S22, S33 or S13, S22, S31 are turned on. At that point, the full unregulated input voltage comes to the load. These states are not used for control. Three states take place when all the three load phases are connected to the same input phase. At that point, the load voltage is zero. In the remaining 18 states, one load phase is connected to some input phase and two other load phases to another input phase.

The MC voltage space vectors are analogous to the system of the VSI space vectors, with the difference that the same vector direction can be provided under different combinations of the turned-on switches and that these vector amplitudes are time dependent and just the vector direction can change for the opposite. The summary of the voltage space vector states is given in Table 4.2 and in Figure 4.74.

Consider the input current vectors. Since the voltage vector determines the load current, the input current is determined by the latter and the switch states; that is, the current I_g is input and the current I_p is output. The positions of the current space vector are also given in Table 4.2 [16].

An analogy of the VSI and MC space vectors gives an opportunity to use, for the choice of the MC states, the same methods that apply to VSI, with necessary alternations. Specifically, the present state of the input voltage space vector has to be taken into account. The input voltages can be divided into six sectors, as shown in Figure 4.75.

TABLE 4.2

MC States

Vk	Turned On	Voltage		Current	
		Module	Phase	Module	Phase
+1	S11, S22, S32	$K_p U_{pab}$	0	$K_i I_a$	$-\pi/6$
−1	S12, S21, S31	$-K_p U_{pab}$	0	$-K_i I_a$	$5\pi/6$
+2	S12, S23, S33	$K_p U_{pbc}$	0	$K_i I_a$	$\pi/2$
−2	S13, S22, S32	$-K_p U_{pbc}$	0	$-K_i I_a$	$-\pi/2$
+3	S13, S21, S31	$K_p U_{pca}$	0	$K_i I_a$	$7\pi/6$
−3	S11, S23, S33	$-K_p U_{pca}$	0	$-K_i I_a$	$\pi/6$
+4	S12, S21, S32	$K_p U_{pab}$	$2\pi/3$	$K_i I_b$	$-\pi/6$
−4	S11, S22, S31	$-K_p U_{pab}$	$-\pi/3$	$-K_i I_b$	$5\pi/6$
+5	S13, S22, S33	$K_p U_{pbc}$	$2\pi/3$	$K_i I_b$	$\pi/2$
−5	S12, S23, S32	$-K_p U_{pbc}$	$-\pi/3$	$-K_i I_b$	$-\pi/2$
+6	S11, S23, S31	$K_p U_{pca}$	$2\pi/3$	$K_i I_b$	$7\pi/6$
−6	S13, S21, S33	$-K_p U_{pca}$	$-\pi/3$	$-K_i I_b$	$\pi/6$
+7	S12, S22, S31	$K_p U_{pab}$	$4\pi/3$	$K_i I_c$	$-\pi/6$
−7	S11, S21, S32	$-K_p U_{pab}$	$\pi/3$	$-K_i I_c$	$5\pi/6$
+8	S13, S23, S32	$K_p U_{pbc}$	$4\pi/3$	$K_i I_c$	$\pi/2$
−8	S12, S22, S33	$-K_p U_{pbc}$	$\pi/3$	$-K_i I_c$	$-\pi/2$
+9	S11, S21, S33	$K_p U_{pca}$	$4\pi/3$	$K_i I_c$	$7\pi/6$
−9	S13, S23, S31	$-K_p U_{pca}$	$\pi/3$	$-K_i I_c$	$\pi/6$
0a	S11, S21, S31	0	—	0	—
0b	S12, S22, S32	0	—	0	—
0c	S13, S23, S33	0	—	0	—

$K_p = 2/3, \; K_i = 2/\sqrt{3}.$

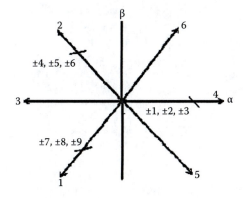

FIGURE 4.74 MC voltage space vectors.

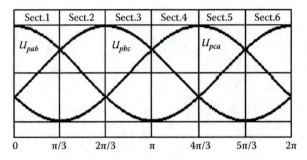

FIGURE 4.75 Voltages supplying MC.

It is obvious that MC states have to be selected in the every sector that correspond to the input phase-to-phase voltages that do not change their signs and are maximum by amplitude.

For example, we want to realize the vector 4 (Figure 4.74) and the voltage input vector is in the first sector (Figure 4.75). The state vectors ±1, ±2, ±3 are directed along the vector 4. The states ±2 are defined by the voltage U_{pbc} that changes its sign in the first sector, the states −1 and +3 give the "negative" value (i.e., the corresponding voltage vectors align with the state vector 3), and the possible states +1 и −3 remain. At that point, for the state +1, input current phase is −π/6, and for the state −3 it is π/6. Choosing these states in turns, the phase shift close to zero can be obtained.

There are several methods for control of the MC output voltage [15]. Furthermore, the space vector modulation is used. Note that the demonstrational model of the three-phase matrix converter is included in SimPowerSystems, in which another control method is used; this model is simpler compared to the described as follows and does not provide for the possibility to show all the MC features. In the method under consideration, unlike VSI, not two but four active MC states are selected and are applied for suitable time intervals within each period T_c. The zero state is used to complete T_c. These states are given in Table 4.3 for the possible sectors of the input U_p and output U_g voltages. The sectors U_g are determined by the position of this vector on the plane of the MC space vectors (рис. 4 74); moreover, the first sector lies between the vectors 4 and 6, and the sectors are counted off in anticlockwise direction; the number of the sector U_p is defined as in Figure 4.75.

Designate as $\Delta\theta$ the angle that determines the position of the vector U_g in the sector and counted off from the sector centre, that is, $-\pi/6 < \Delta\theta < \pi/6$. If the angle that is counted off from the sector center of the input voltage U_p is designated as $\Delta\alpha$ (Figure 4.75), the following relationships are valid for realization of the relative time intervals (duty cycles) of the states given in Table 4.3 [16]:

$$\tau_1 = -1.155\,s\,m\cos\left(\frac{\pi}{3} + \Delta\theta\right)\cos\left(\Delta\alpha + \frac{\pi}{3}\right)$$

$$\tau_2 = 1.155\,s\,m\cos\left(\frac{\pi}{3} + \Delta\theta\right)\cos\left(\Delta\alpha - \frac{\pi}{3}\right)$$

TABLE 4.3

Feasible MC States

$U_g \backslash U_p$	1	2	3	4	5	6
1	+1 −3	+2 −3	−1 +2	−1 +3	−2 +3	+1 −2
	−7 +9	−8 +9	+7 −8	+7 −9	+8 −9	−7 +8
2	−7 +9	−8 +9	+7 −8	+7 −9	+8 −9	−7 +8
	+4 −6	+5 −6	−4 +5	−4 +6	−5 +6	+4 −5
3	+4 −6	+5 −6	−4 +5	−4 +6	−5 +6	+4 −5
	−1 +3	−2 +3	+1 −2	+1 −3	+2 −3	−1 +2
	+5 −6					
4	−1 +3	−2 +3	+1 −2	+1 −3	+2 −3	−1 +2
	+7 −9	+8 −9	−7 +8	−7 +9	−8 +9	+7 −8
5	+7 −9	+8 −9	−7 +8	−7 +9	−8 +9	+7 −8
	−4 +6	−5 +6	+4 −5	+4 −6	+5 −6	−4 +5
6	−4 +6	−5 +6	+4 −5	+4 −6	+5 −6	−4 +5
	+1 −3	+2 −3	−1 +2	−1 +3	−2 +3	+1 −2

$$\tau_3 = 1.155 \, s \, m \cos\left(\frac{\pi}{3} - \Delta\theta\right) \cos\left(\Delta\alpha + \frac{\pi}{3}\right)$$

$$\tau_4 = -1.155 \, s \, m \cos\left(\frac{\pi}{3} - \Delta\theta\right) \cos\left(\Delta\alpha - \frac{\pi}{3}\right)$$

(4.27)

where

　m is the voltage transfer ratio that is equal to the ratio of the prescribed amplitude
　　of the output voltage to the input voltage amplitude, $m \leq 0.866$

　$s = (-1)^{p+g}$

　p, g are the numbers of the U_p and U_g sectors, respectively

Two of the quantities computed by (4.27) are >0, and two of them <0. In this case, the negative states are selected from Table 4.3.

During MC operation, the load voltage is derived from a train of alternating high-frequency input voltage sectors; the high-frequency component of this voltage is smoothed usually by the load inductance. The input current (network current) consists of a train of the alternating with high-frequency load current sectors. The current high-frequency components result in large changes in the MC input voltage, because of the network inductances, and distort essentially the operation of MC and the other connected loads; therefore, the L-C filter is set usually at the MC input.

　The system under consideration is realized in model4_15. Its main part is the subsystem **Matrix** consisting of nine blocks **Ideal Switch** connected according to the scheme in Figure 4.73. The MC is supplied from the source 380 V through the intermediate block **Series RLC Branch** that is used as a part of the input filter. The same block that is connected with the help of the three-phase breaker is used for the capacitance part of the filter. As a load, the block **Series RLC Load** with the active load of 100 kW and the inductive load of 50 kvar is employed.

Three subsystems placed in the subsystem **Control_&_Measurement** are developed for the switch gate pulse generation. The first of them, **Comp_time**, computes the times $\tau_1-\tau_4$ by (4.27). The input and output voltage sectors are calculated the same way: with the measured U_p or as specified in pu, U_g; α and β components are computed as

$$U_\alpha = U_a, \quad U_\beta = \frac{(U_b - U_c)}{1.732} \tag{4.28}$$

Knowing U_α, U_β, with help of the Simulink block **Cart2 Polar**, the angles of the space vectors α and θ and their modules are calculated. This module is m for U_g. The angles change from $-\pi$ to π. At the outputs of the switches **Sw** and **Sw1**, the angles change in the range $0-2\pi$, and in the range $0-6$ after **Gain** and **Gain1**. The whole numbers $0-6$ are formed at the outputs of the blocks **Round** that, according to the characteristic of rounding-off, change in the middle of the rounding sector, that is, for angles $30°$, $90°$ and so on. These signals define the numbers of the sectors of the input and output voltages S_in = p, S_out = g. Note that the value of 0 and 6 define the same sector. The quantities $\Delta\alpha$ and $\Delta\theta$ are calculated as

$$\Delta\alpha(\Delta\theta) = \alpha(\theta) - p(g)\frac{\pi}{3} \tag{4.29}$$

The signal s with the necessary sign is formed at the output of **Sw2**. The blocks **t1–t4** compute the times $\tau_1-\tau_4$ by (4.27).

This subsystem is the triggered one (with an external actuation). Calculations in it are fulfilled once for cycle T_c—at the beginning of each cycle.

The second subsystem **Time/Code** generates the code that determines which MC switches have to be turned on every instant. An order of the times $\tau_1-\tau_4$ has an influence on the load current THD. The following order is accepted: τ_1, τ_3, τ_4, τ_2. The interval number at the given moment is formed at the output of the summer **Sum1**. The Simulink block **Repeating Sequence** designated here as **Triangle** generates a triangular waveform with the amplitude of 1 and the frequency of $1/T_c$. At the start of each period, the subsystem **Comp_time** is triggered, and 1 is set at the output of **Sum1**. After expiration of τ_1, the output of **Sum1** becomes equal to 2, after expiration of $\tau_1 + \tau_3$—equal to 3, after expiration of $\tau_1 + \tau_3 + \tau_4$—equal to 4, and after expiration of $\tau_1 + \tau_3 + \tau_4 + \tau_2$—equal to 5. The present output code depends on the sector numbers p and g and on the number of the interval. It is seen from Table 4.3 that for the sectors 1 and 4, 2 and 5, 3 and 6 the states differ only by sign. If we take this fact into account, the same table can be used for said states; that is, it is sufficient to have three tables. Which of the tables is used at a particular time depends on the position of the multiport switch **Sw2**. Pay attention to the fact that the sector number at the subsystem input increases by 1; that is, it takes values 1–7. The first table is chosen under positions **Sw2** of 1, 4, 7, the second one under positions of 2 and 5, and the third one under positions of 3 and 6.

The available MC states are encoded the following way: The null is added on the right to the number of the "positive" states (i.e., the state +8 transforms in the code 80), and the zero states have the code 101, 102, 103. The row number is defined by g, and the column number is determined by the interval number. If the negative value τ corresponds to the given interval, the logical 1 appears at the output of **OR** gate that switches **Sw1**, and the number 1 is added to the table code with help of the summer **Sum7**, that is, the "negative" states are encoded as 11, 21, ..., 81, 91.

The formed codes are converted into gate signals for MC switches in the subsystem **Code/Pulses**. It consists of nine identical blocks that model the table with one input. The table inputs are all feasible MC states, that is, 10, 11, 20, ..., 91, 101, 102, 103. Besides, simulation showed that there are cases from time to time when, because of unfavorable combination of the delays, the number 1 is also added to the zero code for a short time; that is, the code 104 is formed. For preventing malfunction in this case, the additional input with this code is put into the tables, having the same output as the code 103.

Take, for example, the input code is 81. According to Table 4.2, under the code −8 the switches S12, S22, S33 are turned on. It is seen that under this input code (the 16th position in the tables), in the 16th output position of the tables with the said designations are 1, and in the other tables are 0.

Several scopes can observe the processes in the system. The scope **Source** fixes the supplying voltage and the current in the phase A—both the actual current and the current smoothed by the filter with the time constant of 0.5 ms. The scope **Load** records the load-phase voltage, the filtered load voltage, and the load current. The scope **Code** shows the output codes of the control system and the scope **Scope_Pulses** shows the gate pulses of the MC switches.

Let us begin with simulation. At first, the supply source is supposed to be ideal. For that purpose, the zero inductance and very small resistance have to be set in the blocks **Three_Phase Source** and **L_F**; the initial state of the block **Breaker** is open, and the transition time is 10 s. The amplitude of 0.85 and the frequency of 30 Hz are specified in the block **Reference**. The source current I_p and the filtered current are shown in Figure 4.76. Comparison of the phase of the latter with the phase of the supplying voltage shows that there is the phase shift of 12° that corresponds to the phase shift of **Filt1**. Hence, the source voltage and the current fundamental harmonic are in phase. The load-phase voltage and the load current are shown in Figure 4.77. The former has a lot of harmonics that, however, are filtered easily by **Filt2** with a small time constant. It is seen that the load current is nearly sinusoidal. With the help of the option *Powergui/FFT Analysis* one can find that the load current $THD = 1.57\%$ and the phase voltage fundamental harmonic amplitude is 260.7 V, so that the actual value of the modulation factor is $m_{act} = 260.7 \times 1.73/(1.41 \times 380) = 0.84$.

The processes under the reference frequency of 100 Hz are shown in Figure 4.78. Thus, MC makes it possible to have the output frequency both less and more than the supply frequency.

Now take into consideration the inductance of the supplying network. For that purpose, the interior impedance of the block **Three-Phase source** is specified by using short-circuit level; this level is set equal to 10 MVA at the base voltage of 380 V (the source phase inductance is 46 μH). One can see after simulation that the voltage

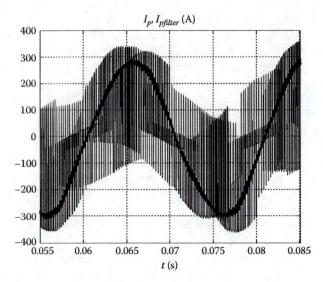

FIGURE 4.76   MC input current without input filter.

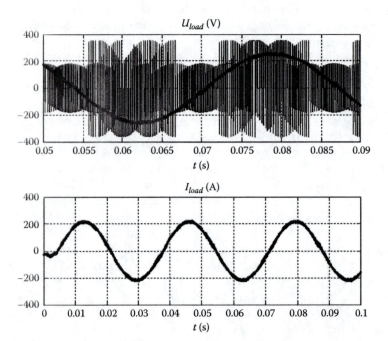

FIGURE 4.77   Load voltage and current in MC under output frequency of 30 Hz.

instantaneous values on the source terminals and across the load is greater by several folds than the rated values, what is obviously impermissible. The other loads connected to this network are exposed to danger too. Therefore, the filter is necessary, usually L-C filter. The inductance of 0.24 mH and the resistance of 0.013 Ω are set in the block **L-F** and the capacitance of 160 μF is set in the block **C-F**. The initial state

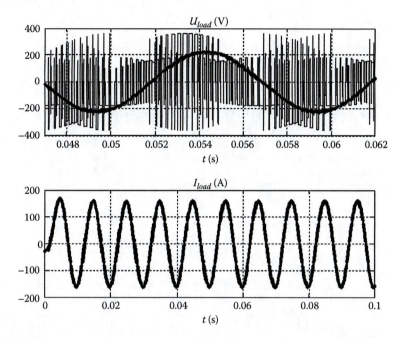

FIGURE 4.78 Load voltage and current in MC under output frequency of 100 Hz.

is "close" for the block **Breaker**. The output-source voltage and its current resulting from simulation under the output frequency of 30 Hz are plotted in Figure 4.79. THD of the voltage is 1.5%, and of the current, 5.1% (under setting in Powergui *Start time* = 0.04 s, *Number of cycles* = 2, *F* = 50 Hz). It is recommended that the reader will trace the influence of changing filter parameters on the network's current shape.

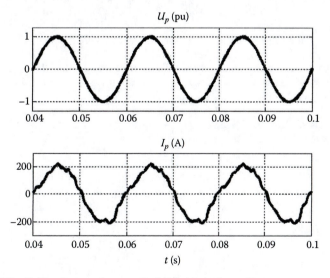

FIGURE 4.79 Grid voltage and current in MC with the input filter.

REFERENCES

1. Mathworks, SimPowerSystems™, User's Guide, 2004–2011.
2. Bose, B.K., *Modern Power Electronics and AC Drives*, Prentice Hall PTR, Upper Saddle River, NJ, 2002.
3. Perelmuter, V.M. and Sidorenco, V.A., *DC Thyristor Electrical Drives Control Systems (in Russian)*, Energoatomizdat, Moscow, Russia, 1988.
4. Hingorani, N.G., High-voltage DC transmission: A power electronics workhorse, *IEEE Spectrum*, 33(4), 63–72, April 1996.
5. Meynard, T.A., Fadel, M., and Aouda, N., Modeling of multilevel converters, *IEEE Transactions on Industrial Electronics*, 44(3), 356–364, June 1997.
6. Gateau, G., Fadel, M., Maussion, P., Bensaid, R., and Meynard, T.A., Multicell converters: Active control and observation of flying-capacitor voltages, *IEEE Transactions on Industrial Electronics*, 49(5), 998–1008, October 2002.
7. Peng, F.Z., Z-source inverter, *IEEE Transactions on Industry Applications*, 39(2), 504–510, March/April 2003.
8. Peng, F.Z., Z-source inverter for motor drives, *Annual IEEE Power Electronics Specialists Conference*, Aachen, Germany, pp. 249–254, 2004.
9. Ding, X., Qian, Z., Yang, S., Cuil, B., and Peng, F.Z., A direct peak DC-link boost voltage control strategy in Z-source inverter, *Proceedings of IEEE Applied Power Electronics Conference*, Sydney, New South Wales, Australia, pp. 648–653, 2007.
10. Ding, X., Qian, Z., Yang, S., Cuil, B., and Peng, F.Z., A PID control strategy for DC-link boost voltage in Z-source inverter, *Proceedings of IEEE Applied Power Electronics Conference*, Sydney, New South Wales, Australia, pp. 1145–1148, 2007.
11. Cuzner, R.M., Abel, J., Luckjiff, G.A., and Wallace, I., Evaluation of actively clamped resonant link inverter for low-output harmonic distortion high-power-density power converters, *IEEE Transactions on Industry Applications*, 37(3), 847–855, May/June 2001.
12. Cuzner, R.M., Nowak, D.J., Bendre, A., Oriti, G., and Julian, A.L., Mitigating circulating common-mode currents between parallel soft-switched drive systems, *IEEE Transactions on Industry Applications*, 43(5), 1284–1294, 2007.
13. Lesnicar, A. and Marquardt, R., An innovative modular multilevel converter topology suitable for a wide power range, *Bologna PowerTech Conference*, Bologna, Italy, 2003.
14. Rohner, S., Bernet, S., Hiller, M., and Sommer, R., Modulation, losses, and semiconductor requirements of modular multilevel converters, *IEEE Transactions on Industrial Electronics*, 57(5), 2633–2642, May 2010.
15. Wheeler, P.W., Rodriguez, J., Clare, J.C., Empringham, L., and Weinstin, A., Matrix converters: A technology review, *IEEE Transactions on Industrial Electronics*, 49(2), 276–288, April 2002.
16. Casadei, D., Serra, G., Tani, A., and Zarri, L., Matrix converter modulation strategies: A new general approach based on space-vector representation of the switch state, *IEEE Transactions on Industrial Electronics*, 49(2), 370–381, April 2002.

5 Electric Machine and Electric Drive Simulation

5.1 DIRECT CURRENT (DC) MOTORS AND DRIVES

5.1.1 DC DRIVES WITH CHOPPER CONTROL

DC electrical machine (DCM) description is accepted in SimPowerSystems™ as follows: An armature circuit equation is

$$U_d = E + R_a I_a + L_a \frac{dI_a}{dt} \tag{5.1}$$

where
U_d is the DCM armature voltage
E is the emf of DCM
R_a is the armature circuit resistance
I_a is the armature current
L_a is the armature inductance

$$E = K_e \omega \tag{5.2}$$

ω is the rotation speed of DCM, rad/s,

$$K_e = L_{af} I_f \tag{5.3}$$

L_{af} is the mutual inductance of the armature and excitation circuits
I_f is the excitation current

DCM torque is

$$T_e = K_e I_a \tag{5.4}$$

Thus, it is supposed that emf and torque are proportional to the excitation current and not to the flux, as it is in reality. It means disregarding saturation, which can often be an essential shortcoming of the accepted model.

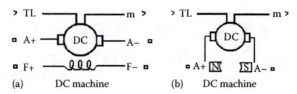

(a)   DC machine   (b)   DC machine

FIGURE 5.1   DC model picture. (a) With excitation winding. (b) With permanent magnets.

The armature motion equation is

$$J\frac{d\omega}{dt} = T_e - \text{sign}(\omega)T_f - B_m\omega - T_l \qquad (5.5)$$

where
   J is the inertia moment of the armature with the connected load
   B_m and T_f are coefficients of the viscous and dry friction, respectively
   T_l is the external moment of the motion resistance

In Figure 5.1 the graphical DCM model representation in SimPowerSystems is shown. One can see that the excitation winding is carried out so that it is possible to connect it in series with the armature winding for a series DCM; the winding has the resistance R_f and the inductance L_f. The model output m is a four-dimensional vector: [ω (rad/s), I_a (A), I_f (A), T_e (N-m)].

Beginning from the version 4.6, the dialog boxes of the electrical machines have three pages that are to relevant to specific types of transformers (see Chapter 2). For DCM, the first page *Configuration* gives the opportunity to select one of the standard DCMs. Beginning from version 4 of SimPowerSystems, an electric machine model can work in two modes: with the specified load torque T_l and with the fixed rotation speed, the mode is selected on the first page too; in the last case, the model's mechanical part is ignored. Furthermore, we shall use only the first option. Beginning from version 5.2.1, it has become possible to model DCM with permanent magnets. When the option *Permanent magnet* is chosen in the line *Field type* of the dialog box, the block picture changes (Figure 5.1b). On the second page (Figure 5.2), the parameters of DCM are given, whose meanings are clear from the description provided previously. The value of the coefficient K_e can be found by using the DCM nameplate data as $\omega_n = \pi n_n/30$, $T_{en} = P_n/\omega_n$, $I_n = P_n/(\eta U_n)$, $K_e = T_{en}/I_n$, where, P_n is the rated power (W), U_n is the rated voltage (V), n_n is the rated speed (rpm), and η is the efficiency. The third page *Advanced* appears by the introduction of a new quality of the electrical machine models in SimPowerSystems, where it is possible to choose different sample times for different model blocks. In field "Sample time" the sample time of the block has to be specified: for -1 the sample time is defined by Powergui, or it can be lengthened a number of times, which gives the opportunity to accelerate the simulation process.

As the first example of use of DCM model, we investigate chopper control of the DC motor D. The schematic diagram of the main circuits is shown in Figure 5.3a. Here K is the force-commutated semiconductor: GTO—thyristor

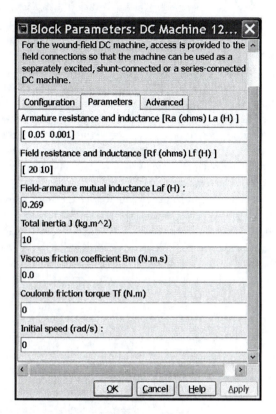

FIGURE 5.2 The second page of the DC dialog box.

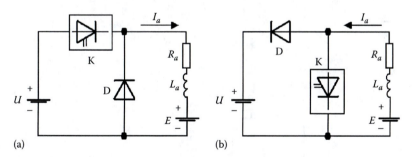

(a)   (b)

FIGURE 5.3 Diagrams in the motor and generator modes. (a) Motor mode. (b) Generator mode.

or IGBT—transistor. When K is closed, the current flows from the source to motor and increases. When K is open, the current, owing to the inductance, flows in the same direction for some time, taking the path through diode D, and decreases in magnitude, because it draws against motor emf. With interchange of the closed and open K states, the mean value of the motor armature current can be controlled.

For the deceleration mode with power recuperation, the schematic diagram in Figure 5.3b is used. When K closed, the DCM armature is short out, and the DCM current increases by the force of emf, being in direction opposite to that observed in the previous mode. With such a current direction, DCM torque acts against rotation direction. When K is open, the current, keeping up the direction, takes the path through D and the source in the direction that is opposite to the source polarity, so that the energy is returned to the source. DCM decelerates with power recuperation.

The circuit that provides both modes can be configured in different ways. Figure 5.4 shows a circuit in which the mode switch-over is accomplished by a contactor. In position 1 of the switch P, there is a motor mode, and in position 2, there is a braking mode; this is easy to understand by comparing the circuits shown in Figure 5.3. In Figure 5.5, the circuit without contacts, but with double number of the semiconductors, is shown. In the motor mode, K1 and D1 are working, K2 is turned off, and in the braking mode K2 and D2 are working and K1 is turned off. This circuit is especially suitable for use with IGBT because the antiparallel diodes that are built into IGBT housing can be used. This circuit is more reliable than the previous one because the contactors are absent, but an additional disconnector has to be set up for protection. Keep in mind that the mode switching-over has to be achieved with zero current in the DCM armature as a rule.

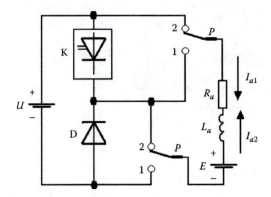

FIGURE 5.4   Mode switch-over with contactors.

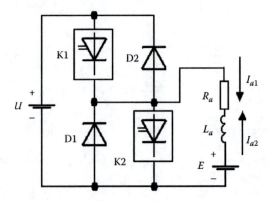

FIGURE 5.5   Static mode switch-over.

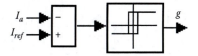

FIGURE 5.6　Relay current controller.

The current controller is an integral part of the system under consideration. Usually either bang-bang or pulse-width controllers are used. A block diagram of the former is shown in Figure 5.6. If the DCM current I_a is less than the reference one I_{ref}, output g is set in 1, and the controlled switch (e.g., K in Figure 5.3) conducts. If $I_a > I_{ref}$, then $g = 0$. Switch-over is carried out with a certain hysteresis. The circuit is very simple, but it has a drawback because the switching frequency depends on parameters specific to the motor operation mode. In the use of pulse-width controller, the switching frequency is constant. The method is based on comparing the control signal u with a saw-tooth signal u_s having the constant period T. At the beginning of the period u_s, the signal $g = 1$ forms, and when $u_s \geq u$ $g = 0$ sets. The relative duration of the state $g = 1$ is $a = u/U_{smax}$ where U_{smax} is the saw-tooth signal amplitude. With increasing u, the relative time, when the switch is closed, is increasing too, and the current is rising. The controller diagram is given in Figure 5.7. The difference between the reference and actual currents is integrated by the integrator with limited output [0, 1]. With the help of the amplifier K, the regulator gain is set. Sawtooth generator produces the saw-tooth pulses of the frequency $f = 1/T$ with amplitude 1. At the beginning of the saw-tooth pulse, its value is less than the integrator output and output limiter is set in 1 and the controlled switch is in a conducting state (switch turned on). When saw-tooth signal rises up to the level of integrator output, the output limiter goes 0 and the controlled switch is turned off. For the controller, the standard Simulink® blocks are used.

The electrical drives under consideration operate often at the rotation speed above the rated ω_{rate}, with field weakening. For this purpose, an additional switch is set in the excitation winding circuit. A possible diagram of the control circuit is shown in Figure 5.8. The excitation current reference is defined as $I_{fref} = UK_1/L_{af}\omega$, the divider is bounded below in order to eliminate division by 0. The amplifier K1 is used for adjusting U and ω correspondence. If the source voltage U can change over a wide range, then this voltage is to be measured.

The limiter at divider output limits the possible references by I_{fmin}, I_{fn}. When $\omega < \omega_{rate}$, the excitation current reference is I_{fn}; The actual current approaches the

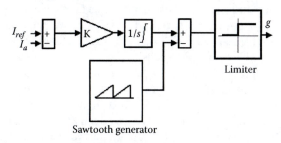

Sawtooth generator

FIGURE 5.7　Pulse-width current controller.

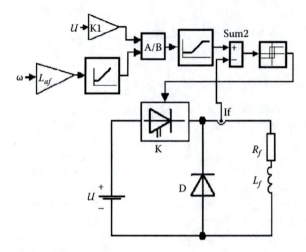

FIGURE 5.8   DC flux control.

reference value, with the help of the switch circuit that is similar to one shown in Figure 5.3. The controller is fulfilled in accordance with the diagrams shown in Figure 5.6 or 5.7. When during acceleration ω begins to exceed the reference value, the limiter output decreases. Operation at the speed mentioned earlier takes place, with the rated emf value remaining constant.

Some variations in the use of the described control methods are shown in model5_1 – model5_3. DCM with a separate excitation 120 kW, 550B, 1760 rpm, $\eta = 0.9$, and $I_{fnom} = 10$ A is used for simulation. Coefficient L_{af} is computed as

$$\omega_n = \pi \frac{1760}{30} = 184.3 \text{ rad/s}, \quad T_{en} = \frac{120{,}000}{184.3} = 651 \text{ N-m},$$

$$I_n = \frac{120{,}000}{(0.9\,550)} = 242 \text{ A}, \quad K_e = \frac{651}{242} = 2.69, \quad L_{af} = \frac{2.69}{10} = 0.269,$$

and the rest of the motor parameters are given in Figure 5.2. The source armature rated voltage is 600 V, and excitation is 220 V. The load torque has a steady component and the component that is proportional to the square of ω, which is typical for vehicle electrical drives (the accepted value is derived as follows: $T_l = 50 + \omega^2/400$ N-m). It is worth noting that simulation showed that for its speeding up it is reasonable to take $T_f = 0$ in (5.5) and to simulate dry friction as given in the present models.

In model5_1 the main circuit is designed according to the diagram shown in Figure 5.4 (subsystem **Arm_contr**). The additional inductance of 3 mH is set in the armature circuit. The blocks **Ideal Switch** are used for models of switching devices. **Switch 1_1** and **Switch 1_2** model the switches marked "1" in Figure 5.5, and **Switch 2_1** and **Switch 2_2** model the switches marked "2" in Figure 5.5. **IGBT/Diode** block is used for switch K, and the controller is carried out according to the diagram in Figure 5.6. Current reference (400 A) is determined by the block **Step** and passes through **Rate Limiter**. **Switch1** defines the time of transfer from the mode *Run* to the mode *Brake*.

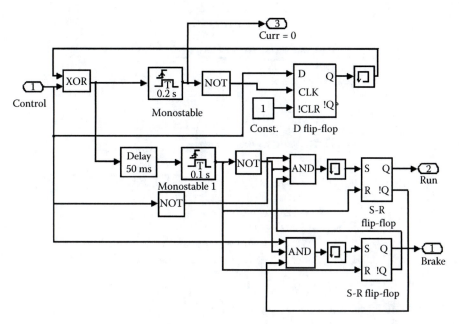

FIGURE 5.9 Simulink diagram of the block **Mode** for mode switch-over.

Block **Mode** whose Simulink diagram is shown in Figure 5.9 controls mode switching. Zero signal in the mode *Run* and signal 1 in the mode *Brake* come to the input *Contr*. The output of D Flip-Flop is set in the state that corresponds to the present mode. During mode change, the disparity signal (logical 1) appears on the output of **XOR** gate, and the block **Monostable** switches for 0.2 s. At output 3 (marked Curr = 0) signal 1 appears that turns over the **Switch** in model5_1 and the current reference drops to 0. With a delay of 50 ms that is sufficient for the current decay, univibrator **Monostable1** is switched on. Both S-R flip flops that control the switches in the main circuit are reset and both switches in the main circuit are open. After the pulse at **Monostable1** output decays, the switch that corresponds to the input *Contr* is switched on, and after the pulse at **Monostable** output decays, the current reference will be switched on.

This model can be used for studying DCM in static and dynamic modes, particularly for studying how the inductances of the armature and excitation windings affect the currents, torque, speed pulsation, and so on. Processes for the next conditions are shown in Figure 5.10: at $t = 0$ an excitation is switched on and at $t = 0.5$ s the command for acceleration with current 400 A is given; DCM speeds up. At about $t = 2.5$ s speed reaches 185 rad/s and the excitation current decreases. DCM torque decreases too and at the same time the load torque increases, so that DCM acceleration goes down close to 0. At $t = 12$ s, the command for deceleration is given. The current decreases to 0, the switches switch over in the main circuit, the current increases up to 400 A again, and under the combined influence of motor and load torques DCM decelerates fast (the residual armature voltage is caused by the current flowing through snubbers because the neutral position of the main circuit

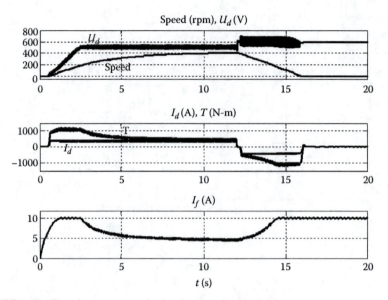

FIGURE 5.10 Transient responses in the chopper electrical drive.

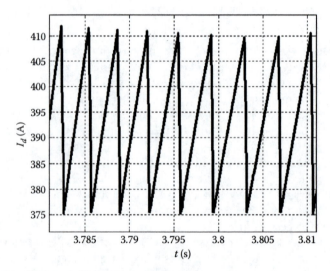

FIGURE 5.11 Current oscillation in the steady state.

switches is absent in the model). Armature current swinging is 37 A in Figure 5.11. One can note that during the general sample time of 10 ms the accepted sample time for DCM is 50 ms.

Model5_2 differs from the previous one, given the use of the current controller, as illustrated in Figure 5.7 (subsystem **Curr_Reg**). The integrator reset to 0 when its output reaches 1 is used in the controller. So the amplitude of saw-tooth voltage is equal to 1, and the frequency is equal numerically to the value of the constant signal at the integrator input (accepted 2000 Hz). At the time of switching over, when the

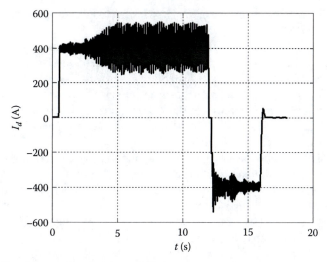

FIGURE 5.12   Current changing in model5_2.

values of the reference and actual currents are equal to 0, the gate pulses have to be prohibited; for this purpose, the integrator of the controller is kept in 0 state with the help of null detectors that are built with the blocks **Relay** and gate **OR**. The processes in this model are rather the same as in the previous one, but current swinging is bigger (see Figure 5.12).

In model5_3 the main circuit is designed according to the diagram in Figure 5.5, and the current controller is made according to Figure 5.6. The processes in this model are close to the processes in model5_1. The DCM current under the same condition is shown in Figure 5.13.

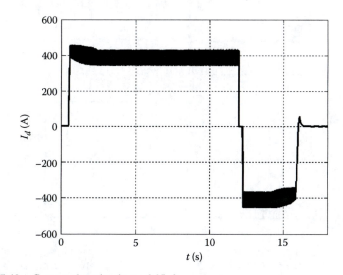

FIGURE 5.13   Current changing in model5_3.

The models of some electrical drives, besides the models of the separate motors, are included in SimPowerSystems too [1]. They are in the folder "Application Libraries/Electric Drives Library." In the section "DC Drives" there are three models of DC electrical drives with chopper control: **One-Quadrant Chopper DC Drive**, **Two-Quadrant Chopper DC Drive**, and **Four-Quadrant Chopper DC Drive**. The main circuits of the first model are the same as in Figure 5.3a, and the second model, as in Figure 5.5; the main circuits of the **Four-Quadrant Chopper DC Drive** are shown in Figure 5.14b. Every electrical drive has a current controller and, as an option, a speed controller. Field control is not provided.

The current controller has a schematic diagram that is close to the one shown in Figure 5.7, with the difference that the controller is PI type but not I type as shown in Figure 5.7. Current reference is formed as a relative value (in pu, as it is accepted in SimPowerSystems), the first-order filter is set at the input of the current feedback. In the diagram of **Two-Quadrant Chopper DC Drive**, the output gate pulses of the controller are sent to both IGBTs in antiphase, and in the diagram of **Four-Quadrant Chopper DC Drive** the output pulses are sent simultaneously to both IGBTs that are placed in the opposite arms and in antiphase to the other pair of IGBTs. The speed controller is PI type also and has the rate limiter at the speed reference input and the 1st-order filter at the speed feedback.

Graphical representation of all three models is the same and as shown in Figure 5.14a, and difference lies in the designation on the blocks: DC5—**One-Quadrant Drive**, DC6—**Two-Quadrant Drive**, and DC7—**Four-Quadrant**.

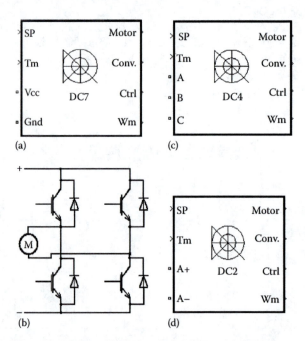

(a)

(b)

(c)

(d)

FIGURE 5.14 Pictures of DC standard electrical drive models. (a) Four-quadrant chopper drive. (b) Main circuit of DC7. (c) Four-quadrant drive with the three-phase rectifier. (d) Four-quadrant drive with the single-phase rectifier.

The models have the following inputs and outputs:

SP—speed reference (rpm) or torque reference (N-m)
Tm—load torque (N-m)
Vcc, Gnd—DC source (+ at Vcc)
Wm—DCM speed (as it was mentioned earlier, the option, for which the motor
   speed is defined from outside, is not used)
Motor—vector of the motor states: armature voltage (V), rotation speed (rpm),
   armature current (A), excitation current (A), torque (N-m)
Conv—voltage on the terminals of the circuit that consists of the motor and
   the smoothing reactor
Ctrl—vector of the controller quantities: current reference, duty cycle, difference between reference and actual values of the speed (or torque, if this
   quantity determines an external controller), and rate limiter output or torque
   reference, if the latter determines the external controller.

A multipage dialog box serves for parameter setting. The third page is shown in
Figure 5.15.

On the first page *DC Machine*, the same DCM parameters are given as in
Figure 5.2. On the second page *Converter*, the parameters of the blocks IGBT and

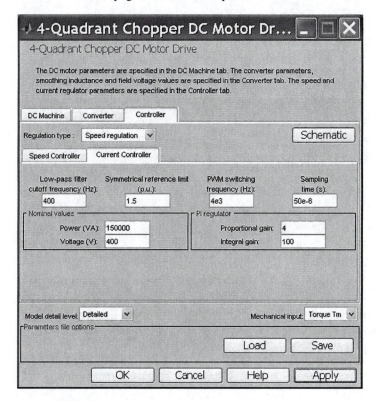

FIGURE 5.15 The third page of the chopper drive dialog box.

Diode are given. In addition, in the upper part of the page, the parameters of the smoothing reactor and the excitation voltage are given.

Let us return to the third page. In the first line "Regulation type" one can select a type of the external regulator: *Speed regulation* or *Torque regulation*. By clicking on the button *Schematic*, the block diagrams of the controllers can be seen. In Figure 5.15, the current controller window is activated. In the spaces of the second line, the cutoff frequency of the filter at current feedback, current limit in pu, the saw-tooth frequency, and current regulator sample time are specified. Furthermore, in the left part the DCM, the rated power and the rated voltage are defined, and in the right—the P and I gains of the current controller.

At the bottom there are two other fields that are on the pages DC *Machine* and *Converter*. In the right field, the mechanical input is chosen. In the left field, the simulation mode can be chosen: detailed or for average values. In the last case the motor model is replaced with the controlled voltage and current blocks whose output values are in certain relations with the duty cycle. Incidentally, a simulation process proceeds much quicker at the expense of loss of most part of the information about real processes in the system.

In Figure 5.16 the speed regulator window is activated. In the lower part of every page there are the buttons *Save* and *Load* that give the opportunity to save the chosen parameters and then to use them again.

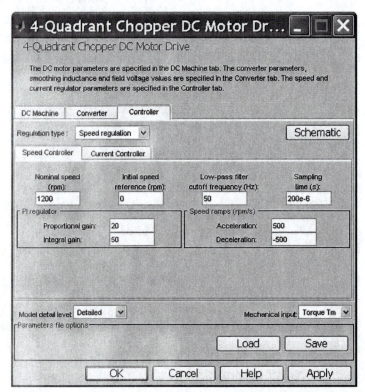

FIGURE 5.16   Parameter window of the chopper drive speed controller.

Use of the electrical drive models developed in SimPowerSystems enables to cut down on elaboration time of the complex systems in which they are used, but, however, it is necessary to bear in mind that it is possible only in cases where the structure of the elaborated or used drives is similar to these models, which is not often the case. In these cases, the user must develop their own models similar to what has been described for the models 5_1–5_3 and the following. For a better study of the chopper models included in SimPowerSystems, the reader can use the given demonstrational models dc5_example—dc7_example.

5.1.2 SATURATION CONSIDERATION

It was mentioned earlier that the DCM model in SimPowerSystems does not take saturation into account. For a machine with a separate excitation, it can be done by that way. Because multiplication by the excitation current is an internal model feature, it means that for saturation consideration the current in the excitation winding has to be equal numerically to the flux. It can be reached by modeling the equation

$$U_f - R_f I_f = \frac{d\Psi}{dt} \tag{5.6}$$

where
U_f is the excitation voltage
I_f is the excitation current
R_f is the winding resistance
Ψ is the machine flux

Usually, the characteristic $\Psi = F(I_f)$ is known, but the inverse relationship $I_f = F^{-1}(\Psi)$ can be used, so instead (5.6) we have

$$U_f - R_f F^{-1}(\Psi) = \frac{d\Psi}{dt} \tag{5.7}$$

that can be solved with using of Simulink blocks. The output Ψ is transformed to the voltage in SimPowerSystems with the block **Controlled Voltage Source** and supplies the machine excitation winding whose resistance is 1 Ω and inductance is zero. The parameter $L_{af} = E_n/\Psi_n \omega_n$ where the index n marks the rated values.

The aforesaid can be explained by the example with the use of the model that is shown in Figure 5.17 and which we recommend to the reader to create on their own. The motor of the rolling mill has the following parameters [2]: $U_n = 880$ V, $P_n = 2000$ kW, $\omega_n = 33$ rad/s, $I_n = 2460$ A, $R_a = 0.018$ Ω, $L_a = 0.42$ mH, $J = 1200$ kg m², $R_f = 4.9$ Ω, $L_{af} = (880 - 0.018 \cdot 2460)/(0.142 \cdot 33) = 178$; the rated excitation current is 40 A, and the dependence flux/excitation current is given in Table 5.1.

In Figure 5.17 the summer input is the excitation voltage. With the help of the block **Lookup Table**, the dependence that is reverse, relatively, magnetization curve is realized, and the block adjustment is shown in Figure 5.18. In the dialog box the said parameters are set including $R_f = 1$, $L_f = 0$. The machine works as a generator. For doing so, the initial speed is set to 33 rad/s, and inertia is equal to a very large

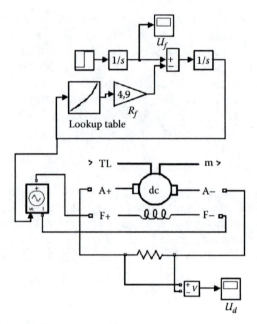

FIGURE 5.17 Magnetization curve model.

TABLE 5.1

Flux Dependence on Excitation Current

I_f, A	0	10	20	30	40	50
Ψ, Wb	0	0.05	0.09	0.12	0.142	0.16

⊟ Function Block Parameters: Lookup Table   ✕

Lookup

Perform 1-D linear interpolation of input values using the specified table.
Extrapolation is performed outside the table boundaries.

| Main | Signal Attributes |

Vector of input values: [0 0.05 0.09 0.12 0.142 0.16]   Edit...

Table data: [0 10 20 30 40 50]

Lookup method: Interpolation-Extrapolation

Sample time (-1 for inherited): -1

OK   Cancel   Help   Apply

FIGURE 5.18 Window of the magnetization curve parameters.

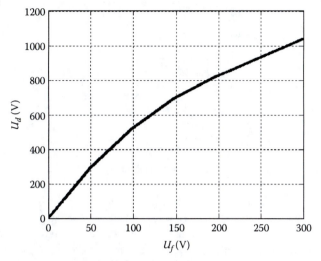

FIGURE 5.19 Actual magnetization curve.

value, for example, $J = 10^{10}$, in order to prevent the speed change during simulation. During the simulation the excitation voltage grows linearly. The simulated dependence of the machine voltage U_d on U_f is shown in Figure 5.19. One can see that the machine characteristic is depicted right. For example, at $U_{frate} = (40A) \cdot (4.9\,\Omega) = 196B$, U_d (no-load emf) is 830 V.

This model of DC machine is used for studying the thyristor drive with two-zone speed control. The drive diagram is shown in Figure 5.20 [2].

The DCM is supplied from the regulated four-quadrant six-pulse rectifier that is supplied from the network 6.3 kV through the transformer. A firing angle α is set by the synchronized six-pulse generator (SPG) and is determined by the current controller CR. CR reference is the output of the speed controller SR. The motor is excited by nonreversible exciter, whose firing angle is set by the flux controller FR that receives the signals of the excitation current and also the motor voltage and current (for emf calculation), of the motor speed. The logic switching unit LSU switches over the rectifier bridges using the signals of the reference and actual motor currents.

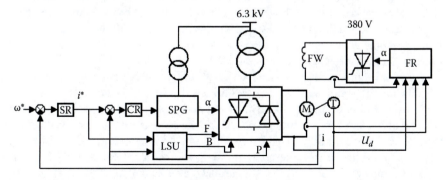

FIGURE 5.20 Block diagram of the reversible DC drive with flux weakening.

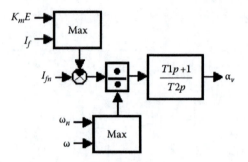

FIGURE 5.21   Block diagram of the excitation controller.

FR block diagram is shown in Figure 5.21 [2]. The controller is PI type. Its gains are decreasing with the help of divider when the speed ω is above the rated ω_n. At the FR input, the reference of the rated excitation current I_{fn} and the largest of two signals—the excitation current I_f and emf $k_m E$—are compared; moreover, the scale of emf is fixed so that under ω_n and I_{fn} the signals I_f and $k_m E$ are numerically equal. While $k_m E < I_{fn}$ (up to ω_n), a current control is active; during an increase in speed $k_m E$ becomes more than I_f and emf control (keeping up rated value emf) is active.

Model5_4 simulates this electrical drive. The motor data are as provided earlier. The saturation is determined as shown in Figure 5.17. The load is permanent, 30 kN-m, and it is modeled in the subsystem **T-load**. The smoothing reactor with parameters $L = 0.64$ mH, $R = 0.8$ mΩ is installed in the armature circuit. DCM is supplied from the thyristor rectifier by using the blocks **Universal Bridge**, with the option *Thyristors*. The reversible rectifier circuit has been considered in model4_2 already. The rectifier is powered from the network 6.3 kV through the 3-phase trans-former $P_n = 3{,}180$ kVA, $U_1 = 6300$ V, $U_2 = 902$ V, $I_2 = 2042$ A, $e_k = 7.2\%$, $P_{\text{s-c}} = 19{,}700$ W, $P_{\text{no-load}} = 6{,}400$ W, and $I_{\text{no-load}} = 1\%$.

It follows from these data $I_1 = 3180/(1.732 \ 6.3) = 292$ A, $R_1 = 19{,}700/(3 \ 292^2) = 0.077 \ \Omega$, $R_b = 6300/(1.732 \ 292) = 12.5 \ \Omega$, $R_1^* = 0.077/12.5 = 0.006$ where R_1, R_1^* is the transformer phase resistance in SI and pu (referred to the primary winding) and R_b is the transformer phase base resistance (referred to the primary winding). Because distribution of this resistance between the windings is not known, we accept that the resistances of the primary and secondary windings are equal in pu.

The inductances of the both windings in pu are equal also and total comes to e_k. The primary no-load current is $(0.01 \ 292) = 2.92$ A, $L_m = 100$ pu; resistance R_m is found from the equation: $P_{\text{no-load}} = 3 \ 2.92^2 \ R_m$, from that $R_m = 250 \ \Omega = 20$ pu. The transformer dialog box is shown in Figure 5.22.

The block **Synchronized 6-Pulse Generator** with the options *frequency* 50 Hz, pulse *width* 10°, and *double pulsing* "on" is used for SPG. The subsystem **LSU** has been considered in Chapter 4 already. The rectifier voltage control subsystem **U-reg** consists of **SR** and **CR**—as in Figure 5.20. If this subsystem is opened, one can see that **SR** has the **Rate Limiter** in the reference circuit, PI-controller, and the **Limiter**, whose output is the current reference. The limit values are set ±5000 A. **CR** is real-ized with using of **PI** subsystem with a parallel connection of the circuits; at the sub-system output the arccosine block is set for characteristic linearization (Chapter 4).

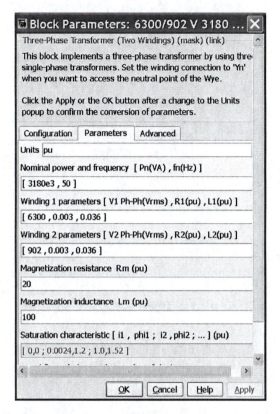

FIGURE 5.22 Dialog box of the power transformer.

In **CR** the input PI signal changes its polarity during changeover of the bridges; that is necessary because for both bridges the same SPG is used and the current direction changes because of the bridge switching-over. During a pause, **CR** is put in the mode that corresponds to inverter in order to prevent the current overshoot. When the reference and actual speeds are equal to zero simultaneously, the logic signal *Block* is formed that removes the pulses from both bridges.

FR subsystem **Field_Contr** is designed according to diagram Figure 5.21. The emf calculation is made as follows. Since

$$E = U_d - I_a R_a (1 + sT_a), \quad T_a = \frac{L_a}{R_a} \tag{5.8}$$

where s is a derivative symbol, then

$$E' = \frac{E}{1 + sT_a} = \frac{U_d}{1 + sT_a} - I_a R_a \tag{5.9}$$

and instead of E, we use E'; thus, a differentiation is avoided. The thyristor exciter is simulated as a first-order link with a time constant of 10 ms and with the output voltage limiter ± 500 V. By scope **Motor** one can observe the motor rotating speed ω, the

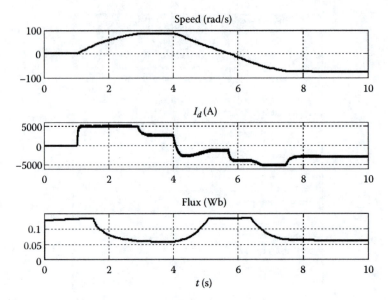

FIGURE 5.23   Transient responses in the DC drive.

armature current I_a, the flux Ψ, the torque T_e, and by scope $\mathbf{U_a}$—the motor voltage. In Figure 5.23 the acceleration to the speed 80 rad/s and the reverse speed −80 rad/s is shown. One can see clearly the areas of a field weakening.

The new DCM model developed takes saturation into account: block **DC_Machine_Sat**. The reader can move the block in his catalogue and use it during the simulation of the DC drive.

The principles of DC machine model that are accepted in SimPowerSystems are used at most under the block development. The structure of the block is shown in Figure 5.24. The signal emf E is formed in the subsystem **Mex** that uses the Simulink blocks and is transferred in SimPowerSystems with the help of the block **Controlled**

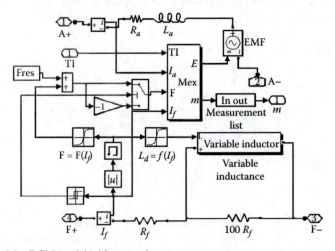

FIGURE 5.24   DCM model with saturation structure.

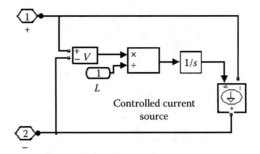

FIGURE 5.25 Variable inductance model.

Voltage Source (EMF); R_a and L_a are the resistance and the inductance of the armature, respectively. The excitation circuit is a series connection of the resistor R_f and of the variable inductance, whose model is borrowed from the Demos [1] and is given in Figure 5.25. The model uses the relationship

$$I_f = \frac{1}{L} \int V \, dt \tag{5.10}$$

where
    I_f is the current in the inductance L
    V is the inductance voltage

I_f is determined by the block **Controlled Current Source** output. Simulink signal that numerically is equal to the inductance value under the given current I_f comes at block **Variable Inductance** input; moreover, the so-called dynamic inductance $L_d = d\Psi/dI_f$ is used. The dependence $L_d = f(I_f)$ is carried out with a table.

The flux Ψ could be calculated by an integration of the value $U_f - R_f I_f$ where U_f is the excitation winding voltage, but the simulation shows that, if the number of the points of the magnetization curve is not sufficiently large, the inductance that is calculated as a difference of the next flux values referred to I_f changing is determined only approximately, and under total saturation, when the value L_d is not much, the flux is computed with errors; therefore, it is more preferable to determine the flux with the help of another table $\Psi = f(I_f)$ that corresponds to the magnetization curve. The constant value F_{res} is a residual magnetization that is important during the simulation of the deceleration of DCM with a series excitation, which, before this, rotated without supply (run-out). The blocks **Switch** and **Relay** serve for changing the flux sign and, hence, of the torque and emf signs during changes in the current direction in the excitation winding.

The motor motion equation (5.5) under $T_f = 0$ is computed in block **Mex**; the torque and emf are calculated as $T_e = K_e \Psi I_a$, $E = K_e \Psi \omega$; at output m five signals are gathered as a vector: $[\omega, I_a, \Psi, T_e, I_f]$.

Model5_5 shows the use of this model for series DCM. The main circuit diagram is given in Figure 5.26. In comparison with Figure 5.4, the excitation winding is added that is connected in series with the DCM armature through the rectifier that provides the invariable emf polarity under transition in the braking mode. Model diagram differs from model5_1 because the circuits that were connected with the parallel excitation

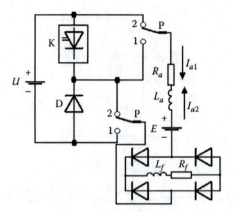

FIGURE 5.26   DC drive with series motor main circuit.

winding are excluded and the rectifier in the series winding circuit that uses the block
Universal Bridge with options «*Number of Bridge Arms =2*» and *Diodes* are included.

Before the start of the model, the program of initialization DC_contr.m has
to be executed, whose example is given below. First, the DCM parameters and the
initial speed are set. Then the points of the magnetization curve are set as a two-
dimensional array: Y (1,:) – I_f points, Y (2,:) – Ψ points where N_{sat} is the number
of the curve points. In order to receive more precise values of the derivative, the
magnetization curve is approximated with a polynomial of the fifth order and using
the MATLAB® function polyfit. With using of the function polyval, the values
of the approximating function for the given excitation current values are computed.
Then the plot of the magnetization curve is built, on which the approximating points
are marked. It is possible to compare the approximation with the reference curve
and, if needed, to increase the polynomial order. Afterwards, the coefficients of the
polynomial derivative are found and its values at the same points I_f are computed;
the calculated dependence of L_d by I_f is shown in figure (2) and then *Fres* is fixed. In
conclusion, the result of calculation is stored in mat-file dc_ser. Of course, one can
give the other name. If the command **load dc_ser** is written in the window *Model
Properties/Callbacks/Model pre-load function* of model5_5, during the model start
the model will be loaded automatically with the requisite parameters, including the
table data. It is possible to write in this window the command **run DC_contr**; then
at start, this program will be executed automatically and will compute parameters as
was done in model5_5 (this command appears in the option *InitFcn*).

Program DC_contr.m

```
Ra=0.05;
La=0.001;
Rf=0.05;
Ke=18.95;
J=10;
Bm=0;
w0=0;
```

```
Nsat=9;%The number points of magnetization curve
Y=zeros(2,Nsat);
Y(1,1)=0;Y(1,2)=60.5;Y(1,3)=121;Y(1,4)=181.5;Y(1,5)=242;
   Y(1,6)=302.5;
Y(1,7)=363.5;Y(1,8)=423.5;Y(1,9)=488;% Excitation current
Y(2,1)=0;Y(2,2)=0.05;Y(2,3)=0.09;Y(2,4)=0.12;Y(2,5)=0.142;
   Y(2,6)=0.15;
Y(2,7)=0.156;Y(2,8)=0.16;Y(2,9)=0.162;%Flux
If=zeros(1,Nsat);F=zeros(1,Nsat);
If=Y(1,:);
F=Y(2,:);
p=polyfit(If,F,5);%approximation
for j=1:Nsat;
s(j)=(j-1)*60.5;
z(j)=polyval(p,s(j));%calculation
end;
plot(If,F,'k',If,z,'.k')%comparison
grid
dp=polyder(p);%output of the polynomial
for j=1:Nsat;
s(j)=(j-1)*60.5;
Ld(j)=polyval(dp,s(j));%inductance calculation
end;
figure(2)
plot(If,Ld,'k')
grid
Fres=0.003;
save dc_ser
```

DCM has the same data as in Figure 5.2, but with series excitation. The rated flux is 0.142 Wb, coefficient K_e = 2.69/0.142 = 18.95. In Figure 5.27 acceleration and deceleration responses are shown as I_{aref} = 242 A (the rated value), and in Figure 5.28 as I_{aref} = 400 A. In Figure 5.27 the torque is 650 N-m. Since without saturation the series DCM torque is proportional to the current in square, when I_{aref} = 400 A it should be expected that the torque is $650(400/242)^2$ = 1776 N-m, whereas, owing to saturation, it is only 1200 N-m.

5.1.3 CONTINUOUS MODELS OF DC ELECTRICAL DRIVES IN SIMPOWERSYSTEMS™

In the folder *Application Libraries/Electric Drives Library/DC Drives* there are four DCM models with continuous (not chopper) control:

    DC1—Two-Quadrant Single-Phase Rectifier DC Drive
    DC2—Four-Quadrant Single-Phase Rectifier DC Drive
    DC3—Two-Quadrant Three-Phase Rectifier DC Drive
    DC4—Four-Quadrant Three-Phase Rectifier DC Drive

The constant flux is assumed common for all the models. DC1 is a single-phase full-wave rectifier with four thyristors. DCM with a smoothing reactor is set in

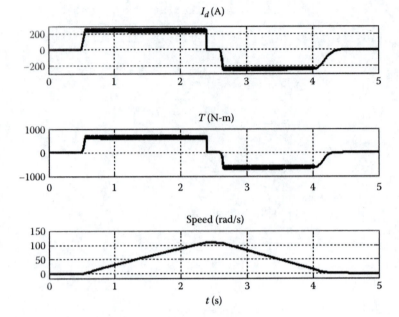

FIGURE 5.27 Transient response under rated current.

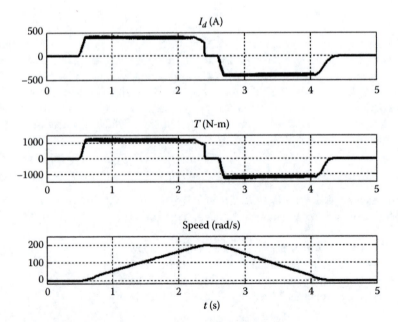

FIGURE 5.28 Transient response under the current of 1.65 rated.

the bridge diagonal. The block SPG is employed for firing pulse generation as in model5_4, the firing angle is determined with PI **CR**, and the loop characteristic is linearized by arccosine transformation. The current reference is generated by **SR** or can be formed outside. **DC3** differs from **DC1** by using the three-phase rectifier instead of the single-phase. **DC2** has two single-phase full-wave rectifiers that are connected to the load antiparallel, as in Figure 5.20 or in model5_4. Each of rectifiers is controlled with separate SPG, whose firing angles α_1 and α_2 are in relation $\alpha_1 + \alpha_2 = 180°$. Since the circulating current flows between the rectifiers, the current-limiting reactors at the rectifiers output are set [2]. Model **DC4,** unlike the previous, has three-phase rectifiers; that is, it differs from model5_4 by the presence of current-limiting reactors and two SPGs.

The model pictures are shown in Figure 5.14c and d. The inputs are the same as for the chopper models, and the terminal for the third phase is added for three-phase models. The output signals are the same as for the chopper drives with the following differences: In the vector *Ctrl* the second component is the firing angle α; for four-quadrant drives the output *Conv* is a four-dimensional vector whose components are equal to the voltages and currents of the rectifier bridges. The dialog boxes are similar to the dialog boxes of the chopper models, but, of course, have a number of differences. It is shown in Figure 5.29.

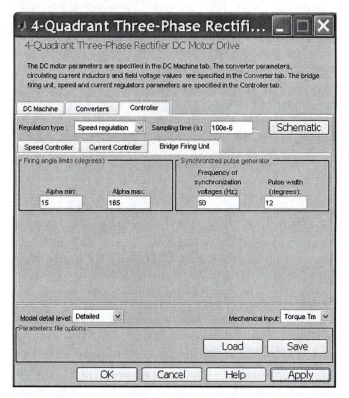

FIGURE 5.29   Dialog box of the continuous DC drive model.

5.2   INDUCTION MOTORS AND ELECTRIC DRIVES

5.2.1   Model Description

Induction motors (IM) are the most widespread type of the motors. The equations of the IM voltages are given as

$$U_{sa} = R_{sa}i_{sa} + \frac{d\Psi_{sa}}{dt}$$

$$U_{sb} = R_{sb}i_{sb} + \frac{d\Psi_{sb}}{dt} \tag{5.11}$$

$$U_{sc} = R_{sc}i_{sc} + \frac{d\Psi_{sc}}{dt}$$

$$U_{ra} = R_{ra}i_{ra} + \frac{d\Psi_{ra}}{dt}$$

$$U_{rb} = R_{rb}i_{rb} + \frac{d\Psi_{rb}}{dt} \tag{5.12}$$

$$U_{rc} = R_{rc}i_{rc} + \frac{d\Psi_{rc}}{dt}$$

Here

U_s, U_r are the stator- and rotor-phase voltages
R_s, R_r are the stator and rotor winding resistances
Ψ_s, Ψ_r are the stator- and rotor-phase flux linkages
i_s, i_r are the stator and rotor winding currents

All the values are referred to a stator.

The equations that link the stator flux linkages with currents are

$$\Psi_{sa} = L_s i_{sa} + l_{sr}\left[i_{ra}\cos x + i_{rb}\cos\left(x + \frac{2\pi}{3} \right) + i_{rc}\cos\left(x + \frac{4\pi}{3} \right) \right]$$

$$\Psi_{sb} = L_s i_{sb} + l_{sr}\left[i_{ra}\cos\left(x + \frac{4\pi}{3} \right) + i_{rb}\cos x + i_{rc}\cos\left(x + \frac{2\pi}{3} \right) \right] \tag{5.13}$$

$$\Psi_{sc} = L_s i_{sc} + l_{sr}\left[i_{ra}\cos\left(x + \frac{2\pi}{3} \right) + i_{rb}\cos\left(x + \frac{4\pi}{3} \right) + i_{rc}\cos x \right]$$

Here

$L_s = L_{1s} + L_m$, L_{ls} is the stator winding leakage inductance
L_m is the mutual inductance of a phase that takes into account the influence of the currents in other phases [3]
$l_{sr} = 2/3L_m$, x is the electrical angle between the phase A stator and the phase A rotor axes in the counterclockwise direction

Analogous for the rotor

$$\Psi_{ra} = L_r i_{ra} + l_{sr} \left[i_{sa} \cos x + i_{sb} \cos\left(-x + \frac{2\pi}{3}\right) + i_{sc} \cos\left(-x + \frac{4\pi}{3}\right) \right]$$

$$\Psi_{rb} = L_r i_{rb} + l_{sr} \left[i_{sa} \cos\left(-x + \frac{4\pi}{3}\right) + i_{sb} \cos x + i_{sc} \cos\left(-x + \frac{2\pi}{3}\right) \right]$$ (5.14)

$$\Psi_{rc} = L_r i_{rc} + l_{sr} \left[i_{sa} \cos\left(x + \frac{2\pi}{3}\right) + i_{sb} \cos\left(x + \frac{4\pi}{3}\right) + i_{sc} \cos x \right]$$

$$L_r = L_{1r} + L_m.$$

When U_s, U_r for currents and flux linkages computation are already known, one must know the time dependence of the rotor position x that can be found from the electromechanical equilibrium equations:

$$J \frac{d\omega_m}{dt} = T_e - f\omega_m - T_m$$ (5.15)

$$\frac{dx}{dt} = Z_p \omega_m$$ (5.16)

Here
  J is the total moment of inertia of IM and of mechanism connected with it
  ω_m is the IM angular rotation speed
  T_e is the IM torque
  f is the viscous friction coefficient
  T_m is the load moment (load torque)
  Z_p is the number of pole pairs

Joint solving of (5.11)–(5.16) is rather difficult, owing to trigonometric functions; therefore, IM simulation, as a rule, is carried out by reducing the three-phase IM to the two-phase IM; such a model is accepted in SimPowerSystems; however, in some cases, the three-phase IM equations have to be used. Such a situation takes place, for instance, when the resistances of the stator windings are different.

For converting to the equivalent two-phase IM, the three-phase values, for example, voltages U_a, U_b, U_c, are replaced with two variables U_q, U_d in the reference frame rotating with speed ω by formulas

$$U_q = \frac{2}{3} \left[U_a \cos\theta + U_b \cos\left(\theta - \frac{2\pi}{3}\right) + U_c \cos\left(\theta + \frac{2\pi}{3}\right) \right]$$

$$U_d = \frac{2}{3} \left[U_a \sin\theta + U_b \sin\left(\theta - \frac{2\pi}{3}\right) + U_c \sin\left(\theta + \frac{2\pi}{3}\right) \right]$$ (5.17)

Here θ is the angle between the phase A axis and the axis q of the reference frame. It is supposed that IM windings are connected in a three-wire Y configuration with an insulated middle point, so that the zero sequence current is null.

Because the line (phase-to-phase) voltages are applied to IM, after substitution of the formulas,

$$U_{ca} = -U_{ab} - U_{bc}, \quad U_a = \frac{U_{ab} - U_{ca}}{3}, \quad U_b = \frac{U_{bc} - U_{ab}}{3}, \quad U_c = \frac{U_{ca} - U_{bc}}{3} \quad (5.18)$$

it follows for the stator:

$$U_{qs} = \frac{1}{3}\left[2\cos\theta U_{abs} + (\cos\theta + \sqrt{3}\sin\theta)U_{bcs}\right]$$

$$(5.19)$$

$$U_{ds} = \frac{1}{3}\left[2\sin\theta U_{abs} + (\sin\theta - \sqrt{3}\cos\theta)U_{bcs}\right]$$

Analogous for the rotor

$$U_{qr} = \frac{1}{3}\left[2\cos\beta U_{abr} + (\cos\beta + \sqrt{3}\sin\beta)U_{bcr}\right]$$

$$(5.20)$$

$$U_{dr} = \frac{1}{3}\left[2\sin\beta U_{abr} + (\sin\beta - \sqrt{3}\cos\beta)U_{bcr}\right]$$

where $\beta = \theta - \theta_r$, θ_r is the rotor angle position. For the squirrel-cage IM $U_{dr} = U_{qr} = 0$.

Usually, one of three reference frames are used: stationary, $\theta = \omega = 0$, and synchronous.

$$\theta = \int \omega_s dt \quad (5.21)$$

ω_s is the synchronous speed that is equal to the stator voltage frequency, $\omega = \omega_s$, and connected with the rotor, $\omega = \omega_r = Z_p \omega_m$

$$\theta = \int \omega_r dt \quad (5.22)$$

The inverse transformation from qd to abc is carried out by formulas:

$$i_{as} = i_{qs}\cos\theta + i_{ds}\sin\theta$$

$$i_{bs} = 0.5\left[i_{qs}(-\cos\theta + \sqrt{3}\sin\theta) - i_{ds}(\sin\theta + \sqrt{3}\cos\theta)\right] \quad (5.23)$$

$$i_{sc} = -i_{as} - i_{bs}$$

and analogous for the rotor with changing θ by β. All the rotor values are referred to the stator.

In axes q-d IM equations are

$$U_{qs} = R_s i_{qs} + \frac{d\Psi_{qs}}{dt} + \omega\Psi_{ds}$$

$$U_{ds} = R_s i_{ds} + \frac{d\Psi_{ds}}{dt} - \omega\Psi_{qs}$$

(5.24)

$$U_{qr} = R_r i_{qr} + \frac{d\Psi_{qr}}{dt} + (\omega - \omega_r)\Psi_{dr}$$

$$U_{dr} = R_r i_{dr} + \frac{d\Psi_{dr}}{dt} - (\omega - \omega_r)\Psi_{qr}$$

(5.25)

$$T_e = 1.5 Z_p \left(\Psi_{ds} i_{qs} - \Psi_{qs} i_{ds} \right)$$

(5.26a)

$$T_e = 1.5 Z_p \left(\frac{L_m}{L_m + L_{lr}} \right) \left(\Psi_{dr} i_{qs} - \Psi_{qr} i_{ds} \right)$$

(5.26b)

$$\Psi_{qs} = (L_m + L_{ls}) i_{qs} + L_m i_{qr}$$

$$\Psi_{ds} = (L_m + L_{ls}) i_{ds} + L_m i_{dr}$$

(5.27)

$$\Psi_{qr} = (L_m + L_{lr}) i_{qr} + L_m i_{qs}$$

$$\Psi_{dr} = (L_m + L_{lr}) i_{dr} + L_m i_{ds}$$

(5.28)

These equations give a generic form of IM differential equations, but they are inconvenient for simulation, because both currents and flux linkages enter (5.24), (5.25). Equations 5.24, 5.25, 5.27, and 5.28 can be converted to the form that is used in SimPowerSystems [4]:

$$\frac{d\Psi_{qs}}{dt} = \omega_{bb} \left[u_{qs} - \omega\Psi_{ds} + R_s / L_{ls} \left(\Psi_{mq} - \Psi_{qs} \right) \right]$$

$$\frac{d\Psi_{ds}}{dt} = \omega_{bb} \left[u_{ds} + \omega\Psi_{qs} + R_s / L_{ls} \left(\Psi_{md} - \Psi_{ds} \right) \right]$$

(5.29)

$$\frac{d\Psi_{qr}}{dt} = \omega_{bb} \left[u_{qr} - (\omega - \omega_r)\Psi_{dr} + \frac{R_r}{L_{lr}} \left(\Psi_{mq} - \Psi_{qr} \right) \right]$$

$$\frac{d\Psi_{dr}}{dt} = \omega_{bb} \left[u_{dr} + (\omega - \omega_r)\Psi_{qr} + \frac{R_r}{L_{lr}} \left(\Psi_{md} - \Psi_{dr} \right) \right]$$

(5.30)

$$\Psi_{mq} = X_{aq}\left(\frac{\Psi_{qs}}{L_{ls}} + \frac{\Psi_{qr}}{L_{lr}}\right)$$

$$\Psi_{md} = X_{ad}\left(\frac{\Psi_{ds}}{L_{ls}} + \frac{\Psi_{dr}}{L_{lr}}\right) \tag{5.31}$$

$$X_{ad} = X_{aq} = \left(\frac{1}{L_m} + \frac{1}{L_{ls}} + \frac{1}{L_{lr}}\right)^{-1} \tag{5.32}$$

$$I_{qs} = \frac{\Psi_{qs} - \Psi_{mq}}{L_{ls}}$$

$$I_{ds} = \frac{\Psi_{ds} - \Psi_{md}}{L_{ls}} \tag{5.33}$$

$$I_{qr} = \frac{\Psi_{qr} - \Psi_{mq}}{L_{lr}},$$

$$I_{dr} = \frac{\Psi_{dr} - \Psi_{md}}{L_{lr}} \tag{5.34}$$

In (5.30) through (5.34) all the reference and calculated values are given in pu; specifically, the speeds are related to the base speed ω_{bb} that usually is equal to the rated stator voltage frequency ω_s. The IM torque in pu is given as

$$T_e = \Psi_{ds}i_{qs} - \Psi_{qs}i_{ds} \tag{5.35}$$

Equation 5.15 acquires a form

$$2H\frac{d\omega_r}{dt} = T_e - f\omega_r - T_m \tag{5.36}$$

$$H = \left(\frac{0.5}{Z_p^2}\right)\left(\frac{J\omega_b^2}{P_b}\right) \tag{5.37}$$

where the load torque is given in pu also, and instead of the moment of inertia J the inertia constant H is introduced, and P_b is the base power. The IM base values are defined in SimPowerSystems as follows. The main parameters are P_b, the base voltage V_b, and the base frequency ω_{bb}. The value P_b is equal to the rated IM power (VA), and the value V_b is equal to the rms of the IM rated phase voltage. The base current value $I_b = 1/3P_b/V_b$, the base torque $T_b = Z_pP_b/\omega_{bb}$, and the base resistance value $R_b = V_b/I_b$. The inductances in pu are equal to the quotient of their reactance for the base frequency and R_b. When the instantaneous values of the voltage, current, flux linkage are considered, their rms values multiplied by $\sqrt{2}$ are accepted for the base values.

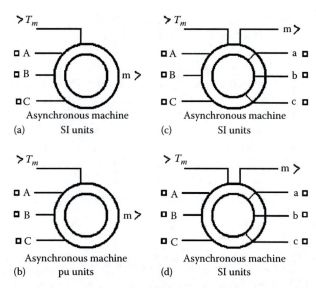

FIGURE 5.30 Pictures of IM models. (a) Squirrel-cage IM, SI parameters. (b) Squirrel-cage IM, pu parameters. (c) Wounded-rotor IM, SI parameters. (d) Wounded-rotor IM, pu parameters.

The IM pictures in SimPowerSystems are shown in Figure 5.30. The models are placed in the folder *Machines*. IM with a wound or a squirrel-cage rotor is picked in the dialog box, first page. IM parameters can be given either as the absolute value (SI) or as the relative value (pu); for that it is necessary to choose in the folder the suitable picture (block icon). Note that simulation is fulfilled in pu always and only the dimensions of the output values and of the load torque are differed. The load torque that is formed in Simulink comes to the input T_m, and the IM variables are the output *m*, which are accessible by the Simulink block (see further).

The first page of the dialog box has the following fields:

"Preset Model". In SimPowerSystems there is information about a number of IM with power from 3.7 to 160 kW. If one of them is investigated, this IM is chosen from the list in falling menu. Otherwise, the text "No" is selected.

"Mechanical input"—we choose "Torque T_m."

"Rotor Type." Either *Wound* or *Squirrel-cage* is put in. With the introduction of version 5.3, it is possible to simulate a double squirrel-cage IM.

"Reference frame." From the menu one of the three reference frames is chosen in which IM is simulated. The general recommendations are as follows [1]: The *Stationary* reference frame is used if the stator voltages are either unbalanced or discontinuous and the rotor voltages are balanced (or 0), the *Rotor* reference frame is used if the rotor voltages are either unbalanced or discontinuous and the stator voltages are balanced, and the *Stationary* or *Synchronous* reference frames are used if all the voltages are balanced and continuous.

On the second page the IM parameters are put in; their meanings are clear from the aforesaid IM equations, including indication about saturation simulation. If it is made, in the next line the $2 \times n$ matrix is put in where the first row contains the values of stator currents and the second contains the values of corresponding stator voltages. The first point corresponds to the point where the effect of saturation begins; n is the number of points of the magnetization curve.

The third page contains information about the sample time for the simulation of this block; this has already been mentioned during the discussions about DCM consideration.

If the IM model is opened with the option *Look Under Mask*, one can see that the model consists of two subsystems. The first one, **Source**, consists of six subsystems in its turn, and one of them computes $\sin \theta_r$, $\cos \theta_r$ and IM slip; the other calculates voltages U_{qs}, $U_{ds} U_{qr} U_{dr}$ by (5.19), (5.20); the third, fourth and fifth realize relations (5.29), (5.30), (5.31); and the sixth one carries out the transformation (5.23). Besides, **Source** computes the torque by (5.35). The second subsystem carries out relations (5.36) and $\theta = \int \omega_r dt$.

The IM variables can be measured with Simulink block **Bus Selector**. This block is connected to the output m; after clicking on the block picture, the dialog box appears, and its left part contains the list of the available output signals; choosing *Select,* the chosen signals are brought over to the right part and appear at the output of the block **Bus Selector**. It is possible to use several such blocks to separate outputs functionally.

In the old versions of SimPowerSystems the block **Machines Measurement Demux** from the folder *Machines* was used for this purpose. Its input must be connected to the output m. Before utilization, its adjustment has to be done. For this purpose, the dialog box has to be open, in which one should select *Machine type: Asynchronous* and further to mark the values that have to be observed.

In the new versions, this block is considered as obsolete; nevertheless, this block is left in the library SimPowerSystems, and its use is often more convenient. Furthermore, both opportunities are used.

Model5_6 simulates IM direct no-load start for the two cases of IM data assignment. The IM has the parameters 5 HP (3730 VA), 460 V, 60 Hz ($\omega_{bb} = 377$ rad/s), $Z_p = 2$. It can be found $V_b = 460/1.73 = 266$ V, $I_b = (1/3)3730/266 = 4.67$ A, and $T_b = 2$ 3730/377 = 19.8 N-m. It is seen after simulation that the forms of the transient response curves are the same and that only the scales that correspond to the said base values are different. In reality, the current amplitude of the first IM in steady-state is 4.7 A, that is, in pu 0.707 4.7/4.67 = 0.71; what one can observe on **Scope1**, the amplitude of the voltages u_{sd} and u_{sq} is 375 V, that is, in pu 375/(266 1.41) = 1, which corresponds to **Scope1** indication; the speed of 185 rad/s corresponds to 1 in pu, and the torque maximum 140 N-m is equal 140/19.8 = 7.07 in pu that corresponds to the last trace of **Scope1** too.

5.2.2 Simulation of IM with Two-Level Voltage-Source Inverter (VSI) and DTC

Furthermore, more complicated simulation tasks for the electrical drives with IM are considered. The first example is the electrical drive with the direct torque control (DTC) [5]. There is the model of such a drive in SimPowerSystems named AC4. The block diagram of the model is shown in Figure 5.31. IM is supplied from VSI that converts the DC voltage of constant value to the AC voltage of variable frequency and amplitude. The stator windings of IM are connected to the inverter outputs.

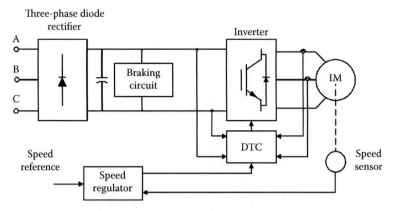

FIGURE 5.31 Block diagram of the electrical drive with IM and DTC.

During deceleration, IM sends energy in DC link, the diode bridge is cut-off, and the capacitor voltage rises up to impermissible level. To avert this, the dynamic braking circuit is used, which consists of the resistor RDT and the switch.

DTC subsystem contains the direct torque controller. The reference values are the required IM torque and the IM stator flux linkage. The reference value Ψ_{sref} is equal to the rated value up to ω_{rate} and decreases inversely proportional when $\omega > \omega_{rate}$. The toque reference is either an output of **SR** (speed controller) or an external signal. DTC contains 3 **Relays**, the torque error e_m is sent to two of them, and the flux linkage error e_φ goes to the third **Relay**.

These relays form the output characteristics shown in Figure 5.32.

With the help of two tables, the first one as $e_\varphi = 1$, the second one as $e_\varphi = 0$, the inverter states are chosen. In total there are 8 states—6 active and 2 zero. Choosing the next state depends on the output of the tables and on the position of the stator flux linkage vector θ. In fact, it is not the precise θ value but the number of 60° sector in which this vector resides is of interest. In Figure 5.33 the vector Ψs is shown in the system of space vectors of VSI voltages.

In this figure the numbers from 1 to 6 are given to each possible state of the voltage space vector, and each of these numbers written as a three-bit code shows the VSI state, as explained in Chapter 4.

Furthermore, if, for instance, the vector Ψ_s is in the first sector, as shown in Figure 5.33, then if $V_k = 6$ and $V_k = 2$ are chosen, the IM torque increases, and when

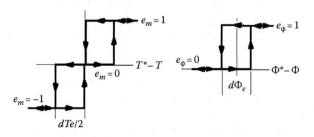

FIGURE 5.32 Relay controller characteristics in the drive with DTC.

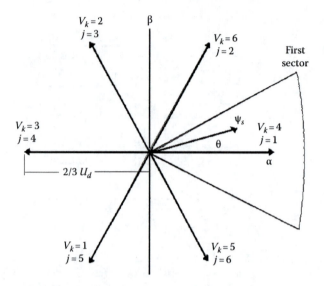

FIGURE 5.33 Vector Ψ_s in the system of VSI voltage space vectors.

$V_k = 1$ and $V_k = 5$, it decreases. Besides, when $V_k = 6$ and $V_k = 5$ the IM flux linkage increases and when $V_k = 2$ and $V_k = 1$ it decreases. If, for example, the flux linkage is small ($e_\varphi = 1$), then when $e_m = 1$, $V_k = 6$ is selected; when $e_m = -1$, $V_k = 5$ and when $e_m = 0$, one of the zero states are selected. If the vector Ψ_s moves to the second sector (during counterclockwise rotation), then the voltage vectors used are displaced counterclockwise as well. We note that in the model described in [1] the other order of the vector designations is used: $V1$ corresponds to $V4$ in Figure 5.33, and afterwards the vectors are numbered in a row in counterclockwise direction; however, we prefer the expounded order that has a physical interpretation.

For speed control, the discrete PI controller is used, a low-pass filter is set at its feedback input, and the rate limiter is set. For computation of the components of the stator flux linkage vector, the relations (5.17), (5.24), (5.26) are used. The starting unit is provided that permits to give the speed reference no sooner than the IM flux linkage reaches some minimum value.

As it was said already, the circuits are provided in the model for the voltage limiting in the DC link. They consist of the series connected switch and resistor. For the switch model, the block **Ideal Switch** is used with the series diode (**Braking chopper** subsystem of the model). The control of the switch is carried out with a width-modulated signal: While DC link voltage U_{bus} is less than the rated value U_{busn}, the switch is open; when $U_{bus} > U_{busn}$ the enabling pulses are given to the switch whose relative duration increases proportionate to the difference $U_{bus} - U_{busn}$. One can consider the drive subsystem diagrams in more detail by carrying out commands Look *Under Mask*.

In Figure 5.34 the picture of the model of the electrical drive with DTC is shown. The system parameters are set in three windows. In the first one, the IM data are put in; in the second one "Converters and DC bus" the parameters of the rectifier, the inverter and the elements in DC link are set (Figure 5.35), and in the third one (Figure 5.36) parameters of SR and DTC are defined.

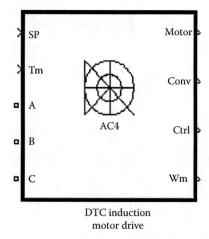

DTC induction
motor drive

FIGURE 5.34   Picture of the electrical drive model with IM and DTC.

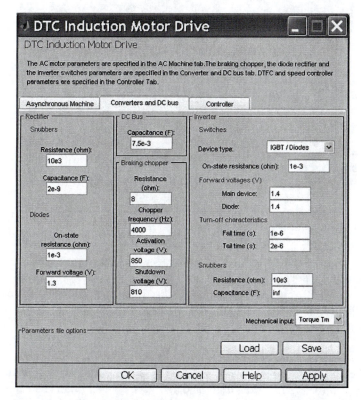

FIGURE 5.35   Parameter window *«Converters and DC bus»* of the electrical drive model with IM and DTC.

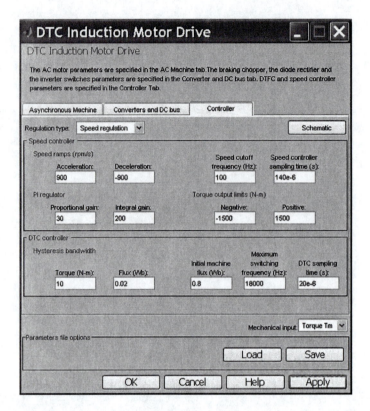

FIGURE 5.36 Parameter window *«Controllers»* of the electrical drive model with IM and DTC.

Модель has the following inputs and outputs:

SPT—reference of the speed (rpm) or of the torque (N-m)
T_m—load torque
A, B, C—supply
W_m—IM rotation speed
Motor—output *m* of IM model
Conv—inverter state vector with components: DC voltage, output current of the rectifier, input current of the inverter
Ctrl—regulator states: torque reference, flux linkage reference, speed error, output of the rate limiter in the circuit of speed reference, or reference torque (depending on the chosen control system type)

Model5_7 shows the use of this model. IM 110 kVA, 400 V is used. One can see the rest of the data in the dialog box (see Figures 5.35 and 5.36 too). The rectifier supply voltage is 570 V. The load torque model of "dry friction" type is achieved with an amplifier having a large gain (set K = 500,000) and Simulink block **Saturation**, whose limit is equal to the load torque (in this case 500 N-m). The scopes fix the

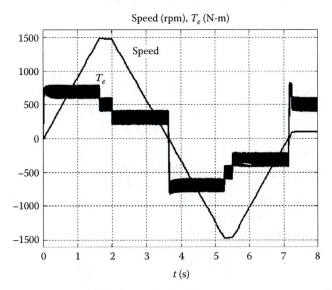

FIGURE 5.37 Transient responses in model AC4.

output signals specified in the model description. The reference speed **Timer** specifies the following operation: speed-up to 1470 rpm, reverse to −1470 rpm, and reverse
to 100 rpm. The speed and torque responses are shown in Figure 5.37.

Use of the said DTC model is limited by the structures of the power and the control
circuits that were incorporated during the design stage itself. But often it is necessary
to investigate the other structures as well; in this case, the user should develop the
model on their own. An example of such a model with DTC is model5_8. The main
differences of this model from the previous one are as follows: The model runs in pu
that simplifies its use when the drive parameters change; the circuit in the dynamic
braking circuit is controlled by the simpler relay circuit; in the control block, (subsystem **DTC_Contr**) modified control algorithm is used; the circuits for estimation of the
stator flux linkage components contain the first-order link with the transfer function

$$W = \frac{T}{Ts+1} \tag{5.38}$$

$T \gg 1$ [6] instead of the integrators, in order to decrease the influence of drifts and
initial values. It is taken $T = 100$ s.

As in the previous model, the reference values for DTC are the required IM torque
and the stator flux linkage. The flux linkage reference is held as $\Psi_{sref}(Phisref) = 1$
because the model in pu is used, and the reference torque is formed by **SR** that uses
the block **Discrete PI Controller**. If the subsystem **DTC_Reg** is opened in the subsystem **DTC_Contr**, we can see that it contains two **Relays** that receive the torque
and flux linkage errors. The hysteresis of the torque **Relay** is ±0.03 and is ±0.01 in
the case of flux linkage **Relay**. The block **Lookup Table1** converts the relay states
in the number 0..3 (Table 5.2).

TABLE 5.2

Coding of the Relay States

DF/DM	0	1
0	0	1
1	2	3

Choosing the next inverter state depends on the output of this table and on the angle position of the flux linkage vector, which is characterized by the number of 60° sector, in which this vector is located (see above). Thus, one of the four inverter states correspond to each input of **Lookup Table1**; the values determined are thus relevant to the sector number of the vector Ψ_s position. When it is necessary to decrease the torque, the zero states are used, which leads to a reduction in the torque pulsation. Such a system works well even at very low speeds when it is reasonable to use only active states. Furthermore, with use of the zero states and with rotation direction changing, it is necessary, for the same sign at output of the block **Relay**, instead of the previously used zero states, to use the active states and vice versa. Therefore, three blocks **Lookup Table0**, +, − are used in the subsystem **DTC_Reg**. They have two inputs: one from the output of **Lookup Table1** and the other from the block **Round** that defines the sector number for the vector Ψ_s. The outputs of the tables are the wanted inverter states. Furthermore, the data of the block **Lookup Table0** appear at the output, if the absolute speed value is less than 0.05; the data of the block **Lookup Table+** appear when $\omega > 0.05$ and the block **Lookup Table** when $\omega < -0.05$. The data of **Lookup Table0** are given as follows (Table 5.3).

The first two columns in the block **Lookup Table+** and the last two columns in the block **Lookup Table**—consist of zero. At the output of these tables, three blocks **Table_a,b,c** are set that convert the number V_k into a binary code that controls the a, b, c phase switches.

The subsystem **Feedback** contains, between its inputs and outputs, the blocks **Zero_order Hold** for imitation of the discrete computation. The hold-time is 25 μs. The block **Memory** (one integration step delay) is put into because in its absence the

TABLE 5.3

Inverter States under Low Speed

Row/Column	0	1	2	3
−3, 3	6	2	5	1
−2	2	3	4	5
−1	3	1	6	4
0	1	5	2	6
1	5	4	3	2
2	4	6	1	3

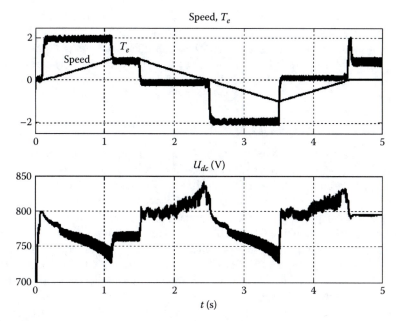

FIGURE 5.38 Transient responses in the second model with DTC.

program informs about the existence of the algebraic loop that is not available in the *Accelerator* mode. It applies to all the following models as well.

The block **Scope** fixes the speed, the stator currents, and the torque of IM, and the scope **Uc** fixes the capacitor voltage.

The block **Timer** schedules the following operations: speeding-up to $\omega = 1$, reverse to $\omega = -1$, and reverse to $\omega = 0.05$. The **Rate Limiter** defines the acceleration time up to $\omega_{rated} = 1$ s. If the dynamic braking circuit is switched off (e.g., to set for the block **Relay** the switch-on point of 8500 V), then the capacitor voltage during deceleration rises to 1800 V. Let us set for the block **Relay** the switch-on point of 850 V. The response is shown in Figure 5.38. Voltage U_c is not as high as 850 V in all the conditions. In the same figure, the speed and torque responses when the load torque is 0.9 are shown.

It is useful to observe the influence of the sample time in the subsystem **Feedback** and the influence of hysteresis values in the blocks **Relay** (the subsystem **DTC_Contr**) on the current shape and on the switching frequency. For measuring the last value, the block **Counter** is used (see model4_6). A time interval during which the number of switching in one phase is computed is set with the blocks **Switch** and **Timer1**. Let us set the time interval T_{iz} from 1.5 s to 2 s (steady state) and the simulation time 2.1 s and carry out the simulation with the different reference speeds in the block **Timer** (from 1 and lower). For $\omega_{ref} = 1$ the number of switching N_a in the phase a is 3057, that is, the switching frequency $f = N_a/(2T_{iz}) = 3057/(2\ 0.5) = 3.06$ kHz (because **Count** fixes both switching-on and switching-off of the switch +). When the reference value decreases, f increases first and then decreases, and reaches for $\omega_{ref} = 0.1 f = 1.8$ kHz. When $\omega_{ref} = 1$, with the help of the option *Powergui/FFT Analysis*, one can find the content of the high harmonics in

IM current. In the window *FFT Analysis*, the *Structure ScopeData1*, input 2, and signal 1 (phase *a*) are selected (*Start time* 1.8 s, *Number of cycles* 10, *Fundamental frequency* 50 Hz, *Max. Frequency* 4000 Hz.) After command Display, one can see THD = 14%. When there is an increase in the flux linkage **Relay1** hysteresis from 0.01 to 0.02, one can see that the switching frequency decreases to 2.7 kHz but THD increases to 16.6%.

5.2.3 MODELS OF THE STANDARD IM DRIVES IN SIMPOWERSYSTEMS™

Besides the previously discussed model AC4, there are three more models of electrical drives with IM in the folder *Application Libraries/Electric Drives Library/AC Drives*, which are described as follows. First, we consider AC2 and AC3 models. The main circuits and the first two pages of the dialog boxes are the same, as for AC4. AC2 uses a space vector modulation and has *V/f* control system in which the IM voltage is proportional to the frequency for retaining the rated IM flux linkage. The control system block diagram is shown in Figure 5.39. The proportional-integral (PI) **SR**, with the rate limiter **RL** at the input and with the filter **F** in the feedback circuit, forms the reference torque; Since the IM torque is proportional to the slip *s*, one can suppose that the controller output is proportional to the requested slip. The summer **Sum** adds this signal to the IM rotation speed; thus, the **Sum** output is proportional to the requested stator frequency, and at the output of the block **Kv** is the requested stator voltage. At the output of the transformation block **Tp1** these values are converted into three-phase voltage system, with amplitude *V* and frequency *f*, and at the output of the transformation block **Tp2** the components α and β of the wanted voltage space vector are computed, and the angle of this vector θ as well. With these data, taking into account the DC link voltage, the on and off times of the transistors are calculated, as was shown in Chapter 4.

The controller page of the dialog box is shown in Figure 5.40. Besides the usual parameters, such as acceleration and deceleration, P and I gains, cutoff filter frequency, sample time, it has fields that indicate the following: *Output limits* for the minimum and maximum slip frequency (field *Controller*), the minimum and maximum IM voltage frequency (field *Frequency*), and the minimum and maximum IM voltage (field *Voltage*). Furthermore, the *Ratio Volts/Hertz* as well as *Zero Speed Crossing Time* are given, and

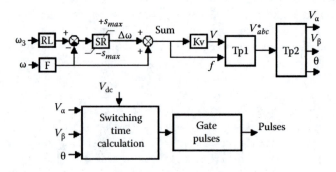

FIGURE 5.39 Block diagram of AC2 model control system.

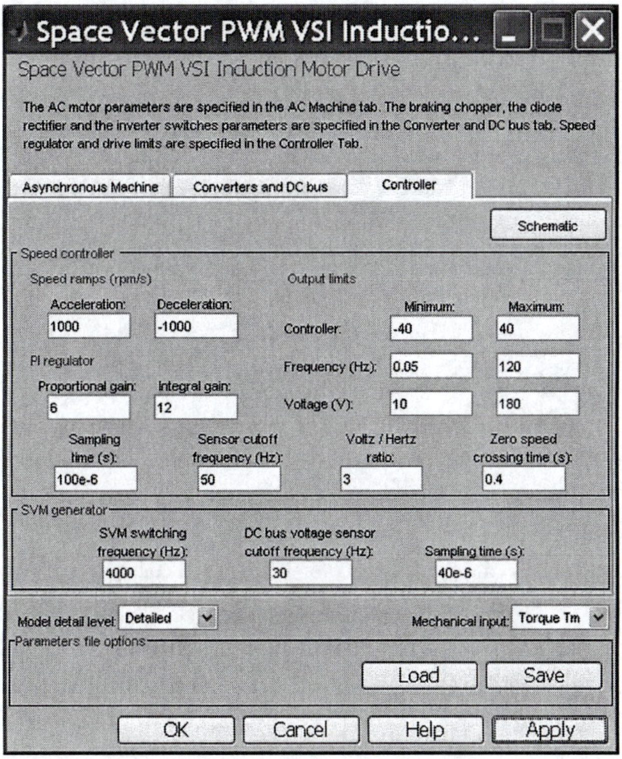

FIGURE 5.40   Parameter window «*Controller*» of AC2 model.

the later is the delay time during zero speed, especially in reverse condition, which is necessary for IM flux linkage decay. In the low part of the page *SVM generator*, SVM switching frequency, the filter cutoff, and the sample period are put in.

The reader can study this model in more detail by following the DEMOS ac2_example. On the whole, this electrical drive does not have the torque and current loops, which limits its speed of response.

Model AC3 uses a principle of the vector control [5] that is based on the control of stator current vector components aligned with and perpendicular to the rotor flux linkage vector. These systems are subdivided into systems of direct and indirection vector orientation. In the former, the position of the rotor vector flux linkage is found by one way or another, and in the latter, an orientation is reached only indirectly, by the corresponding building of the control system. In AC3 model the second method is used.

Inverter control system consists of three units: **SR**, the unit for current references calculation, and the current controller **CR**. SR is the same as in model AC4 and forms the reference signals for the torque T^* and for the rotor flux linkage Ψ_r^* (this value is constant up to the rated speed ω_{mrate} and decreases inverse to ω_m when $\omega_m > \omega_{mrate}$). The block diagram of the second unit is shown in Figure 5.41. The reference value of the current I_d^* is produced by the PI controller of the rotor flux linkage **RFl**; at its input the reference Ψ_r^* and calculated Ψ_r values are compared. The latter

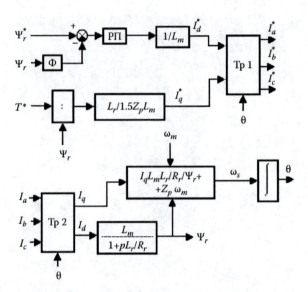

FIGURE 5.41   Block diagram of AC3 model indirect vector control.

is filtered by the filter **F**. The I_q^* reference value is computed from the relation (5.39) that can be received from (5.26), taking into account that during orientation with the flux linkage vector is $\Psi_{qr} = 0$:

$$I_q^* = \frac{2}{3} \frac{T_e^* L_r}{Z_p L_m \Psi_r} \tag{5.39}$$

The reference values I_d^*, I_q^* are converted in the phase current references I_{abc}^* in the stationary reference frame, with the help of the transformation block **TP1**, when the position θ of the vector Ψ_r is known beforehand.

The measured IM phase currents are used for Ψ_r and θ calculation. They are converted in I_d, I_q quantities with the help of the transformation block **TP2**. The value is derived as

$$\Psi_r = \frac{L_m I_d}{1 + s L_r / R_r} \tag{5.40}$$

(s is the Laplace transform symbol), and the Ψ_r rotation speed in the stator reference frame ω_s is equal to the sum of the mechanical speed $\omega_r = Z_p \omega_m$ and of the slip $\Delta\omega$. The latter value can be found from the first equation (5.25) when $U_{qr} = 0$ and from the first equation (5.28) when $\Psi_{qr} = 0$ as

$$\Delta\omega = \frac{R_r L_m I_q}{L_r \Psi_r} \tag{5.41}$$

The integral of ω_s is an estimation of θ.

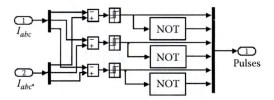

FIGURE 5.42 Scheme of the phase current hysteresis controller.

CR diagram is shown in Figure 5.42. The block carries out a hysteresis regulation. If, for example, the phase *a* current is less than the reference, and both currents are positive, the inverter switch T1 is closed, and in the opposite relation the inverter switch T2 is closed and so on. The regulation method is simple and rapid and does not demand the information about IM parameters. The drawback is an increased distortion of the current shape. At the controller outputs, the flip-flops are set (are not shown in figure), with the help of which the maximum transistor switching frequency is limited.

The unit for the setting of the initial flux linkage is provided in order to make a start with some, usually near to the rated value, rotor flux linkage.

It is worth noting that the rotor reference frame is chosen in the dialog box. The *Controller* page is shown in Figure 5.43.

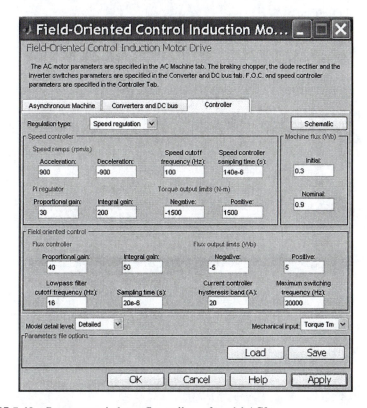

FIGURE 5.43 Parameter window «*Controller*» of model AC3.

Model5_9 simulates the same electrical drive as model5_7, but instead of AC4 (with DTC) AC3 is used. The oscillograms are not given here because they are close to the given ones for AC4 already. We note that the stator current shape is better: THD = 4.26% at the rated speed and with the load torque 500 N-m. It is worth noting that the control system demands, to a considerable extent, precise information about IM parameters that is not known always; thus, the results received are idealized. One of the main aims of simulation is to investigate the influence of having inaccurate IM parameter knowledge on the operation of the electrical drive, which is impossible to ascertain with the model under consideration, and the user has to develop their own model in this case.

Model AC1 uses control *V/f* that has already been used in AC2, but unlike the latter, the system is with open-loop. The reference frequency *f* is converted direct into the gate pulses for IGBT. The control system sets the active states of the inverter that are shown in Figure 5.33 one after the other. Every state is held during the time interval $t = 1/(6f)$. As a result, the line and phase voltages that are shown in Figure 5.44 are formed. Since IM voltage has to change proportional to the frequency, the three-phase thyristor rectifier is used instead of the diode rectifier for changing DC link voltage that defines the amplitude of the IM voltage. The firing rectifier pulses are generated by SPG that has been described earlier, and the firing angle is set by the DC voltage controller. The reference voltage is proportional to the reference frequency.

The circuit is simple enough and was employed much in the past, but to-day, owing to decreasing cost of the power IGBT and the control elements, it is used rarely. Therefore, it is not considered.

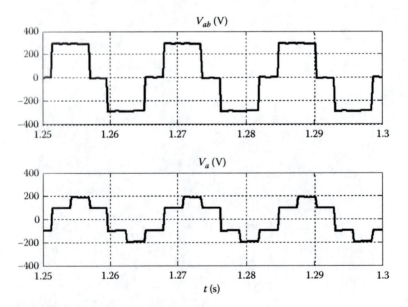

FIGURE 5.44 IM voltages in the model AC1.

5.2.4   IM WITH TWO-LEVEL VSI AND AN ACTIVE FRONT-END RECTIFIER

The circuit with a braking resistor and a switch that is used for reducing of the over-voltage in DC link during braking has a number of drawbacks: during frequent accelerations and decelerations the resistor size turns out too large; the IM energy accumulated is not used effective; with using of the diode front-end rectifier, the supply current is essential non-sinusoidal. For elimination of these shortcomings it is possible to employ an active rectifier; it is a reversible inverter that is connected "back to back" with IM inverter. In the main circuit diagram (Figure 4.30, model5_10) the diode rectifier at the input is replaced with the inverter bridge with the forced-commutated switches. In the control system, control of the components of the supply current space vector that are aligned with the supply voltage space vector (I_{sd}) and perpendicular to it (I_{sq}) is carried out. The reference value $I_{sqref} = 0$, in this case, there is no phase shift between the supply voltage and current; the reference value I_{sdref} defines a value and a direction of the power transferred and keeps the DC link voltage constant. This reference usually is an output of PI voltage controller (**VR**) that keeps up the voltage U_c on the capacitor in DC link. At the motor mode, U_c tries to decrease, and the controller has to support this voltage by increasing of I_{sd} that must be in phase with the supply voltage because the power is transferred from the supply to the motor. At the generator mode, U_c tries to rise, and the controller has to decrease I_{sd} and even to change its sign in order to transfer the power from the motor to the supply. The reference I_{sdref}, I_{sqref} values are converted in the reference phase currents that are controlled with the help of PI controllers and PWM. The controller block diagram is shown in Figure 5.45.

The electrical drive described is simulated in model5_10. IM 37 kW, 400 V is supplied from the inverter bridge **PWM Inverter** that, in turn, is supplied from the rectifier bridge **PWM Rectifier** of the same type. The latter is connected to the grid 380 V through the reactor of 0.5 mH. The subsystem **Control** contains the rectifier control system that was described above and consists of the blocks considered already. The reference value U_c is 900 V. The grid voltage and current are measured with the block **Three-Phase VI Measurement** with options for voltage and current

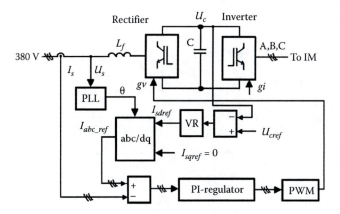

FIGURE 5.45   Block diagram of active front-end rectifier control.

"Use a label". Processes in the grid are fixed by the scope **Source**, the power in the grid is recorded by the scope **Power**, and the capacitor voltage is shown by the scope **Scope_Uc**.

A vector control is provided for IM that uses the same principle as for the model5_9. For this aim, the subsystem **Vector Control** serves that is taken with changes from the one of DEMOS [1] and that is a simplified version of the AC3 control system: the flux linkage controller (it is supposed $I_d^* = \Psi_r^*/L_m$), the initial magnetization and the units for switching frequency limitation are eliminated. In this model, the IM parameters that are used by the control system can be different from the actually IM parameters that allows to use the model for the estimation of the influence of IM parameter inaccurate knowledge. For this purpose, the control system parameters are designated with the additional letter "e"; these parameters are put into the window *File/Model Properties/Callbacks/InitFcn*. It is worth to note that the reference of I_q is calculated in the block Iq* as

$$I_q^* = \frac{2}{3} \frac{T_e^* L_r}{Z_p L_m \Psi_r + b} \tag{5.42}$$

where a small number b (for example, $b = 0.0001 - 0.001$) is put so that not to receive an infinity by start (when $\Psi_r = 0$).

The block **Timer** sets the next schedule: acceleration to the speed 120 rad/s, then reverse to the speed −100 rad/s at $t = 1.5$ s and reverse to the speed 10 rad/s at $t = 3$ s. The load torque is 150 N-m that is equal about 2/3 of the rated one. In Figure 5.46

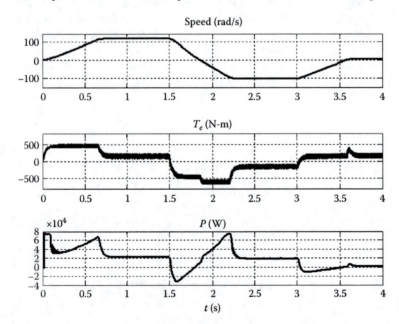

FIGURE 5.46 Transient responses in the electrical drive with VSI and active front-end rectifier.

the speed, IM torque and the supply active power are shown. One can see that during deceleration the power is negative, i.e., an energy returns to the grid. If to look at the supply voltage and current curves, it is seen that during acceleration and at a steady speed, the voltage and the current are in phase; during slow-down, they are in anti-phase. It is seen by the scope **Power** that in all modes the reactive power is about zero. The capacitor voltage is keeping at the reference value 900 V (the deviation is less ±2%), excepting the time of the initial charge. In practice, the special circuits are used for an initial charge.

5.2.5 IM with Three-Level VSI

5.2.5.1 IM with Three-Level VSI and DTC

Model5_11 deals with the three-level inverter that is built with using the block **Three-Level Bridge** (Figure 4.6). There are total $3^3 = 27$ inverter states. If in all the phases the switches connected to "+" or "−" (e.g., T1a, T2a, T3b, T4b, T3c, T4c) are closed, then the inverter voltage space vector diagram does not differ from two-level inverter. These states are named "main"—M-states and, respectively, M-vectors. If the switches connected with the same terminal or with the neutral point are closed (e.g., T1a, T2a, T2b, T3b, T2c, T3c), then the space vectors are aligned with M-vectors but have half amplitude. These states are named H-states. Moreover, each of these vectors can be received by two ways: by connecting to "+" and to the neutral point and to the neutral point and to "−." These states are marked HH and HL, respectively. Besides, there are three zero-states (Z-states), when all the three IM phases are connected together (T1a, T2a, T1b, T2b, T1c, T2c, or T2a, T3a, T2b, T3b, T2c, T3c, or T3a, T4a, T3b, T4b, T3c, T4c). The specific inverter states (I-states) are formed by connecting the three IM phases to three different points (e.g., T1a, T2a, T2b, T3b, T3c, T4c). The space vectors are shown in Figure 5.47, where the states are marked as follows: 1—two top switches are closed, 0—two middle switches are closed, −1—two low switches are closed. The increased number

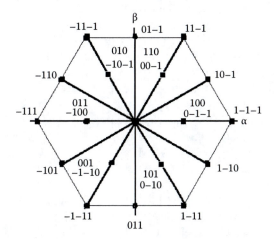

FIGURE 5.47 Voltage space vectors of the three-level VSI.

of the possible inverter states gives the opportunity to receive better characteristics compared to the two-level inverter.

It is reasonable to have different control algorithms at high ($\omega_m > 0.5 - 0.55$) and low speeds. In the first case, M, Z, and I–states are used (also H-states when $T_e > T_{eref}$ and when there is a need to regulate the capacitor voltages [7]). One can see from Figure 5.47 that at Ψ_s location in the beginning of the first sector, by choice $V_k = 2$ (when the torque has to be increased and the flux linkage has to be decreased), IM torque increases very little or there is no decrease at all. If instead of the state −11 −1 that corresponds to $V_k = 2$ I-state 01–1 is selected, the torque rise would be bigger. The same is right for Ψ_s location in other sectors. Really, simulation shows that with the use of I-states for Ψ_s location in the first half of each sector the torque fluctuation and THD of IM current under the same switching frequency decrease by 25%–40% [7].

At low speeds, the active H-states are used, for which the IM voltage is half as much. The torque fluctuation, IM current form distortion, and the switching frequency decrease too. In this case, we can select one of the following states—HH or HL. Since the state selected has an influence on the voltage distribution in the inverter DC link, the state selection is defined by the difference of these voltages and IM mode of operation: In the motor mode ($T_{ref} > 0$) and during IM rotation in the positive direction, the voltage on the capacitor that is connected to the pole "+" decreases when HH-states are used; consequently, the voltage on the other capacitor increases, but in the generator mode situation is opposite. Thus, selecting HH or HL depends on the sign of the product $y = \text{sign}(U_{c1} - U_{c2})\text{sign}(\omega_m)\text{sign}(T_{ref})$.

The model under consideration differs from model5_8 with the use of the **Three-level IGBT bridge**; the capacitor in the DC link is divided into two sections connected in series and their common point is connected to the bridge point N. The resistors that are set in parallel to the capacitors have large resistance and do not influence the control process.

The controller scheme **DTC_REG1** in the subsystem **Control** is modified essentially in comparison with the diagram **DTC_REG** in model5_8. At the output of **Table1**, just as in **DTC_REG**, the numbers 0, 1, 2, or 3 are formed, depending on the signs of the differences between the referent and the measured torque and flux linkage values. Owing to a shift when there is the negative value of the angle, at the output of the block **Round** the numbers 0–6 are formed, and at the output of **Table0** the sector natural numeration 1…6 takes place. With **Sum** and **Relay1** blocks, it is determined, in which half of the sector the vector Ψ_s locates.

Selection of the requested inverter state is carried out by six blocks **Lookup Table**. For that purpose, the inverter states receive definite numbers. M-states have the same numbers as that of the two-level inverter: 1…6; the H-state numeration is carried out by adding 1 for HH and 2 for HL to the M-state number, in whose direction the vector of H-state is aligned (e.g., the state 110 receives the number 61, and the state 00–1—the number 62); and I-states have the two-digit number that consists of the numbers of the adjacent M-states: 26, 23, 13, 15, 54, 46. The blocks **Tablemax+1** and **Tablemax+2** are active at high speeds and when the IM rotation is in the positive direction. Moreover, the first block is active in the second half of the sector and the second block is active in the first one. The first block has the same output states as the block **Lookup Table+** in model5_8 (in the other order because the sector numeration is changed), and in the second block, if there is a need to decrease the flux linkage and to simultaneously increase the torque, M-states are replaced by I-states. **Switch2** selects one of the two blocks **Tablemax+1 or Tablemax+2**, depending on the **Relay1** state. During the IM reverse rotation, the blocks **Tablemax−1** and **Tablemax−2** are active, and the choice is carried out by **Switch3**. The selection of one or the other group of blocks is fulfilled by **Switch 6**, depending on the speed sign. The blocks **Tablemin1** and **Tablemin2** are active under the speed $\omega_m < 0.5 - 0.55$ and are switched on by **Switch5**. These blocks have the same states as the **Lookup Table0** in model5_8 (in the other order), but the second digit 1 is added for the first block and 2 for the second one. The block choice is made by **Switch4**, depending on the said value y. **Switch7** allows to generate the gate pulses some time later (for charging of the capacitors, it is set as 0.1 s), then simulation starts.

The block **Code/Pulses** converts the state codes into gate signals for the switches of **Three- Level Bridge**. This block consists of 12 single-input blocks **Lookup Table**. The input vector for all the tables is the same and contains the list of the possible states. Only one of three possible zero-states is used when all the switches that are connected to the point N are closed. For example, for code zero digit 1 is set in blocks 2, 3, 6, 7, 10, 11, and for code 62 (the last position, code 00-1) digit 1 is set in blocks 2, 3, 6, 7, 11, 12.

The scope **Motor** fixes the IM speed, current, torque, and the scope **U** records the voltage in DC link and the difference between the capacitor voltages.

The simulation results are shown in Figure 5.48. Block **Timer** schedules are as follows: at $t = 0.001$ s, $\omega_{ref} = 1$, at $t = 1.2$ s, $\omega_{ref} = -1$, and at $t = 3.5$ s, $\omega_{ref} = 0.1$.

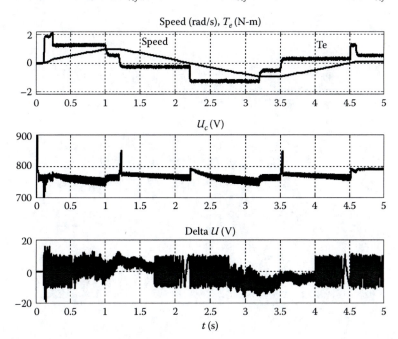

FIGURE 5.48 Transient processes in the electrical drive with the tree-level inverter and DTC.

One can see that voltage imbalance does not exceed 15 V. With the help of Powergui, the content of harmonics in IM current can be studied under different system parameters.

5.2.5.2 IM with Three-Level Inverter and *L-C* Filter

L-C filters are often installed at the inverter output before IM, in order to diminish the overvoltage on the IM windings that arise because of inverter commutation and to decrease the current harmonics. The resonance is possible in this case; the active damping (AD) is used to put it out. One of the possible AD methods is the use of cascaded control method. In this case, the outermost controllers are IM speed and flux linkage controllers, they govern IM currents I_{ad}, I_{aq}, the controllers of the filter capacitor voltages V_{cfd}, V_{cfq} are submitted to this current controllers, and in the innermost control loop there are the controllers of the inverter currents I_{invd}, I_{invq} that are controlled by PWM [8].

The block diagram of the control system is given in Figure 5.49. The reference frame angle θ is calculated just as described for model AC3 earlier. The reference I_{ad}^* is defined by the flux linkage controller output, and the reference I_{aq}^* is defined by the speed controller output, as shown in the Figure 5.41 (not shown in Figure 5.49). In order to diminish the mutual influence of *d* and *q* circuits, the decoupling signals are added to the regulator outputs.

This structure demands information about capacitor voltages, currents of IM and inverter; since use of additional sensors is not desirable, the estimations of the demanded quantities are used. Usually, an electrical drive has DC-link voltage U_c sensor and the inverter current I_{inv} sensors, with the help of which the drive protection is realized.

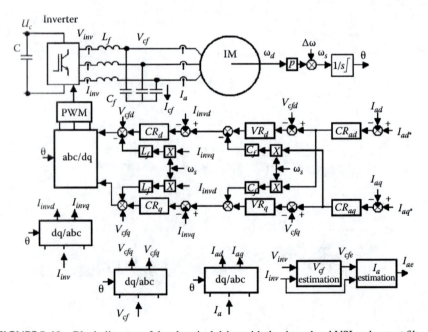

FIGURE 5.49 Block diagram of the electrical drive with the three-level VSI and output filter.

By using the inverter switching functions and voltage U_c, it is possible to compute inverter voltage U_{inv}. Then V_{cf} voltages can be estimated as $V_{cfe} = U_{inv} - R_f I_{inv} - L_f dI_{inv}/dt$. In order to reduce an influence of the differentiation noise, all the terms on the right are put through a filter with the time constant T_1 (it is taken $T_1 = 50\,\mu s$). The IM current I_{ae} can be computed as $I_{ae} = I_{inv} - I_{cfe}$, $I_{cfe} = C_f V_{cfe} s/(1 + R_c C_f s)$ (it is taken $R_c = 0.01\,\Omega$). These signals put through the filter with the same time constant T_1 too. In the control system (Figure 5.49), instead of V_{cf}, I_a, their estimations V_{cfe}, I_{ae} are sent to the inputs of the transformation blocks dq/abc.

In model5_12 IM 1100 kVA, 1200 V is supplied from the three-level inverter with PWM with the switching frequency of 1320 Hz. DC inverter voltage is 2×1000 V, and the circuits for these voltage fabrication are not simulated to speed up simulation. The inductance of the filter phase is 0.4 mH (0.096 pu), resistance is 0.01 Ω, and the phase capacitance is 160 μF. A part of the drive parameters are given in the window *File/Model Properties/Callbacks/InitFcn* where they can be changed. The load torque is 0.9 of the rated value. With the help of the subsystems **Line/Phase**, two phase-to-phase voltages are transformed into three phase-to-ground voltages that are shown in the scope **Voltages**. By the sensors **Iinv**, the inverter current (inductance currents) are measured that are referred to pu afterwards. The display **Frequency** shows the stator voltage frequency, and this value is used in Fourier transformation.

The subsystem **AD_Control** contains the circuits that correspond to the structure shown in Figure 5.49. The subsystems **IM_Control** and **IM_Control1** carry out an indirect vector control as given in model5_10 and are not described here. The difference between these subsystems is that **IM_Control1** is not intended for AD and is used for studying the system characteristics without AD only and during the subsequent simulation it can be removed. In **AD_Control**, the controllers control the filter capacitor voltage components; the subsystem **Inv_Control** contains the controllers of the inverter current components and the circuits for IM current and the capacitor voltage components estimation. The block **Timer** schedules go up to a speed of 0.95 and in reverse to the speed of −0.5 at $t = 1.5$ s.

During the simulation without C_f (in the block C_f a branch type R is selected where R is large enough, for example, 1000 Ω; the tumbler **Switch1** is in the upper position, and simulation time is 1.5 s), the phase voltages at the reactor input and output are formed as shown in Figure 5.50. It is seen that although the reactor reduces IM voltage jumps, they still remain rather big. By choosing the option *Powergui/FFT* one can find that IM current THD = 6.32%, which is not much. If the simulation is repeated with return to a branch type C in the block C_f, we can see that IM current is distorted essentially under both the inverter current feedback (tumbler **Switch2** in the subsystem **IM_Control1** is in the lower position) and the IM current feedback especially, so use of AD is necessary. The subsequent simulation is carried out with the tumbler **Switch1** in the lower position.

At first, we suppose that filter capacitor voltages and IM currents can be measured. For that, the tumbler **Switch** in the subsystem **AD_Control** and the tumbler **Switch3** in the subsystem **IM_Control** are set in the upper positions. After simulation it can be seen that IM voltage and current are nearly sinusoidal: current THD = 2.43%. In order to verify the possibility to use the estimations described earlier, one can repeat a simulation at first with the setting of tumbler **Switch3**

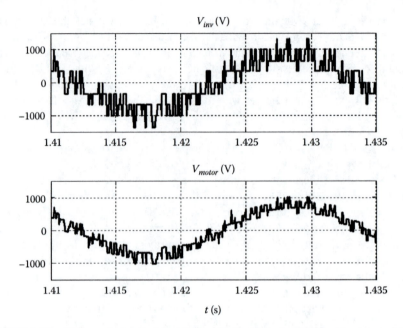

FIGURE 5.50 Voltages in the electrical drive with the three-level VSI without capacitance filter.

(IM current estimation) and afterwards with the setting of tumbler **Switch** (voltage estimation in addition) in lower positions. In the first case, IM current THD = 2.4%, and in the second, THD = 2.22%. We note that THD was calculated by 5 periods of the frequency that was fixed by the block **Frequency**, beginning from $t = 1.35$ s. In Figure 5.51, the waveforms of the voltage and the IM current for this case are shown.

The readers can observe the IM speeding-up and reverse themselves by an increase in the simulation time up to 3.2 s.

5.2.6 SIMULATION OF IM SUPPLIED FROM CHB INVERTER

Simulation of the cascaded H-bridge multilevel (CHB) inverter was considered in Chapter 4.

Here such an inverter is used for IM supplying and control.

In model5_13, IM 2.4 MVA, 3.6 kV is supplied from the CHB inverter that is made according to the diagrams in Figures 4.37, 4.38a with $n = 3$. Nine blocks Zigzag Phase-Shifting Transformer are used, the primary windings are connected in parallel and are supplied from the source 6.3 kV, and the secondary transformer voltage is 1.1 kV.

The subsystem **Cell_Phase** consists of the three foregoing transformers and the subsystem **Inv_phase** that in turn contains three units connected as shown in Figure 4.38a: the rectifier, the smoothing capacitor, the single-phase inverter, and the block PWM too.

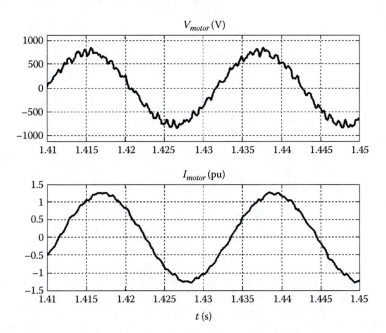

FIGURE 5.51 IM voltage and current in the electrical drive with the three-level inverter, capacitance filter, and AD.

PWM uses single-polar modulation as discussed in Chapter 4, Figure 4.41. The modulation relations are as follows:

if $V_{ref} \geq 0$ and $V_{ref} \geq V_{cp}$, conduct T1 and T4, that is, $V_k = E$
if $V_{ref} \geq 0$ and $V_{ref} < V_{cp}$, conduct T2 and T4, that is, $V_k = 0$
if $V_{ref} < 0$ and $V_{ref} > V_{cp}$, conduct T2 and T4, that is, $V_k = 0$
if $V_{ref} < 0$ and $V_{ref} \leq V_{cp}$, conduct T2 and T3, that is, $V_k = -E$

The triangular carrier waves of each sell in the same phase are shifted by about 120°.

The voltage sensor and the scope are set in the phase A subsystem.

Two modes of IM control are provided that are selected with the tumbler switch **S**. In its lower position, the three-phase signal V_{ref} sets the requested amplitude and frequency of the voltage at the inverter output that allows to study the peculiarity of the inverter output voltages and input current forming; in the upper position, the **Speed_ref** defines IM speed. Speed control system is similar to the one used in AC3, Figure 5.41, and differs, in principle, by I_d, I_q current regulation instead of the phase currents I_a, I_b, I_c. Since in this model a high-speed response is not intended, for its simplification the circuits for decoupling of the d and q regulators are eliminated [5]. In the window *Model Properties/Callbacks/InitFcn* the IM parameters are repeated because the regulator parameters are given in the alphabetical symbols and during simulation their values are to be in the Simulink workspace.

The scope **Motor** fixes IM stator currents, its speed, and the torque in pu. The scope **Inverter** shows the phase inverter voltages, and the scope **Network** records the grid voltage and current.

This system does not assume power recuperation (an inversion mode). At the start of operation, the voltages on the capacitors are seen to rise sharply. Use of braking resistors with the series switches in parallel to capacitors, as in model5_8, is undesirable because of the larger number of these circuits. It is possible to reduce the probability of the origin of the inversion mode essentially by setting either of the low decelerating rate in the **Rate Limiter** or the low braking torque at the **SR** output. However, during deceleration with a big load torque, when IM does not go over to power recuperation, duration of braking would be increased unjustifiably. To avoid this, it is possible to lower the IM deceleration rate as a function of the voltage on one of the capacitors: If this voltage exceeds the permissible value, the deceleration rate decreases. Such a method is used in the model under consideration in the block **Control**: When the voltage on the capacitor of the first phase A inverter cell exceeds 1500 V, at the output of **Gain1** the signal appears, and output of **Sum3** decreases, which in turn leads to a reduction of the reference changing rate at the block ZI output. It is supposed that the voltages on the smoothing capacitors do not differ much.

Speeding up to the rated speed at $t = 0.5$ s and reverse to the speed -0.5 at $t = 2$ s are scheduled. The response curves when $T_m = 1$ are shown in Figure 5.52. It is seen that during deceleration IM torque remains positive and the IM does not go over to power recuperation, so the capacitor voltage U_{dc} is in permissible boundary; one

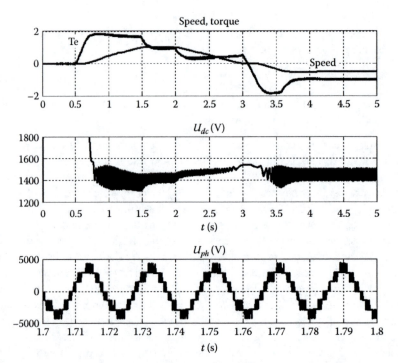

FIGURE 5.52 Processes in the electrical drive with the cascade inverter without recuperation when there is the big load torque.

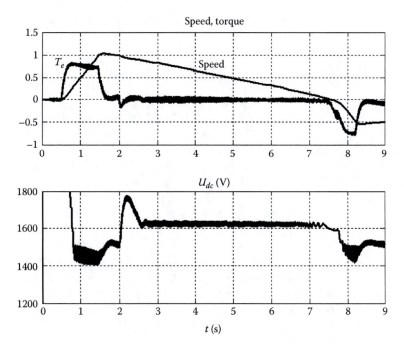

FIGURE 5.53 Processes in the electrical drive with the cascade inverter without recuperation when there is a small load torque.

can see a harmonic with the frequency of 100 Hz and with the swing of 80 V in this voltage when $\omega = 1$. In the phase voltage U_{ph} one can see seven inverter voltage levels. In the grid current THD = 3.2%, with predominance of the fifth and seventh harmonics.

Let $T_m = 0.1$, **Gain1** = 0 in the subsystem **Control**; after simulation it is seen that at $t = 2.25$ s the capacitor voltage reaches 2.6 kV, which is inadmissible. The response curves that are received with Gain1=1/150 are shown in Figure 5.53. The capacitor voltage remains limited, but the braking time increases. Thus, the capacitor voltages are keeping up automatically during deceleration with different load torque by their influence on the braking rate.

In model5_14 the inverter provides the inversion mode, and its cells are made according to the diagram in Figure 4.38b. For speeding-up of simulation that goes on for a rather long time, in each phase only two sells are set, so the phase voltage has only five levels. Accordingly, the IM rated voltage decreases to 2.3 kV. Since the inverter input current is formed by active rectifiers, there is no need to use complex transformers; all the transformer windings are connected in delta with the secondary voltage 725 V. The three-phase transformer model uses three blocks **Multi-Winding Transformer**. The subsystem **R_Control** that is analogous to the one used for this aim in model5_10 controls by each of the six active rectifiers. The capacitor voltage is taken equal to 1300 V.

The inverter control is the same as in the previous model, but the triangular carrier waves of the sells in the same phase are shifted by 180°. Moreover, the signal

ΔV is added to the three-phase **AD_Control** subsystem output **V** (the same for each phase) where

$$\Delta V = -\frac{\max(\mathbf{V}) + \min(\mathbf{V})}{2}$$

that has a triple frequency and the amplitude 0.25 (when the magnitude **V** is 1). The linearity PWM range rises by 15.5%. The circuits for changing of **Rate Limiter** parameter depending on U_{dc} are excluded.

The scope **Motor** fixes IM stator currents, the speed, and the torque, the scope **V_inv** shows the phase inverter voltages, the scope **Network** records the grid voltage and current, and the scope **Udc** fixes the capacitor voltages for one cell of each phase. The scope **Uabc_ref** records the input PWM signals.

The next process is simulated: speeding up to the speed 0.98 at $t = 0.5$ s and reverse to the speed -0.5 at $t = 2$ s. In Figure 5.54 the response curves of the speed and the torque under the load torque $T_m = 0.1$ (Figure 5.54a) and $T_m = 1$ (Figure 5.54b) are shown. One can see that the speed plots are the same and in the first case IM works in the power recuperative mode during braking. By scope **Network** it is seen that the grid voltage and current are in-phase, except the braking area under $T_m = 0.1$ when they are in antiphase, that is, when IM gives its energy in the grid. The grid current is

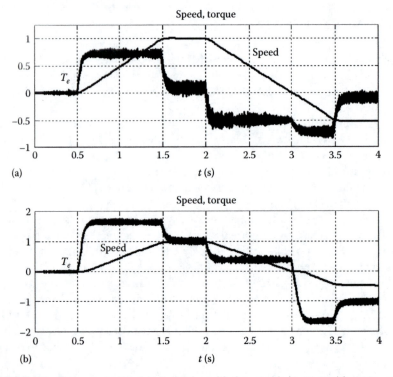

(a)

(b)

FIGURE 5.54 Processes in the electrical drive with the cascade inverter with recuperation. (a) Small load. (b) Big load.

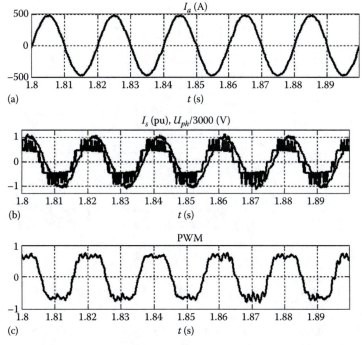

FIGURE 5.55 Currents and voltages in the electrical drive with the cascade inverter with recuperation. (a) Mains current. (b) IM phase voltage and current. (c) PWM input.

nearly sinusoidal (Figure 5.55a), THD < 1%. The scope **Udc** shows that the capacitor voltages keep up constant sufficiently precisely; the oscillations with the frequency of 100 Hz and amplitude of 100 V are observed in these voltages. The IM phase voltage has five levels, and the 13th harmonic with the amplitude about 2% is seen in IM current (Figure 5.55b). In Figure 5.55c the input signal PWM is shown; it is seen that it does not reach the PWM linearity bound (equal to 1), which gives the opportunity to reduce the capacitor voltage and, respectively, the transformer secondary voltage. If ΔV circuits are excluded, the PWM input signal will be about 1.

5.2.7 IM Supplied from the Four-Level Inverter with "Flying" Capacitors

In Chapter 4 the circuits and functioning of the inverter with "flying" capacitors with a passive load were considered. Model5_15 illustrates use of this inverter for IM supplying. IM 1.1 MVA, 1200 V connects to the inverter with the help of the breaker **Br2**. The breaker **Br3** shorts out IM when **Br2** is open; at that point, simulation of the initial capacitor charge is carried out faster; naturally, in the reality it is not needed. The capacitance of the inverter capacitor is 1 mF. The indirect vector control from model5_13 is used. A sequence of actions take place: At first, **Br2** is open and the breakers, through which the capacitors are charged, are closed, as it was described in Chapter 4. At $t = 1.2$ s these breakers are open and **Br2** closes

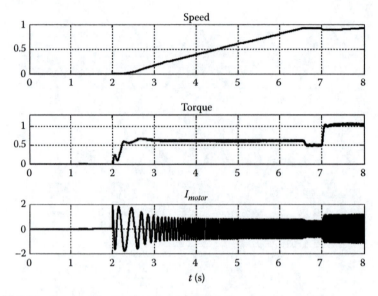

FIGURE 5.56 Processes in the electrical drive with the «flying» capacitors.

(**Br3** is getting open). At $t = 1.4$ s the gate pulses are sent to the inverter switches, and at $t = 2$ s the speed reference is switched on that is equal to 0.95 of the rated value. IM is speeding up with the rated load torque of 0.5. At $t = 7$ s the load torque rises to the rated value. In Figure 5.56 the IM speed, its torque, and the current are shown, and in Figure 5.57 and Figure 5.58 the capacitor voltages and the IM stator voltage and the

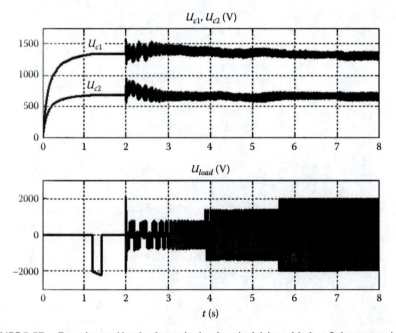

FIGURE 5.57 Capacitor and load voltages in the electrical drive with the «flying» capacitors.

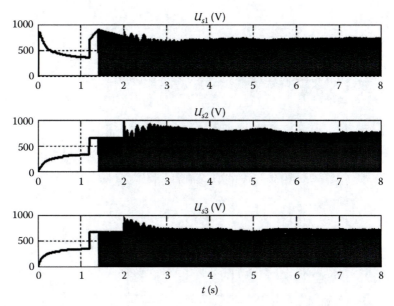

FIGURE 5.58 Voltages across the switches in the VSI with the «flying» capacitors.

inverter switch voltages are shown, respectively. In order to investigate the influence of the auxiliary R-L-C-circuit, one can remove it from the model and repeat simulation. It will be seen that the voltage fluctuation on the switches and the capacitors increase dramatically.

5.2.8 SIMULATION OF THE FIVE-LEVEL H-BRIDGE NEUTRAL-POINT CLAMPED INVERTER (5L-HNPC) SUPPLYING IM

The diagram of a phase of the 5L-HNPC inverter that evolves from the three-level NPC inverter is shown in Figure 5.59. The phase is composed of two NPC legs in the form of H-bridge (Figure 4.6). In Table 5.4 the available switch states, the output voltages, and the active capacitors under these states are given.

Each phase provides five voltage levels (0, $\pm V_d/2$, $\pm V_d$). When three phases are connected as three-phase Y-source, the phase-to-phase voltage has nine levels: 0, $\pm V_d/2$, $\pm V_d$, $\pm 3V_d/2$, $\pm 2V_d$. In this inverter the switch voltage is $V_d/2$ and the output voltage maximum is $2V_d$. In the three-level NPC inverter under the same switch voltage the output voltage maximum is only V_d. The voltages $\pm V_d/2$ can be obtained with different states, when different capacitors are active. It gives the opportunity to use this redundancy for capacitor voltage balancing.

The phases are supplied from the separated sources. Since this inverter is used usually for the power electrical drives, when it is important to reduce the harmonics in the supply current, the multipulse rectifiers are used for V_d fabrication, for example, 18-pulse rectifier that consists of 3 three-phase bridges in series supplied from three windings of the three-phase transformer having a relative phase shift of 20°, as shown in Figure 4.37, or 36-pulse rectifier, when the two aforementioned circuits

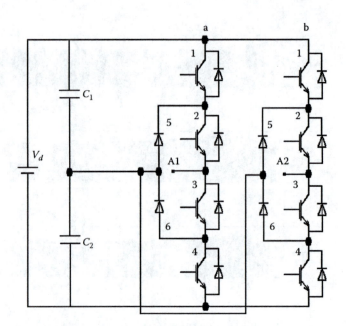

FIGURE 5.59 One phase of the five-level HNPC VSI scheme.

TABLE 5.4

Voltages at the Output of a Phase of the 5L-HNPC Inverter

1a	2a	3a	4a	1b	2b	3b	4b	V_{out}	C
0	0	1	1	0	0	1	1	0	—
0	0	1	1	0	1	1	0	$-V_d/2$	C_2
0	0	1	1	1	1	0	0	$-V_d$	$C_1 + C_2$
0	1	1	0	1	1	0	0	$-V_d/2$	C_1
0	1	1	0	0	1	1	0	0	—
0	1	1	0	0	0	1	1	$-V_d/2$	C_2
1	1	0	0	0	0	1	1	V_d	$C_1 + C_2$
1	1	0	0	0	1	1	0	$V_d/2$	C_1
1	1	0	0	1	1	0	0	0	—

are connected in series and the primary transformer windings have the phase shift of 30°. Such electrical drives are produced for the power of 22 MVA and more, the voltage of 6.9–7.2 kV; both DTC and PWM are used in the control system.

In PWM it is possible to compare two modulating waves U_{ref+} and U_{ref-} that have the same frequency and amplitude, but are 180° out of phase, with two triangular waves with frequency of F_c that are vertically disposed; one of these waves changes in the range 0…1 and the second one in the range −1…0. From the Table 5.4 it follows both the states of the switches 1 and 3 and the states of the switches 2 and 4 are

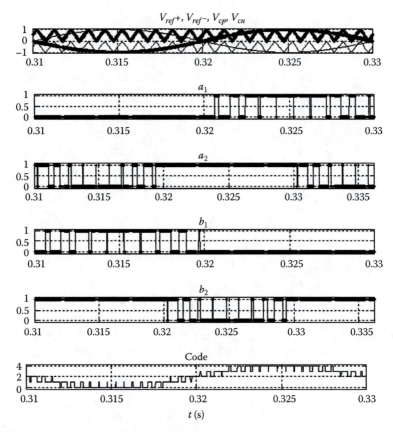

FIGURE 5.60 PWM modulation in 5L-HNPC.

always opposite. Thus, it is sufficient to generate only four gate signals, for example, for $a1$, $a2$, $b1$, $b2$. In Figure 5.60 and in the left area of Figure 5.61 gate pulse generation is shown. The average switching frequency of a switch is $F_c/2$, and the voltage harmonics have sidebands centered around $2F_c$ and its multiples. However, such a control is possible when there is no problem to balance the capacitor voltages, i.e., when each capacitor receives the independent supply that takes place far from always. In general, the control system has to provide such balancing. The possible schematic diagram is shown in Figure 5.61. The outputs of the comparison circuits are summed up by the block Sum. Its output is a 5-level step signal *Code* (Figure 5.60), whose levels correspond to the demanded levels of the inverter-phase voltage (in Figures 5.60 and 5.61 the signal *Code* is shifted by 2 unity up so that it changes from 0 to 4). The blocks **Table** are used for converting the signal *Code* into inverter gate signals; moreover, for each switch one of two tables can be used, depending on the signal that controls the voltage capacitor difference dU_c. For example, it follows from Table 5.4 that for the switch $a1$, during a change in the signal *Code* from 0 to 4, the output of the corresponding tables has to be [0 0 0 1 1] or [0 0 0 0 1]. The contents of the other tables are not given here and can be gathered by studying model5_16.

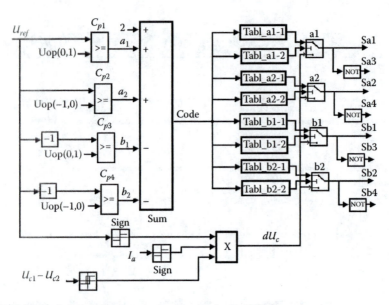

FIGURE 5.61   Scheme of the gate pulse forming in 5L-HNPC.

The signal dU_c depends on the sign of the difference $U_{c1} - U_{c2}$ and on the sign of load power, and this sign is equal to the logical product of the signs of the voltage and the current; instead of the output voltage, the modulating wave can be used. If, for example, $U_{c1} > U_{c2}$ and the power is positive, the pulse sequence should be used, with which the capacitor C_1 is loaded, and if the power is negative, the pulse sequence should be used, with which the capacitor C_2 is loaded because in this case its voltage is increasing (the instantaneous power comes to the supply) and, correspondingly, C_1 voltage is decreasing.

It is worth noting that the relationship between the switching frequencies of the different switches depends on the used states that correspond to the voltage $\pm V_d/2$ because output voltage fabrication occurs with switching-over either between the levels 0 and $\pm V_d/2$ or between the levels $\pm V_d/2$ and $\pm V_d$. In the first variant of PWM mentioned earlier, the on times of each of the four possible states corresponding to $\pm V_d/2$ are about equal, which provides nearly equal switching frequency for every switch. In the second variant of PWM, in contrast, the switches can switch over with different frequencies. Since the capacitor voltage difference rises rather slowly for the first scheme, it is possible to propose the method, in which the most of the times the first PWM scheme is used, and from the origin of a rather large voltage difference, the second scheme is used for a short time.

The system described is shown in model5_16. The load is 8-pole IM, 6 MVA, 5600 V.

Each of the inverter phase models (5L_PhaseA, 5L_PhaseB, 5L_PhaseC) contains the inverter switches and the capacitor according to Figure 5.59 and PWM according to Figure 5.61 (the subsystem **Phase_Control** consists of several subsystems). The subsystem **Compare** contains the two triangular waveform generators, the comparators for comparison with the input-modulating signals, and the summer. The subsystem **Tables** contains the blocks **Table** for converting the signal *Code* into the gate signals for the switches a1, a2, b1, b2, depending on the signal dU_c. The subsystem

Pulses1 takes four switch-control signals from either the subsystem **Compare** or the subsystem **Tables** and transforms them in the direct and inverse gate signals. Besides, the subsystem **Phase_Control** has the circuits for forming the signal dU_c according to Figure 5.61 and the circuits for any specific PWM variant to be chosen. For this aim, **Switch2** and the tumbler **Switch1** are used. If the latter is in the lower position and the constant **Contr_Mode** =1, the first PWM variant is realized, and if this constant is 0, the second variant (with capacitor voltage balancing) is realized. When the tumbler **Switch1** is in the upper position, the system operates in the following way: When the capacitor voltage difference (absolute value) rises and reaches the value that is set up in the block **Relay1**, the system, which used the first variant before that, goes over to the second variant, until the said difference decreases and **Relay1** turns off.

The subsystem **Count** computes the switching frequency of the switches a1, a2, b1, b2 and the average value. Duration of the measurement (set 0.5 s) and the time when a computation begins (set 0.8 s) are specified by the block **Timer**.

The scope **Scope_Phase** fixes the inverter-phase voltage and current, capacitor voltages, and their difference.

The inverter is supplied from the subsystem **Power**. Each of the phases receives the DC voltage from three diode bridges connected in series that are supplied from 3 three-phase transformers 800 kVA that are connected zigzag with the voltage-phase shifts 0 and ±20°. The primary voltage is 6.3 kV, and the secondary one 1100 V. The scope **Power** records the supply phase-to-phase voltage and the total current supplied.

There are two versions of the control systems that are placed in the subsystem **AD_Control** and selected by the tumbler **Switch**. In the first case, the three-phase harmonic signal of the reference amplitude and frequency is applied to the PWM input (perhaps, through the scheme for the third harmonic addition), and in the second case, the output of the IM vector control system that is taken from model5_10 is applied to the PWM input.

With the help of the scope **Scope_Load**, one can observe the load phase-to-phase voltage and the load-phase current.

Let us begin with simulation. In the subsystem **AD_Control** the tumbler **Switch** is in the upper position. It is set in the block **3-phase Programmable Source**: amplitude of 1, frequency of 50 Hz. By setting **Gain1** = 0.5, the third harmonic is added. The triangular wave frequency F_c = 1000 Hz is set in the option *File/Model Properties/Callbacks*. In the subsystem **Phase_Control** the tumbler **Switch1** is set in the lower position, and the constant **Contr_Mode** is equal to 1. The initial slip is set as 0.02 in the IM model. The simulation time is set as 1.3 s. After simulation one can see that the switching frequencies of the switches are nearly the same and equal to 500 Hz. Observing the 4th trace of the scope **Scope_Phase**, it is seen that the difference of the capacitor voltages is rising continuously and at t = 1.3 s reaches of 200 V. If the simulation is repeated with **Contr_Mode** = 0, it can be seen that, with the average frequency of 500 Hz, the switching frequencies of the individual switches are different essentially but the difference of the capacitor voltages is not more than 50 V. If the simulation is repeated with the tumbler **Switch1** (in the subsystem **Phase_Control**) in the upper position, the average switching frequency is equal to 502.5 Hz and the individual switching frequencies differ insignificantly (492–517 Hz); the capacitor voltage difference fluctuates in the range of 10–40 V.

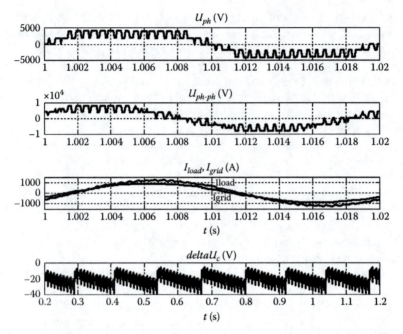

FIGURE 5.62 Phase and line load voltages, load and supply currents, capacitor voltage difference in 5L-HNPC.

The phase- and line-load voltages, the load and supply currents, and the capacitor voltage difference are shown in Figure 5.62. The first voltage has five levels and the second one nine; both currents are nearly sinusoids (THD = 3.6% and 2.6%, respectively). The rms value of the line voltage is 4960 V and the mean value of the DC inverter voltage is 4050 B, which correspond to the theoretical value: $V_{leff} = 0.612\ 2\ 4050 = 4957$ V. With the help of the option *FFT Analysis*, one can find that harmonics are centered about the frequency of 2000 Hz and multiple of it.

Let us make simulation using the following conditions: in the subsystem **AD_Control** the tumbler **Switch** is in the lower position, the timer **Speed_Ref** schedules speeding-up to the speed 0.2 and afterwards to 0.98, the load torque is equal to 1, the initial IM slip is 1, and the simulation time is 2.5 s. In Figure 5.63 the speed and the torque responses are shown, and in Figure 5.64 the capacitor voltages are plotted. Now the displays show the double switching number for the time interval [0.8–2.5] s, that is, in transient.

Consider DTC for this type of the inverter. At first, it is necessary to receive a picture of the inverter space vectors. If the phase voltages are V_a, V_b, V_c, then the components of the voltage space vector are

$$V_\alpha = \frac{2V_a - V_b - V_c}{3}$$

$$V_\beta = \frac{V_b - V_c}{\sqrt{3}}$$

(5.43)

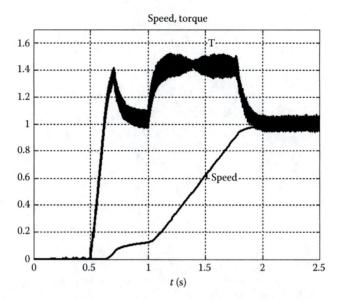

FIGURE 5.63 Transient processes in the electrical drive with 5L-HNPC.

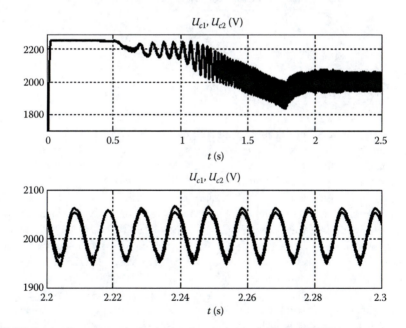

FIGURE 5.64 Capacitor voltages in the electrical drive with 5L-HNPC.

If $V_a + V_b + V_c = 0$, $V_\alpha = V_a$. In order to build the system of the space vectors, it is necessary to give to each of the phase voltages its five possible values. These space vectors are shown in Figure 5.65 [9]. The letters at the triangle corners show inverter states, under which these vectors are realized. In this figure decoding of the accepted letter notation is given. The states of the switches 1a, 2a, 1b, 2b are given. It is seen

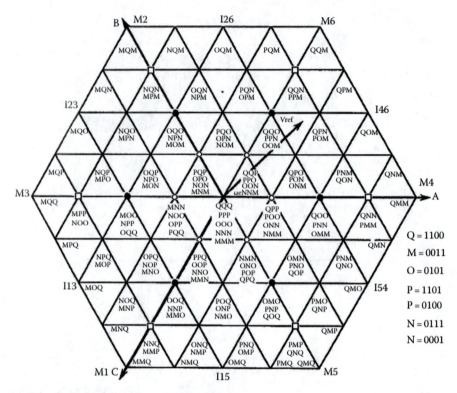

FIGURE 5.65 5L-HNPC inverter voltage space vectors.

that zero vector can be received in five ways, and while approaching the polygon periphery, the number of feasible variants reduces. The vectors that are terminated on the perimeter can be received in one way only. The vectors that correspond to P and N states can be obtained in two ways; moreover, the upper states load the upper capacitor and the lower ones the low capacitor (Figure 5.59).

Theoretically, the increase in the number of the feasible space vectors gives an opportunity to improve control quality, but the system complexity grows as a result. The system considered as follows is a development of the DTC with three-level inverter (model 5_11). Remember that in the latter one of the main vectors **M1–M6** and 0 are used and also the intermediate vectors **I13, I15, I23, I26, I46, I54** and "halb" vectors (H-vectors) that are marked in Figure 5.67 with the points. **M** and **I** vectors are used at $|\omega| > \omega_{0.5}$ (the index here and further points to the approximate speed, under which a switch-over from one of the vector systems to the other takes place). When the torque is to be reduced, the zero states are used; owing to this, while rotation direction is changing, the order of the vector selection changes too. The vectors **I** are selected in order to increase the torque and to decrease the flux linkage when the stator flux linkage vector resides in the first half of the sector. The vectors that are selected when $|\omega| < \omega_{0.5}$ use only the active states, but with their use the inverter capacitors are loading in different ways; therefore, depending on the sign of the capacitor voltage difference, the sign of the power, and the IM rotation

and $\omega > 0$, and **Tablemax–1** и **Table max–2** with $|\omega| > \omega_{0.75}$ and $\omega < 0$; these tables are similar to the ones used in model5_11.

The four vectors that marked in Figure 5.65 with the small squares are realized by using H-states (states P or N). Since with the use of these states the capacitors are loaded unevenly, when $\omega_{0.75} > |\omega| > \omega_{0.5}$, for each rotation direction, two tables are used that correspond to two available ways these voltages are formed (Figure 5.65). The tables **Table_+ 0.75_1,2** are formed from **Tablemax+1** by adding the digits 12 and 22, respectively; analogously the tables **Table_–0.75_1,2** are formed from **Tablemax–1** by adding the digits 12 and 22.

The tables **Table_0.5_1,2** are the same as **Tablemin1,2** in model5_11, and the tables **Table_0.25_1,2** differ from them with the use of second and third digits 11 or 12. The outputs of the last three groups of the tables go to the outputs Out 1,2 through multiposition switches **Multiport1,2**, depending on the speed value. The speeds of the transition from one vector system to another are tuned with the help of **Relay_0.25, Relay 0.5**.

The subsystem **dUContr** has three similar circuits for selection of the inverter-phase states that provide the same phase voltage but load different capacitors. The scheme for selection is the same as in model5_11, that is, depending on the logical product of the following signs: the phase capacitor voltage difference (with hysteresis), the reference torque, and the IM speed. Besides, the logic signal $|\omega| > \omega_{0.75}$ is formed.

In the subsystem **DTC_REG**, selection of the switches **Switch4, Switch9**, and **Switch10** for each of the phases the state code demanded is a function of the capacitor voltage difference; the switches **Switch 5,7,8** select the effective code demanded, taking into account the IM speed. This code goes to the subsystem **Code/Pulses**. One can observe the capacitor voltage difference for each of the phase by the scope **dUCond**.

The subsystem **Code/Pulses** transforms the state code into the gate pulses that these states fabricate.

Since the same space vector can be formed in different ways, in the Table 5.5 the accepted model phase states are given, depending on the direction of the M-vector.

The subsystem contains 12 converters: code—switch on (1) or switch off (0), whose data are given in the Table 5.6. Numeration of the switches of 1a, 2a, 1b, 2b for the phases A, B, C correspond to Figure 5.61.

In the subsystem **Tables**, the input and output values are given as the vectors. In order to put these vectors into the model workspace, the program `Table.m` is to be fulfilled; it is made automatically because the command to run this program is written

---

TABLE 5.5
Selected Phase States as a Function of M

M-Vector	1	5	4	6	2	3
$|\omega| < \omega_{0.25}$	00P	0N0	P00	00N	0P0	N00
$\omega_{0.25} < |\omega| < \omega_{0.5}$	NNP	PNP	PNN	PPN	NPN	NPP
$\omega_{0.5} < |\omega| < \omega_{0.75}$	NNQ	PMP	QNN	PPM	NQN	MPP

direction, one of two possible **H**-vectors is selected. A notation of **H**-vectors consists of two digits; the first one corresponds to **M**-state vector, in whose direction **H**-vector is aligned; and the second digit is 1 or 2, depending on which of the capacitors are active in this state (e.g., 61 and 62).

For the inverter under consideration, the space vectors of four values of the module can be used: marked in Figure 5.65 with small squares, points, small circles, and also **M** and **I** vectors. The smallest vectors (with small circles) are marked by the addition of 1 as the third digit (e.g., 611 and 621), and the vectors designated with small squares are marked by the addition of 2 as the third digit (e.g., 612, 622). Then with $|\omega| < \omega_{0.25}$ the vectors 611, 621, and so on; $\omega_{0.5} > |\omega| > \omega_{0.25}$, the vectors 61, 62, and so on; $\omega_{0.75} > |\omega| > \omega_{0.5}$, the vectors 612, 622, and so on; and at last with $|\omega| > \omega_{0.75}$, **M** or **I** vectors are selected. As the first three vector systems load the capacitors unevenly, the one of two feasible states (611 or 612, 621 or 622, etc.) is selected, depending on these conditions. Besides, with $\omega_{0.75} > |\omega| > \omega_{0.5}$ and in order to reduce the torque, zero states are selected; in this connection, it is necessary to change the order of the vector choice when a change in the rotation direction takes place. With that approach, the control algorithm that was used in the electrical drives with DTC and three-level inverter are maintained on the whole. The main differences are: There are more tables for inverter state choice, and the tables for converting of the inverter state code into gate pulses are modified (subsystem **Code/Pulses**). Besides, in this case there are three pairs of the capacitors, whose voltages have to be balanced; so when $\omega < \omega_{0.75}$, for each the phase there are the separate blocks. Moreover, of all feasible **H**-vectors, those ones are selected that are obtained with the help of P and N states because in this case there is a possibility to control the capacitor voltage.

Model5_17 is a simplified model of the electrical drive with 5L-HNPC and DTC. The simplifications consist in following: The inverter is supplied from DC sources 4400 V, though in reality, the transformers are to be used, as in model5_16; there are no dynamic brake circuits; the stator flux linkage and the IM torque signals are taken from IM model, although in reality it is necessary to use an observer, as in model5_11.

L-C filter with an inductance of 1 mH and a capacitance of 0.16 μF reduces IM stator current harmonics. The block **Space** fixes the dependence of the voltage space vector amplitude on IM speed rotation.

The subsystem **DTC_REG** in the subsystem **Control** consists of several subsystems in turn. The subsystem **DTC** produces the sector number signal, the signal that shows, in which half of the sector vector Ψ_s resides, the error codes of the torque and the stator flux linkage. It is a part of the subsystem **DTC_Reg** from model5_11. The difference is that changes in the hysteresis bands, depending on IM speed, are provided. It is carried out with the help of the dividers and the tables that define these dependencies. Since the operation bands of **Relay** and **Relay2** are set ±1, they will set and reset, when the errors will be larger than the hysteresis bands in the tables.

Subsystem **Table** contains the tables and the switches that form the code of the required inverter state at the subsystem outputs, without taking into account the capacitor voltage difference. **Tablemax+1** and **Tablemax+2** are active with $|\omega| > \omega_{0.75}$

TABLE 5.6
Control Code Decoding

Code	1aA	2aA	1bA	2bA	1aB	2aB	1bB	2bB	1aC	2aC	1bC	2bC
0	0	1	0	1	0	1	0	1	0	1	0	1
1	0	0	1	1	0	0	1	1	1	1	0	0
2	0	0	1	1	1	1	0	0	0	0	1	1
3	0	0	1	1	1	1	0	0	1	1	0	0
4	1	1	0	0	0	0	1	1	0	0	1	1
5	1	1	0	0	0	0	1	1	1	1	0	0
6	1	1	0	0	1	1	0	0	0	0	1	1
11	0	1	1	1	0	1	1	1	1	1	0	1
12	0	0	0	1	0	0	0	1	0	1	0	0
13	0	0	1	1	1	0	0	1	1	1	0	0
15	1	0	0	1	0	0	1	1	1	1	0	0
21	0	1	1	1	1	1	0	1	0	1	1	1
22	0	0	0	1	0	1	0	0	0	0	0	1
23	0	0	1	1	1	1	0	0	1	0	0	1
26	1	0	0	1	1	1	0	0	0	0	1	1
31	0	1	1	1	1	1	0	1	1	1	0	1
32	0	0	0	1	0	1	0	0	0	1	0	0
41	1	1	0	1	0	1	1	1	0	1	1	1
42	0	1	0	0	0	0	0	1	0	0	0	1
46	1	1	0	0	1	0	0	1	0	0	1	1
51	1	1	0	1	0	1	1	1	1	1	0	1
52	0	1	0	0	0	0	0	1	0	1	0	0
54	1	1	0	0	0	0	1	1	1	0	0	1
61	1	1	0	1	1	1	0	1	0	1	1	1
62	0	1	0	0	0	1	0	0	0	0	0	1
111	0	1	0	1	0	1	0	1	1	1	0	1
112	0	1	1	1	0	1	1	1	1	1	0	0
121	0	1	0	1	0	1	0	1	0	1	0	0
122	0	0	0	1	0	0	0	1	1	1	0	0
211	0	1	0	1	1	1	0	1	0	1	0	1
212	0	1	1	1	1	1	0	0	0	1	1	1
221	0	1	0	1	0	1	0	0	0	1	0	1
222	0	0	0	1	1	1	0	0	0	0	0	1
311	0	1	1	1	0	1	0	1	0	1	0	1
312	0	0	1	1	1	1	0	1	1	1	0	1
321	0	0	0	1	0	1	0	1	0	1	0	1
322	0	0	1	1	0	1	0	0	1	1	0	0
411	1	1	0	1	0	1	0	1	0	1	0	1
412	1	1	0	0	0	1	1	1	0	1	1	1
421	0	1	0	0	0	1	0	1	0	1	0	1
422	1	1	0	0	0	0	0	1	0	0	0	1
511	0	1	0	1	0	1	1	1	0	1	0	1

(*continued*)

TABLE 5.6 (continued)
Control Code Decoding

Code	1aA	2aA	1bA	2bA	1aB	2aB	1bB	2bB	1aC	2aC	1bC	2bC
512	1	1	0	1	0	0	1	1	1	1	0	1
521	0	1	0	1	0	0	0	1	0	1	0	1
522	0	1	0	0	0	0	1	1	0	1	0	0
611	0	1	0	1	0	1	0	1	0	1	1	1
612	1	1	0	1	1	1	0	1	0	0	1	1
621	0	1	0	1	0	1	0	1	0	0	0	1
622	0	1	0	0	0	1	0	0	0	0	1	1

in the option *Callbacks*. It is worth noting that instead of the code 111, the code 101 is used, because the model with code 111 does not work properly (the reason is not known to the author). In this subsystem, the scope **Count** is set that every 0.1 s shows the average switching frequency (the amplitude of the saw-tooth signal).

Two variants for the speed reference are provided. For the first one, **Timer** assigns the speed that rises by steps; it gives the opportunity to estimate the average switching frequency at different speeds. For the second variant, **Timer1** schedules speed up to speed 0.1, afterwards to 0.97, later with the reverse speed to −0.97 and repeated reverse to 0.6. The sample time is set as 35 μs for **Zero-Order Holds** in the subsystem **Feedback**.

In Figure 5.66 IM speed, the amplitude of the voltage space vector, and the output of the average switching frequency count are shown for the speed reference from the block **Timer**. One can observe an alteration of the space vector amplitude while speed changes; moreover, beginning from the speed of 0.5, the zero states are used, so that

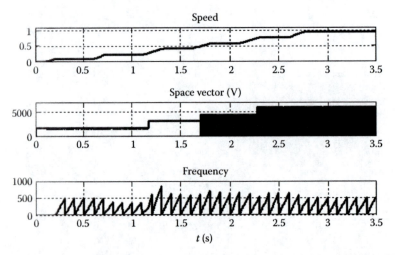

FIGURE 5.66 IM speed, voltage space vector amplitude, and switching frequency in the electrical drive with 5L-HNPC.

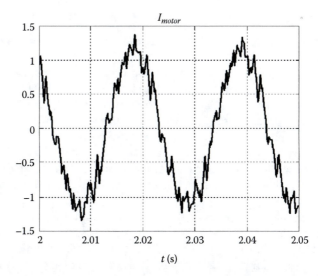

FIGURE 5.67   IM current at the speed of 0.97 in the electrical drive with 5L-HNPC.

the amplitude fluctuates between certain values and zero. At small and big speeds, the switching frequency is about 500 Hz, and sometimes higher during the middle speed.

In Figure 5.67 IM current at the speed 0.97 is shown, THD = 13%. In Figure 5.68 IM speed, its torque, the capacitor voltage difference for the phase A, and the switching frequency for the speed reference from **Timer1** are plotted. The capacitor voltage difference is not more than 1% of the rated DC voltage.

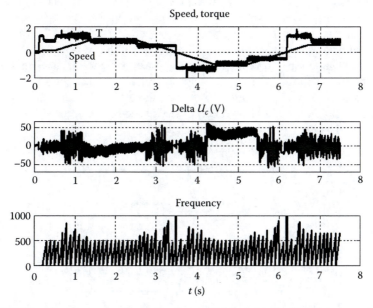

FIGURE 5.68   Transient processes in the electrical drive with 5L-HNPC during speeding-up and reverse.

5.2.9 SIMULATION OF THE IM WITH PHASE-WOUND ROTOR

The next example demonstrates the simulation of the IM with phase-wound rotor (model5_18).

The rotor-resistance start is investigated, when the resistors are inserted into the rotor circuit in order to reduce the starting current. IM has $P = 160$ kVA, $V = 380$ V, $f = 50$ Hz, and its data are taken from SimPowerSystems. It is supposed that the starting resistors are switched over by a servomotor with the constant speed that begins to rotate simultaneously with the switching of power voltage. At first, the model of the controlled resistor should be made. This model is shown in Figure 5.69 [1]. The resistor is modeled as the current source. At the input R Simulink signal comes that is numerically equal to the present resistance value. The scheme of the start model is given in model5_18.

Step switching takes place every 0.1 s. Resistance spreading in steps is determined by the block **Table**. The following resistance steps are accepted: 0.67 Ω, 0.6 Ω, 0.54 Ω, 0.34 Ω, 0.2 Ω, 0.15 Ω, 0.05 Ω, 0.01 Ω. The parallel resistors do not influence the starting process. They are put in because the controlled resistors are modeled as the current sources, and their series connection is impermissible. The bypass contactors are closed at $t = 0.7$ s. The start occurs without load.

During the direct start of IM, the starting current reaches 4500 A (rated current 243 A). With the starting resistors, the starting current is not more than 700 A (Figure 5.70).

The next example simulates IM speed control in Scherbius cascade scheme. In this scheme, IM stator is supplied direct from the three-phase AC grid, and the frequency converter of bidirectional energy transmission is set in the rotor circuit. The converter input voltage and frequency can change, and the output frequency is equal to the grid frequency. As such for a converter, two "back-to-back" VSIs with decoupling capacitor can be used, see Figure 5.71 [5]. The VSI functions that are pointed out in figure are symbolic because with the change in the energy transmission direction, their functions change too. Owing to the independent supply of the rotor winding, the current with the required value and frequency can be generated for any slip, in order to provide the torque demanded. IM can run at speeds both lower and higher than the synchronous speed, in both motor and generator modes.

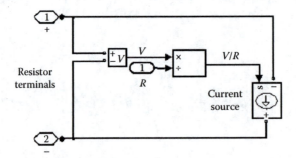

FIGURE 5.69 Variable resistor model.

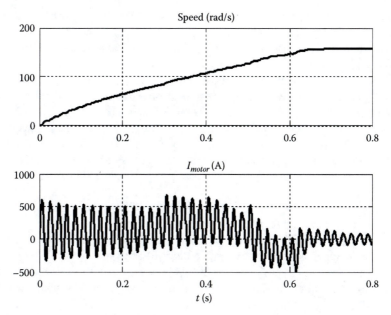

FIGURE 5.70 Rotor resistance IM start.

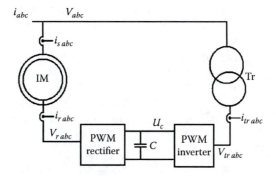

FIGURE 5.71 Scherbius cascade scheme.

This cascade is economical for speed control ranges ±30% of the synchronous. The PWM rectifier controls the IM speed and torque, and the PWM inverter controls the power flow, keeping the capacitor C voltage constant. Power of the devices in the rotor circuit is proportional to the control speed limits and in the said range is about 30%. The transformer Tr serves for matching the rotor and grid voltages. It is necessary to keep in mind that the actual parameters of the rotor circuit devices can vastly differ from the model parameters because the rotor parameters are referred to stator in the model.

The scheme described is realized in model5_19. IM 1.6 MW, 1200 V is supplied from the grid with the frequency of 50 Hz. In the rotor circuit two "back-to-back" VSI using the blocks **Universal Bridge** with IGBT are set. The capacitor in the DC link has a capacitance of 10 mF. The rotor circuit is connected to the grid through the transformer 0.5 MVA, 1200/450 V. The bridge **PWM_Rect** is controlled by the subsystem **Rect_Contr** that realizes DTC, which is similar to the one used in model5_8. The zero states are not used, so only one block **Lookup Table0** is used. A main difference is in the models considered earlier, the stator flux linkage Ψ_s is controlled. Therefore, in order to increase the torque, it is necessary to accelerate rotation of Ψ_s vector. For this purpose, when it resides in the first sector, the states $V_k = 6$ or $V_k = 2$ are selected. In the system under consideration, the rotor flux linkage Ψ_r is controlled during the uniform rotation of Ψ_s; in order to increase the torque, it is necessary to slow down Ψ_r rotation, that is, when Ψ_r resides in the first sector, to use the states $V_k = 1$ or $V_k = 5$. Such a change of algorithm can be achieved by changing the block **Relay** in the subsystem **DTC_REG**: the output "On"= 0, and not 1, and the output "Off" = 1 and not 0.

The subsystem **Contr_Inv** is used for the bridge **PWM_Inv** control; this subsystem is similar to the subsystem **Control** in model5_10. The capacitor voltage is 1500 V. Two variants of the current feedback is provided: At the lower position of the tumbler switch **Sw**, the feedback is carried out by the transformer secondary winding current; in the upper position—by the supply current. In the second case, it is possible to have null-phase shift between the supply voltage and current, but at the expense of oversize of the devices in the rotor circuit.

Several scopes are set to observe the processes in the system. The scope **Motor** fixes IM variables: the speed, the torque, the stator current, and the rotor flux linkage. The scope **Uc** shows the capacitor voltage. The scope **Supply** records the supply voltage and current and the transformer secondary current, and the scope **Power** fixes the active and reactive supply power.

A number of simplifications have taken place in the model. At first, IM is to be speeded up to the synchronous speed or close to it by using the methods that are considered in model5_6, or in model5_18 or in model 5_21. In model5_19 it is admitted that IM rotates at the synchronous speed already. Furthermore, it is accepted that IM speed and torque and the components of Ψ_r are measured. In reality, only the stator and rotor currents are measured (the speed/position sensor is set often also), and Ψ_r is to be calculated. It is supposed that during 2 s after simulation start the no-load mode of IM is achieved and the capacitor is charging, the speed controller switches at $t = 2$ s, and the gate pulses can be sent on the bridge **PWM_Rect**; the rated load is applied at $t = 2.2$ s.

The value of Ψ_r depends on the criterion accepted for its choice. If to accept

$$\Psi_{r\,ref} = \left(\frac{L_r}{L_m}\right) \quad \Psi_s = \frac{\Psi_s}{K_r} \tag{5.44}$$

the stator current will be minimum, and by

$$\Psi_{r\,ref} = \left(\frac{L_m}{L_s}\right) \quad \Psi_s = K_s \Psi_s \tag{5.45}$$

the rotor current is minimum. In the first case, there is no phase shift between the stator voltage and current. For IM under investigation $K_s = K_r = 0.96$. The $\Psi_{r\,ref}$ that gives a minimum loss in the IM windings is determined as [10]

$$\Psi_{r\,ref} = \Psi_s \frac{K_s}{K_r} \frac{R_s K_s + R_r K_r^2}{R_s K_s^2 + R_r K_r} \tag{5.46}$$

For this IM, $\Psi_{r\,ref} = 1.01$ in pu with $\Psi_s = 1$.

The block **W_ref** (Timer) schedules at $t = 2\,s$, $\omega_{ref} = 1.3$, and at $t = 5\,s$, $\omega_{ref} = 0.7$. Let Flux_ref = 1.01 and switch **SW** be in the lower position. In Figure 5.72 the IM speed, the IM current, and the capacitor voltage are shown. With the help of Powergui it can be found that at $t = 4.5\,s$ (steady state) the amplitude of the supply current first harmonic is 1482 A, THD = 2.02%, and the amplitude of the transformer secondary current is $I_{tr} = 936$ A, THD = 4.93%. The stator current amplitude in pu is 1.042, and the rotor current (frequency of 15 Hz) is 1.09. The scope **Power** shows that in this moment $P = 2.17$ MW, $Q = 0.28$ Mvar.

Simulation with **Sw** in the upper position shows that at $t = 4.5\,s$ the amplitude of the supply current first harmonic is 1471 A, THD = 1.58%, and the amplitude of the transformer secondary current is $I_{tr} = 1070$ A, THD = 7.45%; P is the same and $Q = 0$. If simulation is carried out with **Sw** in the lower position and with *Flux_ref* = 1.04, it can be seen that there is no phase displacement between supply voltage and the supply and the transformer currents, Q is reduced to 35 kvar, but the rotor current amplitude increases to 1.134 pu.

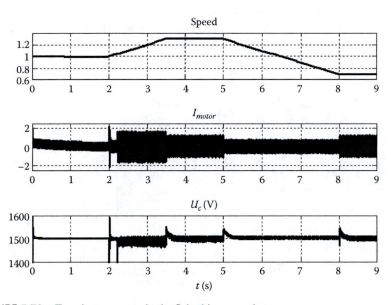

FIGURE 5.72 Transient response in the Scherbius cascade.

5.2.10 IM with Current Source Inverter

The electrical drive with IM and current source inverter (CSI) is shown in Figure 5.73. The rectifier function is to regulate the current in DC link with the help of the current regulator CR; an inductor of rather large inductance is set in DC link. The rectifier uses thyristors. The inverter is a commutator with the forced-commutated switches. These switches must have the capacity to withstand reverse voltage, so for the inverter model the block **Universal Bridge** cannot be used, given the availability of antiparallel diodes. If in reality the switches used do not have such capability, diodes are set in series. The inverter produces six-pulses load current. The phase current consists of positive and negative half-waves; each one has a duration of 120° separated by an interval of 60°. At every moment, one switch connected to the positive terminal and one switch connected to the negative terminal conduct. Switching frequency determines IM stator current frequency.

The control system contains the speed controller SR, whose output determines the absolute slip $\Delta\omega$ that causes IM torque. Sum of the rotation speed and the slip is the desirable IM current frequency ω_s. With help of the pulse former, the gate pulses with the assigned frequency are generated. In order to keep up the specified Ψ_r, DC current I_d needs to change with slip changing as

$$I_d = 0.906 \frac{\Psi_r}{L_m}\sqrt{1+\left(\frac{\Delta\omega L_r}{R_r}\right)^2} \tag{5.47}$$

This dependence of the current reference on $\Delta\omega$ is due to the function generator.

The CSI positive features are as follows: four-quadrant operation with power recuperation, reliability, and dispensing with capacitor with limited service life. The negative features are as follows: a big inductor, and the switches with symmetric blocking ability.

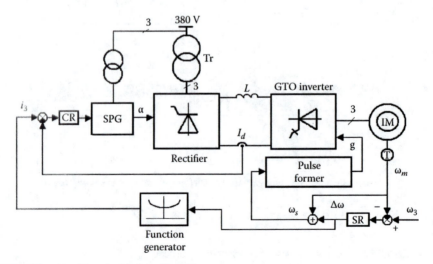

FIGURE 5.73   Electrical drive with CSI.

In model5_20 IM 1600 kVA, 1200 V is supplied from the grid with the voltage of 6 kV through the transformer 6/1.5 kV, the thyristor rectifier, and GTO CSI. The block **Universal Bridge** with option *Thyristors* is used for the rectifier; CSI model circuits is placed in the subsystem **GTO**. The inductor with the inductance of 10 mH is set between the rectifier and CSI. The rectifier current controller is the same as in model5_4. The block **Fcn** forms the nonlinear function (5.47) in the corresponding scale. The speed controller **Speed_contr** is similar to the one used in model5_4 also, but its output determines not the current but the slip $\Delta\omega$. The signal $\omega_s = \omega_m + \Delta\omega$ goes to the subsystem **Inv_Reg**, where it converts into the gate pulses of the same frequency. This subsystem contains a saw-tooth oscillation generator (by using an integrator) and a pulse shaper **Fr_Pulse**. External reset is used for the integrator when its output reaches 360. **Gain4** is chosen so that with the input signal 1, the oscillation frequency is equal to 50 Hz. The gate pulses are formed in the subsystem **Fr_Pulse** with the help of 6 blocks **Relay**. At the output of the block **Divide** the signal x changes from 0 to 6. For example, at the output of the block **Relay** there is the signal 1 at $x \geq 2$. Since it goes to the output A+ through NOT-circuit, at this output is the signal 1 at $0 < x < 2$, that is, with $0 < \omega_s t < 120°$. At the output of the block **Relay4** the signal 1 exists at $x \geq 3$, and at the output of the block **Relay5** at $x \geq 5$. As the latter goes to the element **AND1** through **NOT**-circuit, at the output A− the signal 1 takes place with $3 < x < 5$, that is, when $180 < \omega_s t < 300°$, and so on.

One can observe the current, the speed, and the torque of the IM by the scope **Motor** and the rectified current by the scope **Id** and the rotor flux linkage by scope **Flux**.

The timer **Wref** schedules, at first, speeding up to the speed 0.95 and slowing down to 0.1 at $t = 4$ s. The load torque is rated 1. The responses are shown in Figure 5.74. It is seen that after initial magnetization, the rotor flux F remains constant. Pay attention to the fact that simulation is carried out in the continuous mode, with the exclusion of

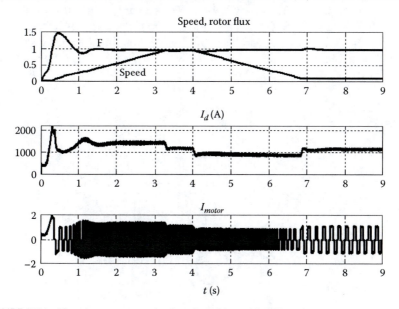

FIGURE 5.74 Transient responses in electrical drive with CSI.

snubbers, the inner resistances of the electronic components, and the voltage drop in them. The ballast resistors with big resistance are to be set at the outputs of the rectifier and the CSI, noting that simulation is made noticeably quicker.

5.2.11 SIMULATION OF IM SOFT-START

During direct IM start (by connected IM to the grid), the big current inrush (6–10 times of the rated current) in the IM windings and in the grid and considerable torque oscillations appear that affect badly the mechanical part of an electric drive. The soft-start systems are intended for reducing these phenomena. One of the often used schemes for the swift start is shown in Figure 5.75. Two antiparallel thyristors are inserted in each IM phase. The thyristor firing signals are applied with a displacement of $t = \alpha/\omega_0$ relative to the moment, when the thyristor anode voltage becomes positive, so that the definite sections are removed from the IM voltage curve. The amplitude of the first voltage harmonic reduces, and the current inrush and torque pulsation reduce too. The control angle α can be constant or can decrease by the definite program at the start or can be the output of the current controller (CR). Only the latter option will be considered here. After the completion of the start-up, the contactor K closes.

Model5_21 simulates soft-start IM 1600 kVA, 6 kV. IM is supplied from the grid 6.3 kV, 50 Hz. The subsystem **Start_Power** contains the thyristors and the contactor according to Figure 5.75. Since the inductances of the source and the IM are connected in series, a small resistor must be set (e.g., 100 W), as it is shown in the model. Change in the load torque as a function of the speed is carried out in the subsystem **T_load**; a steady component and a component that is proportional to square of the speed are provided.

The subsystem **Control** governs start. It consists of the units: SPG described in Chapter 4, **CR**, and the circuit for **K** control. The scheme of connection of the synchronization voltages for SPG is altered in relation to the one used in model5_4. The fact is that these voltages are distorted by the voltage drop in the source impedance. For smoothing of these distortions, the filters with the time constant 5.5 ms are used; these filters gives the phase shift 60° under the frequency of 50 Hz. So, at the filters

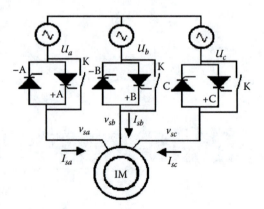

FIGURE 5.75 Scheme of IM soft start.

inputs, the voltages are given that leads demanded ones by 60°. For example, the line-to-line voltage U_{ab} crosses zero 60° earlier as U_{ca} but in the opposite direction. Therefore, instead of U_{ca}, the filtered voltage U_{ab} with opposite sign is used.

The **CR** output sets the firing angle α that is limited by 20°–120°. As a feedback, the subsystem **Rect** output is used that simulates six-pulse rectification of the three-phase current. It is necessary to have in mind that with an increase in the reference, α has to decrease. The block **Is_Reference** sets the starting current. When IM reaches the speed close to the synchronous one, **CR** cannot keep up the current, α decreases to the low limit, this moment is fixed as a start completion, and the by-pass contactor **K** closes.

One can observe the current, the speed, and the torque of IM by the scope **Motor**, and the firing angle α and IM voltage by **Scope**. In Figure 5.76 the IM current, the speed, and the torque are shown for speeding up with $I_{s\_ref} = 3I_{s\_rated}$ and for the load torque $T_m = 0.05 + 0.2\omega^2$. At the same Figure the section of the current curve is shown in large scale. It is worth noting that with the use of this start method, a main drawback of the IM starting characteristics remains: the large slip at the beginning of the start and, correspondingly, the small rotor flux linkage value that causes the small start torque value. The IM torque in pu T_e^* for the slip s and with reference current in pu I_s^* can be computed as

$$T_e^* = 1.1 \frac{K_s^2 s s_k \sigma L_s I_s^{*2}}{\sigma^2 s_k^2 + s^2} \tag{5.48}$$

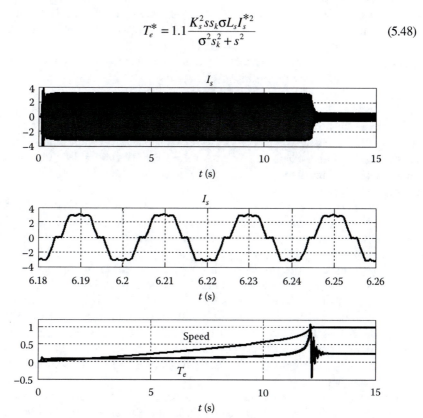

FIGURE 5.76 Transient responses during IM soft start.

where $s_k = R_r/\sigma L_r$ is a critical slip, $\sigma = 1 - K_s K_r$, the resistances and inductances are given in pu, and the coefficient 1.1 takes into account an approximate relationship between the current reference and the first harmonic amplitude. In our case

$$K_s = K_r = 2.416/(2.416 + 0.09) = 0.964, \quad \sigma = 1 - 0.964^2 = 0.071,$$

$$s_k = 0.00773/(0.071\,2.506) = 0.043, \quad T_e^* = 0.0078s, \quad I_s^* 2/(0.93\,10-5+s^2),$$

and at the beginning of the start with $I_s^* = 3$, $T_e^* = 0.07$, which agrees well with simulation results. Thus, this starting method can be used during the start with very small steady load.

5.2.12 IM MODEL WITH SIX TERMINALS

IM model in SimPowerSystems has a connection of the stator windings as star with isolated neutral. There are cases when the neutral terminal is used. Moreover, at the last time, owing to an increase in the electric drive power, the schemes are developed, in which all six terminals of three-phase winding are used [11]. So, the IM model with six terminals was developed. This model uses the subsystems of IM model in SimPowerSystems to a considerable extent. The scheme of the model is given in Figure 5.77. Only the model of the squirrel-cage IM in the stationary reference frame is shown, the results are given in pu.

The block **Rotor** realizes the relations (5.30), (5.34) when $U_r = 0$, $\omega = 0$, and the block **Flux** calculates relations (5.31). In the block **Stator**, the flux linkages Ψ_{ds}, Ψ_{qs}, Ψ_{0s} from the phase flux linkage Ψ_{as}, Ψ_{bs}, Ψ_{cs} are computed by formulas of (5.17) with $\theta = 0$, then by (5.33), the components I_{qs}, I_{ds}, and also the zero-component $I_{0s} = \Psi_{0s}/L_{ls}$ are calculated. With the help of the blocks **Selector and Product**, IM torque is computed by (5.35). The stator-phase currents are calculated by (5.23) with $\theta = 0$, and I_{0s} is added afterwards.

The flux linkages Ψ_{as}, Ψ_{bs}, Ψ_{cs} are calculated in the block **Uabc-Fabc** by (5.11) from the measured phase voltages U_a, U_b, U_c and from the computed values of the block Iqd-Iabc stator phase currents. At the output m1, the output vector is formed in pu: $[I_{qr}, I_{dr}, \Psi_{qr}, \Psi_{dr}, I_{sa}, I_{sb}, I_{sc}, I_{qs}, I_{ds}, \Psi_{qs}, \Psi_{ds}]$.

The mechanical part of IM is simulated with the block **Mex** that solves the equations (5.15), (5.16) in pu. The values of ω_r, T_e, θ come to the output m2.

With the help of the voltage sensors SVA, SVB, SVC, the values of the phase voltages are put into model, and with the help of the current sources SCA, SCB, SCC, a flow of the IM phase currents is provided in the external circuits. Since simulation is fulfilled in pu, at the model input the measured voltages are divided by the voltage base value, and at output of the phase currents, their computed values are multiplied by the current base value.

Before the start of simulation, the program IM_Model_6term, in which IM parameters are set, has to be fulfilled. The program example is given as follows.

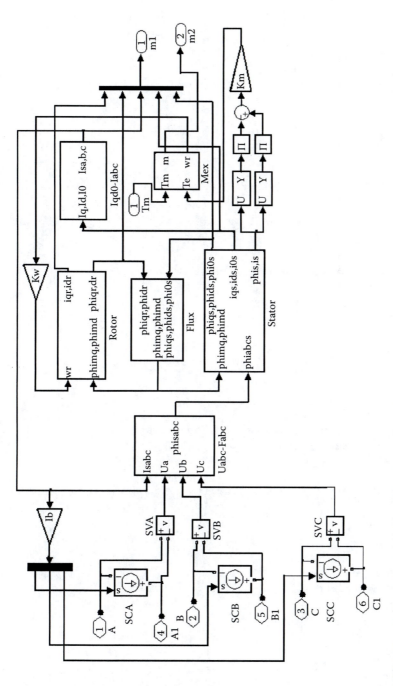

FIGURE 5.77 Model of the three-phase IM with six terminals.

Program IM_MODEL_6term:

```
%IM model with 6 terminal initialization
clear
clc
Unleff=5500%V, rms line-to-line voltage
Pn=3500000%VA, rated power
Zp=4%number of pole pairs
f=50%Hz
%Base values computation:
Vb=1.4142*Unleff/1.732;
Ib=2*Pn/(3*Vb);
W0=2*pi*f;
wb_b=W0;
Wb=W0/Zp;%rad/s
Mb=Pn/Wb;
Rb=Vb/Ib;
%------------------------
pu=1% parameters are given in pu
%------------------------
%if parameters will be given in SI
%set pu =0
%------------------------
if pu==1;
Rs=0.0138;%pu
Rr=0.0077;%pu.
Lls=0.09;% stator leakage inductance, pu
Llr=0.09;% rotor leakage inductance,pu.
Lm=2.416;% main IM inductance,pu
H=0.33;% inertia constant, s
F=0;% viscous friction coefficient
wmo=0;tho=0;% initial speed and rotor position
phirdo=0;phirqo=0;%initial rotor flux linkage
phisqo=0;phisdo=0;% initial stator flux linkage
end;
%------------------------
if pu==0
%IM parameters
Rs=0.12;%Ω
Rr=0.0606;%Ω
Llr=0.00248;% H
Lm=0.0666;% H
J=374;% kgm^2 % moment inertia of the rotor, including
 mechanism
%PU computation
Rs=Rs/Rb;%pu.
Rr=Rr/Rb;%pu.
Lls=Lls*W0/Rb;%pu.
Llr=Llr*W0/Rb;%pu.
Lm=Lm*W0/Rb;%pu.
H=0.5*Wb^2*J/Pn;%s
```

```
F=0;
phirdo=0;phirqo=0;
phisqo=0;phisdo=0;
wmo=0;tho=0;
end;
Laq=(1/Llr+1/Lls+1/Lm)^-1;
Lad=Laq;
Km=1;Kw=1;
save IM_MODEL
```

At first, the rated values of the voltage, the power, the frequency and the number of pole pairs are specified. Then the base values are computed. If the IM parameters are given in pu, pu = 1 has to be put; if they are given in SI, pu = 0 has to be put. In the first case, the parameters are used directly, and in the second one, a computation of the parameters in pu is carried out. As a result, the parameter values are formed in the workspace, and simulation can be carried out. It is reasonable to write down the command run IM_MODEL_6term in the window *Model Propeties/Callbacks/Initialization*, then, by model start, this program is fulfilled automatically.

In model5_22, IM phases are connected to the same terminals of the two three-level VSIs that have the common DC voltage from two rectifiers; these rectifiers are supplied from two transformer secondary windings that have connection wye and delta. The output rectifier voltage is about 3.2 kV. The IM parameters are given in the program IM_MODEL_6term. The inverters are controlled by the block **Discrete 3-phase PWM Generator**; moreover, the gate pulses for both bridges are in antiphase. The controller scheme is similar to the one used in model5_15 and differs by the addition of I_0 current regulator with zero reference.

The scopes show: **PhaseVoltage**: IM phase voltage; **IM_Mex**: IM speed and torque; **Motor_Current&Flux**: the phase currents, I_d, I_q, I_0 stator currents, the rotor flux linkages; and **Bridges**: the input and output VSI voltages.

In this scheme, the amplitude of the voltage space vector is $4/3U_{dc}$ where U_{dc} is the full DC voltage (in this case $U_{dc} \approx 6400$ V) that is twice as much in value gained using one VSI.

The responses are shown in Figure 5.78. Transient performance is rather good, and the zero sequence current is about null.

5.2.13 MODEL OF SIX-PHASE IM

Since during the supply of an alternative current motor from the inverters the former can have any number of phases, there is increased interest in the use of polyphase motors. These motors have several merits: smoother rotor flux linkage changing that reduces loss; more reliability because with an open phase, a motor can continue to operate; the inverters can use the electronic devices with less rated currents. The latter feature is especially important for the big power electrical drives because it is possible to avoid parallel device connection.

Among polyphase motors, six-phase IM that have six stator windings take the special place.

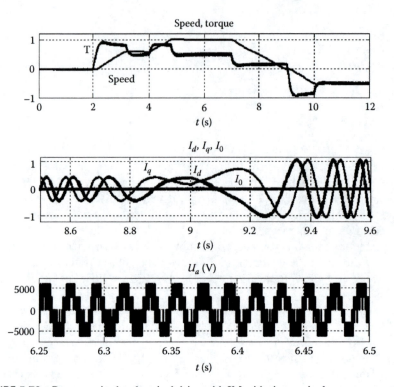

FIGURE 5.78 Processes in the electrical drive with IM with six terminals.

One can differ the symmetric six-phase IM with phase displacement between windings of 60°, and the so-called dual three-phase IM that has two three-phase windings with mutual phase displacement of 30° (electric). Reason for using such an IM is that when supplied from an inverter IM current contains often the 5th and the 7th harmonics, which by interaction with the rotor flux linkage produce a torque fluctuation of sixfold frequency. These fluctuations result in wear and tear of the equipment and are a source of acoustic noise. As the displacement of 30° gives for the 6th harmonic the shift of 180°, the torque fluctuations for both windings are in antiphase, and they can be excluded. Furthermore, such a type of IM is modeled.

The diagrams of arrangement of the IM windings and of their connection are shown in Figures 5.79 and 5.80. Each of three-phase systems has an insulated neutral.

The model equations are derived as follows [12,13]:

If 6-dimensional vectors are introduced $\mathbf{V_s} = [v_{sa}, v_{sb}, v_{sc}, v_{sx}, v_{sy}, v_{sz}]^T$, $\mathbf{I_s} = [i_{sa}, i_{sb}, i_{sc}, i_{sx}, i_{sy}, i_{sz}]^T$, $\mathbf{I_r} = [i_{ra}, i_{rb}, i_{rc}, i_{rx}, i_{ry}, i_{rz}]^T$, the stator equation can be written as

$$\mathbf{V_s} = \mathbf{R_s I_s} + s(\mathbf{L_{ss} I_s} + \mathbf{L_{sr} I_r}) \qquad (5.49)$$

where
    s is Laplace operator
    $\mathbf{R_s}$ is the diagonal matrix of the stator winding resistances

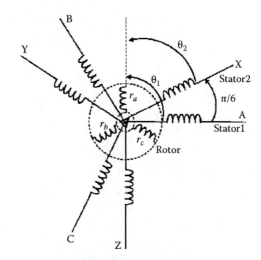

FIGURE 5.79   Arrangement of the six-phase IM windings.

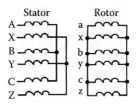

FIGURE 5.80   Connection of the six-phase IM windings.

$\mathbf{L_{ss}} = \mathbf{L_{ls}} + \mathbf{L_{ms}}$, $\mathbf{L_{ls}}$ is the diagonal matrix with elements L_{ls} (the stator winding leakage inductance)

$\mathbf{L_{ms}}$ is the matrix of the mutual inductances that takes into account a reciprocal influence of the phase currents

$\mathbf{L_{sr}} = \mathbf{M}_{sr}[\theta_{sr}]$, $\mathbf{M}_{sr}$ is the maximum of the stator and rotor mutual inductance

$[\theta_{sr}]$ is the matrix, whose elements depend on the angle between the axes of the windings of the stator and rotor phase A

The expressions for these matrixes are not given here because they are rather awkward; the reader can acquaint themselves with these by appropriate references. In general, these formulas are an extension of the expressions (5.13).

Analogous for the rotor

$$\mathbf{R_r I_r} + s(\mathbf{L_{rr} I_r} + \mathbf{L_{rs} I_s}) = 0 \qquad (5.50)$$

where $\mathbf{R_r}$ is the diagonal matrix of the rotor winding resistances, $\mathbf{L_{rr}} = \mathbf{L_{lr}} + \mathbf{L_{mr}}$, $\mathbf{L_{lr}}$ is the diagonal matrix with elements, L_{lr} is the rotor winding leakage inductances, $\mathbf{L_{mr}}$ is the matrix of the rotor winding mutual inductances that take into account the reciprocal influence of the rotor phase currents, and $\mathbf{L_{rs}} = \mathbf{M}_{sr}[\theta_{rs}]$, $[\theta_{rs}]$ is the matrix, whose elements depend on the angle between the axes of the windings of the rotor and stator phase A. These formulas are an extension of the expressions (5.14).

In order to make the analysis, simulation, and synthesis of the control system simpler, 6-dimension system of the equations (5.49), (5.50) can be changed by three 2-dimension subsystems (α, β), (μ_1, μ_2), and (z_1, z_2) with the use of the transformation matrix:

$$
T = k \begin{bmatrix}
1 & -0.5 & -0.5 & \sqrt{3}/2 & -\sqrt{3}/2 & 0 \\
0 & \sqrt{3}/2 & -\sqrt{3}/2 & 0.5 & 0.5 & -1 \\
1 & -0.5 & -0.5 & -\sqrt{3}/2 & \sqrt{3}/2 & 0 \\
0 & -\sqrt{3}/2 & \sqrt{3}/2 & 0.5 & 0.5 & -1 \\
1 & 1 & 1 & 0 & 0 & 0 \\
0 & 0 & 0 & 1 & 1 & 1
\end{bmatrix}
\tag{5.51}
$$

$k = 1/\sqrt{3}$. Thus, instead of the vector V_s, the vector $V_{st} = TV_s = [v_{s\alpha}, v_{s\beta}, v_{s\mu1}, v_{s\mu2}, v_{sz1}, v_{sz2}]^T$ is obtained, and it is analogous to the vectors I_{st}, I_{rt}. If the rotor equations are referred to the stator reference frame, three independent sets of equations are derived:

$$
v_{s\alpha} = (R_s + sL_s)i_{s\alpha} + sM\,i_{r\alpha}
$$

$$
v_{s\beta} = (R_s + sL_s)i_{s\beta} + sM\,i_{r\beta}
$$

$$
0 = sM\,i_{s\alpha} + \omega_r M\,i_{s\beta} + (R_r + sL_r)i_{r\alpha} + \omega_r L_r i_{r\beta}
\tag{5.52}
$$

$$
0 = -\omega_r M\,i_{s\alpha} + sM\,i_{s\beta} - \omega_r L_r i_{r\alpha} + (R_r + sL_r)i_{r\beta}
$$

where s-Laplace symbol, ω_r is the rotor rotation speed (electric) and $L_s = L_{ls} + M$, $L_r = L_{lr} + M$, $M = 3L_{ms}$ is the total inductance (for the usual IM $M = 1.5L_{ms}$, formula (5.13)),

$$
v_{s\mu1} = (R_s + sL_{ls})i_{s\mu1}
$$

$$
v_{s\mu2} = (R_s + sL_{ls})i_{s\mu2}
\tag{5.53}
$$

$$
v_{sz1} = (R_s + sL_{ls})i_{sz1}
$$

$$
v_{sz2} = (R_s + sL_{ls})i_{sz2}
\tag{5.54}
$$

The equations for rotor currents $i_{r\mu1}$, $i_{r\mu2}$, i_{rz1}, i_{sr2} for the squirrel-cage IM do not have the exciting terms (stator currents, fluxes), so these currents are equal to zero. The equations for the flux linkages are

$$
\Psi_{s\alpha} = (M + L_{ls})i_{s\alpha} + Mi_{r\alpha}
$$

$$
\Psi_{s\beta} = (M + L_{ls})i_{s\beta} + Mi_{r\beta}
\tag{5.55}
$$

$$
\Psi_{r\alpha} = (M + L_{lr})i_{r\alpha} + Mi_{s\alpha}
$$

$$
\Psi_{r\beta} = (M + L_{lr})i_{r\beta} + Mi_{s\beta}
\tag{5.56}
$$

IM torque is

$$T_e = 3Z_p(\Psi_{s\alpha}i_{s\beta} - \Psi_{s\beta}i_{s\alpha}) \tag{5.57}$$

The following conclusions can be made from these equations: Only the variables in (5.52) determine IM flux and its torque; hence, only they must be taken into consideration for system synthesis; these equations are the same as the ones applicable for the usual IM. The variables of this system have harmonics of the order of $k = 12n + 1$, $n = 0, 1, \ldots$. The variables in (5.53) have harmonics of the order of $6n + 1$, $n = 1, 3, 5, \ldots$. The currents of these frequencies increase the IM loss; hence, for control system synthesis, steps have to be made to reduced loss in IM. Moreover, when the neutrals of the 2 three-phase systems are not connected, the currents i_{sz1}, i_{sz2} are equal to 0.

Model5_23 contains the model of the dual three-phase IM. The terminals A, B, C and X, Y, Z are connected to six regulated voltage sources from SimPowerSystems; one can change the phase displacements between the source voltage outputs. The load torque signal comes to input T_m. Five groups of the signals are collected at the output $m1$: three phase currents A, B, C, three phase currents X, Y, Z, two stator flux linkage in the axis α, β, rotor flux linkage and rotor currents. The IM speed, its torque, and its rotation angle are collected at the output $m2$. All the values, except the latter, are in pu. The base values of the usual IM of the same power are taken as the base ones. The program IM_MODEL_6Ph has to be executed before simulation start; this program is indicated in the option *File/Model Properties/Callbacks/InitFcn* and is analogous to the program IM_MODEL_6end.

Model **IM_6Ph** consists of the input part that is connected to SimPowerSystems, the subsystem **Voltages** that fabricates the vector of 6 IM phase voltages $\mathbf{V_s}$, the unit for multiplication of $\mathbf{V_s}$ by the transformation matrix $\mathbf{T}$ (the components v_{sz1}, v_{sz2} are not computed), and the subsystem **IM** that calculates IM quantities). $\mathbf{V_s}$ is computed using the two phase-to-phase voltages for each three-phase winding. The blocks from the usual IM model with slight modifications are used in the subsystem **IM**. With the help of the elements with the transfer functions of the first order, the currents $i_{s\mu1}$, $i_{s\mu2}i_{s\mu1}$, $i_{s\mu2}$ are calculated by (5.53); these currents together with the currents $i_{s\alpha}$, $i_{s\beta}$ go to the inverse transformation unit that is made on the basis of the matrix multiplication block; at its outputs the values of six phase currents are formed in pu. After multiplication by the base value, the signals I_a, I_b, I_x, I_y come to the control inputs of the blocks **Controlled Current Source**, whose outputs are connected between the outputs model terminals A-B, A-C, X-Y, X-Z, providing this way a current flow in the external circuits; moreover, C and Z phase currents are fabricated as a sum of the other two phase currents with an opposite sign.

Let us come back to model5_23. The parameters of the equivalent IM are taken as a standard from [3]: $R_s = R_r = 0.03$, $L_{ls} = L_{lr} = 0.1$, $L_m = 2.4$, $L_s = L_r = 2.5$, $K_s = K_r = L_m/(L_m + L_{ls}) = 0.96$, $L'_s = L'_r = 0.2$, $\sigma = L'_s/L_s = 0.08$, s_k (critical slip) $= 0.15$. This information is put down in the program IM_MODEL_6Ph; as the windings are connected in parallel, their active resistance (stator and rotor) double. At the output of the block **Gain**, the rectangle voltage frequency of 50 Hz is formed as shown in Figure 5.81a. This voltage has the fifth and the seventh harmonics with the amplitudes 19.8% and 14.4%, respectively. With the help of the delay blocks D1–D5, two

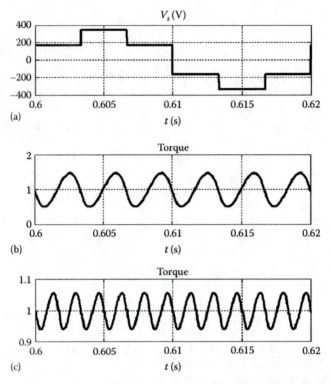

FIGURE 5.81 Processes in the six-phase IM during the supply of step voltage. (a) Voltage shape. (b) Torque oscillation during the supply of in-phase voltages. (c) Torque oscillation during the supply of voltages of 30° displacement.

systems of the three-phase voltages are formed; in addition, there is the opportunity to make the additional phase shift Δ for the second system.

If the simulation is carried out with $\Delta = 0$, the torque shown in Figure 5.81b is received. The pulsation can be seen with the frequency of 300 Hz. If an additional phase shift of 30° is set (additional delay is 0.02/12 s) for the voltage of the second system, the torque response shown in Figure 5.81c is obtained. The torque pulsation reduces drastically; its frequency is 600 Hz. So indeed, during the supply of windings from the voltages with the mutual phase displacement of 30°, the torque pulsation of sixfold frequency disappears.

The six-phase IM under consideration is supplied usually from 2 three-phase inverters with common or separated DC sources (Figure 5.82). Since each of inverters has 8 states, the scheme can realize total 64 states; of them 16 are zero states. PWM must control α and β variables; however, in order to reduce the phase current harmonics, it is preferable to have simultaneously the low average values of the voltages v_{sz1} and v_{sz2} for the modulation period that makes the system more complex. In [13] the simple scheme is suggested with the tooth-sage PWM and with the zero-sequence-voltage additional action.

The control system can use the same principle of operation as the usual IM, in particular the indirect vector control [14]. Such a method is used in model5_24. The control system is shown in Figure 5.83. The formulas (5.39), (5.40), (5.41) are used.

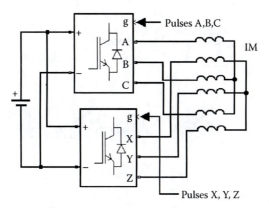

FIGURE 5.82   Six-phase IM supplied from two VSIs.

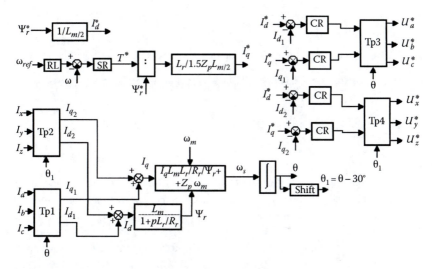

FIGURE 5.83   Control of the electrical drive with six-phase IM.

The integral of ω_s defines the position θ of the synchronous rotating reference frame. The second reference frame lags behind the first one by 30°. The necessary values of $\sin\theta_1$, $\cos\theta_1$, $\theta_1 = \theta - 30°$ are found by values of $\sin\theta$, $\cos\theta$. These values are used for the components I_{d1}, I_{q1} of the space vector I_{sabc} and for the components I_{d2}, I_{q2} of the space vector I_{sxyz} by (3.2), (3.3) computation. These components come as feedback at the inputs of four current controllers **CR**, whose outputs are transformed in two systems of three-phase signals by (3.5)…(3.7) by using of θ, θ_1. These signals go to the inputs of the blocks **PWM Generator** that generate gate pulses.

In model5_24, the dual three-phase IM 400 kVA, 400 V (for each of the windings) is supplied from 2 two-level inverters with the common DC source 650 V. The block **Saturation** defines the rated load torque. The control system is as described earlier. **Timer** sets a small speed of 0.02, then increases the speed to 0.94 and decreases the speed to −0.94 in the reverse direction. **Scope** records the IM speed, its torque, and the angle of the rotor rotation; **Scope1** shows the currents of the first and the second

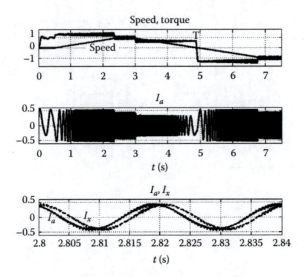

FIGURE 5.84   Processes in the electrical drive with six-phase IM.

three-phase windings and the components of the space vectors of stator flux linkage, rotor flux linkage, and rotor current.

In Figure 5.84 the IM responses are shown. The torque and current pulsations are small. The current of the first winding leads the current of the second one by 30°.

5.2.14   SIMULATION OF THE SPECIAL OPERATION MODES OF THE LINE-FED IM

Simulation of IM with controlled speed has been considered in the previous sections; however, most of IM in industry are supplied from the grid with a constant voltage. The large inrush currents can be observed in some of the operation modes for such an IM, and having knowledge about them is necessary for the correct choice of the IM and the switching and protection devices. Computation of such operation modes demands complex calculations, acceptance of a number of assumptions, and is often only approximate. Simulation of such operation modes using of SimPowerSystems reduces the computation time essentially. These operation conditions are [3]

1. IM direct start
2. Short-circuit in the supply line, two- or three-phase
3. IM switch-over to the reserve power source
4. IM reverse
5. Periodic oscillations of the mechanical load

Model5_6 can be used for the first regime investigation. In model5_25 the regimes 2, 3, 5 are simulated. IM 1600 kVA, 6 kV is supplied from the grid 6.3 kV, 50 Hz. The initial IM speed is synchronous. The single-pole switches **Breaker, Breaker 1** serve for short-circuit simulation, and the three-pole switches **Three-Phase Breaker, Three-Phase Breaker 1** switch over a supply from the main source to the reserve source that is analogous to the first one but has another phase. As a short-circuit happens usually not direct at IM terminal but in the switchgear or in the supply line, a sector of the line from the possible place of the fault to IM is simulated by the circuit *R-L* **Line**.

For simulation of the three-phase short-circuit, the parameter *Time variation of* is set to *None* in the **Three-Phase Programmable Voltage Source** (main), the first switching time (closing) is set as 1 s, and the second switching time (opening) is set as 1.05 s for the single-pole switches. In the block **Relay**, the value of *Switch off* point is negative, for example, −0.85. The tumbler **Switch** is in the lower position. It can be seen after simulation that during both closing of the short-circuit and its opening the phase currents reach 11–13 times the rated current and IM speed decreases by 15%. It is obvious that both IM and switching devices must be sized accordingly. By increasing the winding inductance, one can be convinced that the inrush currents reduce essentially.

For simulation of the two-phase short-circuit, the switching times increase by 10 times for one of the single-phase breakers. It can be seen after simulation that with the closing of the short-circuit the phase currents in the shorted phases reach a 12-fold value of the rated current, the drop in speed is less, and when switching off the fault, the current bumps are reduced too.

For simulation of switching over to the reserve source, the switching times of the single-phase breakers increase by 10 times, the *switch on* point of the block **Relay** is 1.15, and *switch off* point is 0.85. In **Three-Phase Programmable Voltage Source**, the parameter *Time variation of* is set *Amplitude, Type of variation: Ramp, Rate of change* = −10 (i.e., decreasing from the rated value to zero for 100 ms) and *Variation timing* from 1 s to 1.1 s. The block **Three-Phase Measurement** measures the phase-to-phase voltages in pu in reference to $V_b = \sqrt{3}/\sqrt{2}$ 6300 = 8909 V, that is, the block voltage 6.3 kV corresponds to the measured value $\sqrt{3}/\sqrt{2} = 1.22$. The measured voltages are rectified with the block **Rect** and are smoothed by the first-order filter. During a decrease in supply-voltage, down to 0.85 that happens at $t = 1.035$ s, **Three-Phase Breaker** is switched off, and after a delay of 20 ms **Three-Phase Breaker 1** is switched on.

After the simulation is concluded, it can be seen by the scope **Voltage** that IM voltage, which is usually 8800 V, reduces to 6400 V and afterward recovers quickly. By the scope **Motor**, IM current reaches a fourfold value, and the speed deviation is 4%. It is interesting to observe an influence of the reserve source breaker delay and IM inertia on the process of switching-over.

IM are often used in conditions, for example, load torque changes occurring periodically. Usually, the frequency of load oscillation is equal to the load rotation speed that is *i* times less than the IM rotation speed where *i* is a gear ratio. In such conditions, the torque and current oscillations that appear can exceed the rated values. For simulation of this regime, the subsystem **Tm** is used. It produces the load torque by the formula $T_m = T_a + T_b \sin(\omega_s t / Z_p i)$. For activation of the system, the tumbler **Switch** sets in the upper position. Let us set the model in the same state that it was under short-circuit simulation; in this case, the switching times of the single-pole breakers increase by 10 times; let $T_a = 0.5$, $T_b = 0.4$, $i = 2.5$. In Figure 5.85 the IM current, the speed, and the torque resulting from the simulation are shown. Although the load torque changes in the range of 0.1–0.9, the IM torque can take negative values.

Model5_26 simulates IM reverse by phase sequence changing. IM 37 kVA, 380 V is supplied from the grid 400 V (for *B* and *C* phases through the contacts of **Breaker** and

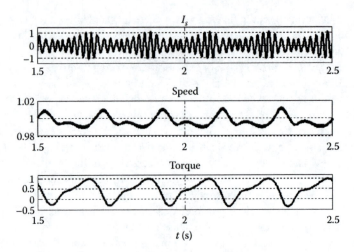

FIGURE 5.85 Processes in IM during load torque oscillation.

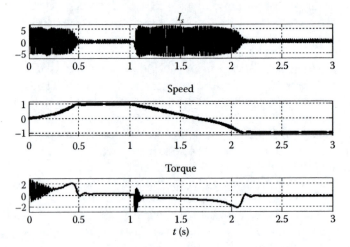

FIGURE 5.86 IM reverse by phase sequence changing.

Breaker 2) and speeds up to the rated speed with a load of 0.2 as rated. At $t = 1$ s these breakers turn off, and 0.05 s later **Breaker1** and **Breaker 3** turn on; moreover, the grid phase B is connected to the IM phase C, and the grid phase C is connected to the IM phase B. IM reverse takes place. The process is shown in Figure 5.86. It is seen that this regime is accompanied with the large overloads of the toque and of the current.

5.3 SYNCHRONOUS MOTORS (SM) AND ELECTRIC DRIVES

5.3.1 SM Model

It is accepted in SimPowerSystems that salient-pole SM, besides the excitation winding, have two damper windings on the rotor that are aligned along the direct and the quadrature axes.

The resistance and the leakage inductance of these windings are designated as R_{kd}, L_{lkd}, R_{kq}, L_{lkq}, respectively. For round-rotor SM, it is accepted that it has two damper windings in the quadrature axis whose parameters are denoted as R_{kq1}, L_{lkq1}, R_{kq2}, L_{lkq2}. In what follows, only equations for the salient-pole SM are given.

If $\mathbf{V}_{abcs}$ is the phase voltage vector, $\mathbf{I}_{abcs}$ is the current vector, and $\mathbf{\Psi}_{abcs}$ is the flux linkage vector—all in the stator reference frame; with $\mathbf{R}_s$ set as the diagonal matrix of the stator resistances, stator voltage equations are written as (5.11)

$$\mathbf{V}_{abcs} = \mathbf{R}_s \mathbf{I}_{abcs} + s \mathbf{\Psi}_{abcs} \tag{5.58}$$

There are three single-phase windings on the rotor: if the q-aligned damper winding is set with parameters R_{kq} and L_{lkq}, then I_{kq} is its current, Ψ_{kq} is its flux; in the case of d-aligned damper winding with parameters R_{kd} and L_{lkd}, I_{kd} is its current, Ψ_{kd} is its flux; and for the excitation winding with parameters R_f and L_{lf}, Ψ_f and I_f are its flux and current, respectively. If to denote as $\mathbf{R}_r$—the resistance matrix, $\mathbf{V}_{qdr}$, $\mathbf{I}_{qdr}$, $\mathbf{\Psi}_{qdr}$—the rotor voltage, current, and flux linkage matrixes, respectively, are arranged as follows: $[\Psi_{kq}, \Psi_f, \Psi_{kd}]$ and so on, where $\mathbf{V}_{qdr} = [0, U_f, 0]$, U_f is the excitation voltage; then the rotor voltage equation is

$$\mathbf{V}_{qdr} = \mathbf{R}_r \mathbf{I}_{qdr} + s \mathbf{\Psi}_{qdr} \tag{5.59}$$

The equations that link currents and flux linkages can be written as

$$\mathbf{\Psi}_{abcs} = \mathbf{L}_s \mathbf{I}_{abcs} + \mathbf{L}_{sr} \mathbf{I}_{qdr}, \quad \mathbf{\Psi}_{qdr} = \mathbf{L}_{sr}^T \mathbf{I}_{abcs} + \mathbf{L}_r \mathbf{I}_{qdr} \tag{5.60}$$

Where $\mathbf{L}_s$, $\mathbf{L}_{sr}$, $\mathbf{L}_r$ are the matrixes of the internal and mutual inductances having the time-independent quantities and also the quantities that depend on the rotor position angle about the stator θ and on the double angle 2θ. Such dependence makes SM modeling in the stationary reference frame very difficult; therefore, modeling in the rotating reference frame that is aligned along rotor position with transformation to the equivalent two-phase SM is usually used. In this case, the inductances are transformed into the constant quantities that do not depend on θ. Such an approach is used in SimPowerSystems.

For transition to two-phase SM, three-phase quantities are transformed with the help of (5.17), (5.19). The stator voltage equations have a form of (5.24). The flux linkage equations are

$$\Psi_{qs}^r = L_q I_{qs}^r + L_{mq} I_{kq}, \quad L_q = L_{sl} + L_{mq}$$

$$\Psi_{ds}^r = L_d I_{ds}^r + L_{md} I_{kd} + L_{md} I_f, \quad L_d = L_{sl} + L_{md}$$

$$\Psi_{kq} = L_{kq} I_{kq} + L_{mq} I_{qs}^r, \quad L_{kq} = L_{lkq} + L_{mq} \tag{5.61}$$

$$\Psi_f = L_{fd} I_f + L_{md} I_{kd} + L_{md} I_{ds}^r, \quad L_{fd} = L_{lf} + L_{md}$$

$$\Psi_{kd} = L_{md} I_f + L_{kd} I_{kd} + L_{md} I_{ds}^r, \quad L_{kd} = L_{lkd} + L_{md}$$

The first and the third equations form the system that does not depend on the system that is formed by the second, the fourth, and the fifth ones, so these systems can be solved for the currents in explicit form; afterwards these currents are substituted into voltage equations (5.59). As a result, the differential equations for the flux linkages are derived as

$$s\Psi_{qs}^r = U_{qs} - \omega_r \Psi_{ds}^r + \frac{R_s}{L_{ls}} \left(\Psi_{mq} - \Psi_{qs}^r \right) \tag{5.62}$$

$$s\Psi_{ds}^r = U_{ds} + \omega_r \Psi_{qs}^r + \frac{R_s}{L_{ls}} \left(\Psi_{md} - \Psi_{ds}^r \right) \tag{5.63}$$

$$s\Psi_{kq} = \frac{R_{kq}}{L_{lkq}} \left(\Psi_{mq} - \Psi_{kq} \right) \tag{5.64}$$

$$s\Psi_f = \frac{R_f}{L_{lf}} \left(\Psi_{md} - \Psi_f \right) + U_f \tag{5.65}$$

$$s\Psi_{kd} = \frac{R_{kd}}{L_{lkd}} \left(\Psi_{md} - \Psi_{kd} \right) \tag{5.66}$$

$$\Psi_{mq} = X_{aq} \left(\frac{\Psi_{qs}^r}{L_{sl}} + \frac{\Psi_{kq}}{L_{lkq}} \right) \tag{5.67}$$

$$\Psi_{md} = X_{ad} \left(\frac{\Psi_{ds}^r}{L_{sl}} + \frac{\Psi_{kd}}{L_{lkd}} + \frac{\Psi_f}{L_{lf}} \right) \tag{5.68}$$

$$X_{aq} = \left(\frac{1}{L_{mq}} + \frac{1}{L_{sl}} + \frac{1}{L_{lkq}} \right)^{-1} \tag{5.69}$$

$$X_{ad} = \left(\frac{1}{L_{md}} + \frac{1}{L_{sl}} + \frac{1}{L_{lkd}} + \frac{1}{L_{lf}} \right)^{-1} \tag{5.70}$$

After calculation of the flux linkages, the currents are computed as

$$I_{qs}^r = \frac{\Psi_{qs}^r - \Psi_{mq}}{L_{sl}} \tag{5.71}$$

$$I_{ds}^r = \frac{\Psi_{ds}^r - \Psi_{md}}{L_{sl}} \tag{5.72}$$

$$I_{kq} = \frac{\Psi_{kq} - \Psi_{mq}}{L_{lkq}} \tag{5.73}$$

$$I_f = \frac{\Psi_f - \Psi_{md}}{L_{lf}} \tag{5.74}$$

$$I_{kd} = \frac{\Psi_{kd} - \Psi_{md}}{L_{lkd}} \tag{5.75}$$

SM torque (in pu)

$$T_e = \Psi_{ds}^r I_{qs}^r - \Psi_{qs}^r I_{ds}^r \tag{5.76}$$

The stator phase currents are calculated by the transformation of (5.23) after finding I_{qs}^r, I_{ds}^r.

The SM rotation speed and angle θ are computed by (5.15), (5.16).

The available SM model pictures in SimPowerSystems are shown in Figure 5.87. They are in the folder *Machines*. SM parameters can be given in SI, in pu, or in so-called standard form (see the following); for that purpose, the corresponding picture has to be selected. Simulation is carried out in pu in any case; only the dimensions of the output variables and the load torque are different. Unlike IM, it is not the load torque but the load power comes at the input *Pm* that is formed in Simulink and is equal to $T_m \omega_m$; such a condition is more typical for SM used as a generator where P_m is the prime mover power. Besides, the negative power corresponds to the SM motor mode and the positive power corresponds to the generator mode. The SM variables that are available for Simulink blocks (see the following) are gathered at the output *m*.

On the first page of the dialog box, one of the existent SM can be selected and also the type of the input mechanical signal, as for IM (furthermore, "Mechanical power P_m" will be chosen). In the field *Rotor Type*, "Salient-pole" or "Round" are put into from the falling menu; in the latter case, the data of the second damper

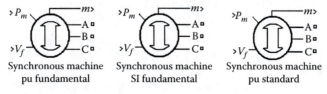

FIGURE 5.87 Pictures of SM models.

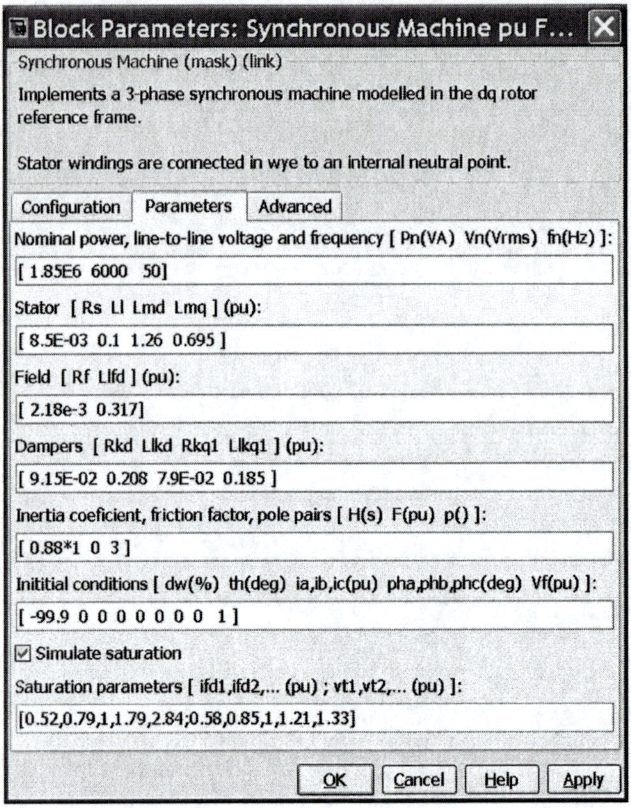

FIGURE 5.88 SM dialog box.

winding appear. On the second page, the information is entered in that is analogous to the information for IM to a great extent, along with the values of parameters whose notations are given earlier (Figure 5.88). The values of the rotor winding resistances and inductances are marked with a stroke, in order to emphasize that these values are referred to the stator. The initial speed is specified in percent in relation to the synchronous speed. Starting from the null point, one can put $dw(\%) = -99.9$. On the third page, besides the sample time, the option "Display V_{fd} which product nominal V_t" can be marked.

When it is made, the excitation voltage that gives the rated excitation current (which produces the rated SM voltage in no-load condition) is determined. If this voltage is sent at the input V_f from system Simulink, the SM excitation current will be the rated one. But it is not the voltage that has to feed the rotor winding because the shown value V_{fd} is referred to the stator. If the rated excitation current is known from SM data, it can be set as the fourth parameter in the field of the SM rated data. Then activation of this option gives the actual voltage that has to be applied to the rotor; the same voltage is used for simulation. If the rated excitation current is not known, the fourth parameter in the field of the SM rated data is omitted.

Consider an example. For the standard SM 8100 VA, 400 V, 1500 rpm, the excitation winding resistance referred to the stator is $R'_f = 1.208\ \Omega$. If the rated excitation current is not given, in the manner described, it can be found $V'_{fd} = 11.56$ V; therefore, the excitation current referred to the stator is $I'_{fd} = 11.56/1.208 = 9.57$ A. During simulation, V'_{fd} is used. Let us suppose it is known that rated excitation current is $I_{fd} = 4.785$ A. The relation is thus

$$I'_{fd} = \frac{2}{3}\left(\frac{N_f}{N_s}\right) I_{fd} \tag{5.77}$$

Where N_f and N_s are the turn numbers of the excitation and stator windings from which the following is derived: $N_f/N_s = 3$. Because the voltages are referred by the relationship

$$V'_{fd} = \left(\frac{N_s}{N_f}\right) V_{fd} \tag{5.78}$$

then $V_{fd} = 3V'_{fd} = 34.7$ V. The actual excitation winding resistance is $R_f = R'_f/1.5$ $(N_s/N_f)^2 = 1.208/(1.5 \times 0.33^2) = 7.25\ \Omega$, from which the result $34.7/7.25 = 4.78$ A is derived, as it should be. If rated data 4.78 A is put in the field of the SM as the fourth parameter and to mark the considered option, $V_{fd} = 34.7$ V appears.

Let us discuss modeling of saturation. No-load SM is modeled. If the option *Simulate saturation* is selected, the additional field *Saturation parameters* appears in which a number of excitation current values and their corresponding rms values of the SM phase-to-phase voltages in the no-load condition are put in. At first, the excitation currents are given separated with commas; then, after semicolon, the same number of the corresponding line voltages are put in. The first value is the point in which a saturation effect begins to manifest itself.

For example, for the SM 1850 kVA, 6000 V, 1000 rpm (Ukraine), the magnetization dependence is given in the Table 5.7.

In the field *Saturation parameters*, it is necessary to put in the following: 73,110,139,249,395; 3480,5100,6000,7260,7980.

If SM in pu is selected, as shown in Figure 5.88, the saturation parameters are given in pu too, and the excitation voltage has to be taken of 1, in order to receive in the no-load condition the rated stator voltage.

The second page of the dialog box when the parameters are given in the so-called standard form is shown in Figure 5.89.

TABLE 5.7
SM Magnetization Dependence

Voltage, pu.	0.58U	0.85U	1U	1.21U	1.33U
Excitation current, A	73	110	139	249	395

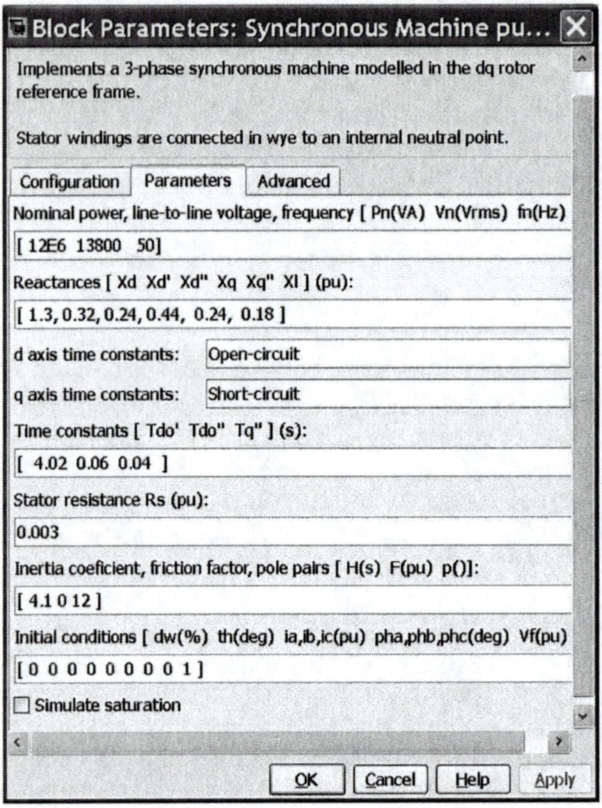

⊡ **Block Parameters: Synchronous Machine pu...** ✕

Implements a 3-phase synchronous machine modelled in the dq rotor reference frame.

Stator windings are connected in wye to an internal neutral point.

| Configuration | Parameters | Advanced |

Nominal power, line-to-line voltage, frequency [Pn(VA) Vn(Vrms) fn(Hz)

[12E6 13800   50]

Reactances [Xd Xd' Xd" Xq Xq" Xl] (pu):

[1.3, 0.32, 0.24, 0.44, 0.24, 0.18]

d axis time constants: Open-circuit

q axis time constants: Short-circuit

Time constants [Tdo' Tdo" Tq"] (s):

[4.02 0.06 0.04]

Stator resistance Rs (pu):

0.003

Inertia coeficient, friction factor, pole pairs [H(s) F(pu) p()]:

[4.1 0 12]

Initial conditions [dw(%) th(deg) ia,ib,ic(pu) pha,phb,phc(deg) Vf(pu)

[0 0 0 0 0 0 0 0 1]

☐ Simulate saturation

[OK] [Cancel] [Help]   Apply

FIGURE 5.89 SM standard dialog box.

Instead of resistances and inductances, the reactances in pu and the time constants are entered. SM reactances are determined as the product of the corresponding inductances by the synchronous frequency ω_e divided by the base resistance and are denoted as X with corresponding indexes. SM is characterized by the following reactances:

The d-axis synchronous reactance $X_d = X_{ls} + X_{md}$, the transient reactance

$$X'_d = X_{ls} + X_{md}X_{lfd}/(X_{md} + X_{lfd}) \qquad (5.79)$$

the subtransient reactance

$$X''_d = X_{ls} + X_{md}X_{lfd}X_{lkd}/(X_{md}X_{lfd} + X_{lfd}X_{lkd} + X_{md}X_{lkd}) \qquad (5.80)$$

the q-axis synchronous reactance

$$X_q = X_{ls} + X_{mq} \qquad (5.81)$$

the transient reactance

$$X_q'' = X_{ls} + \frac{X_{mq}X_{lkq1}}{X_{mq} + X_{lkq1}} \tag{5.82}$$

(in Figure 5.89 denoted as X_q''). For the round-rotor SM, the subtransient reactance appears for q-axis.

The values of the time constants depend on how they are determined: with open or with short stator circuit. Time constants for the axes can be specified independently for either short-circuit or for open circuit. For the open circuit

$$T_{d0}' = \frac{X_{md} + X_{lfd}}{\omega_e R_{fd}} \tag{5.83}$$

$$T_{d0}'' = \frac{X_{lkd} + X_{md}X_{lfd}/(X_{md} + X_{lfd})}{\omega_e R_{kd}} \tag{5.84}$$

$$T_{q0}' = \frac{X_{mq} + X_{lkq1}}{\omega_e R_{kq1}} \tag{5.85}$$

For the short-circuit

$$T_d' = \frac{X_{md}X_{ls}/(X_{md} + X_{ls}) + X_{lfd}}{\omega_e R_{fd}} \tag{5.86}$$

$$T_d'' = \frac{X_{lkd} + X_{ls}X_{md}X_{lfd}/(X_{md}X_{ls} + X_{lfd}X_{md} + X_{ls}X_{lfd})}{\omega_e R_{kd}} \tag{5.87}$$

$$T_q' = \frac{X_{mq}X_{ls}/(X_{mq} + X_{ls}) + X_{lkq1}}{\omega_e R_{kq1}} \tag{5.88}$$

For the round-rotor SM, the second time constant appears for q-axis.

The SM quantities can be measured by Simulink blocks with the help of either the block **Bus selector** as shown in model5_20 or the block **Machines Measurement Demux** from the same folder, *Machines*. The block input is connected to the output *m*. Before we use it, it is necessary to tune the block by selecting *Machine type: Synchronous* and afterwards marking the required quantities. In Figure 5.90 the following are marked: the stator phase currents i_{sa}, i_{sb}, i_{sc}, the excitation current i_{fd}, the main flux linkages Ψ_{mq}, Ψ_{md}, the rotor rotation speed ω_m, the rotor angle θ, degree, the electromagnetic torque T_e, and the load angle δ that is equal to arc tan (v_{sd}/v_{sq}). The rest of the quantities available for the measurement are: the stator currents

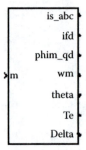

FIGURE 5.90 Measuring block for SM.

i_q, i_d; the damper windings currents; the stator voltages v_{sq}, v_{sd}; the electric power $P_e = T_e\omega_m$; the active and reactive powers that are equal, respectively, $P_{eo} = v_{sq}i_q + v_{sd}i_d$, $Q_{eo} = v_{sq}i_d - v_{sd}i_q$; the deviation of the rotation speed from the synchronous one $\delta\omega$; and the rotor deviation angle $\delta\theta$ that is the integral of $\delta\omega$. It has been said already that this block is considered obsolete, but with this block, the model schemes come out more compactly. It is necessary to have in mind that the described model suppose that the stator currents flow into the windings, but the measured currents i_{sa}, i_{sb}, i_{sc}, i_q, i_d are returned by the model as flowing out of SM.

5.3.2 SIMULATION OF THE ELECTRICAL DRIVE WITH SM AND LOAD-COMMUTATED CONVERTERS

The set of the electrical drive with SM and load-commutated converter (Figure 5.91) consists of the controlled input thyristor rectifier TV1, the smoothing reactor with rather large inductance L (the function of the thyristor T will be explained later), and the thyristor inverter TV2 connected to SM. The scheme of this set is different from the current inverter with IM (Figure 5.73) with the use of the thyristors instead of the forced-controlled switches in TV2. The use of the thyristor inverter is possible in the case under consideration because, with an appropriate SM excitation control, SM can

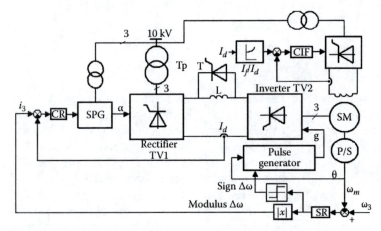

FIGURE 5.91 Electrical drive with SM and load-commutated converters.

operate with leading power factor and provide the thyristor commutation by the SM voltage. The rectifier provides the DC current at the TV2 input. During acceleration and the steady-state operation, the power from the AC mains transfers through TV2 to SM that runs in the motor mode. TV2 operates in the inverter mode when the power from DC link is transferred to the AC circuit of SM. For power economy, it is desirable to have the firing angle by inverting α_2 about 180°, but in order to have the necessary margins for inversion, α_2 is limited by 135°–150°. If α_2 is chosen, it necessary to have in mind that the commutating inductance and, hence, the overlap angle, are defined by the reactances X_d'', X_q' (Equations 5.80 and 5.82) that can be rather large. If the firing pulses are formed by the rotor position, the rotor lags behind the stator voltage during SM loading, so that the firing angle approaches the dangerous area.

During deceleration, TV2 runs in the rectifier mode and TV1 in the inverter one so that the braking energy returns to the network. In this mode, it is reasonable to have the angle α_2 minimum, but the TV2 voltage does not have to be more than the maximum TV1 voltage for inverting.

The control system contains the speed controller SR, whose absolute value of the output is the reference for the current controller CR. The sign of the CR output determines TV2 operation mode—rectifier or inverter. The constant firing angles are expected in the inverter and rectifier modes in regard to the position of the rotor or the flux linkage vector behind subtransient inductance (see further), but there are and more complex schemes [15]. Besides, there is the excitation current controller CIF, whose reference increases with an increase in DC current; such a controller is often submitted to the stator flux linkage controller.

The flux linkage vector is used for estimation of the voltage vector position because the former is not so much exposed to the influence of commutation distortion. The next relations are used for computation of the linkage flux vector behind subtransient inductance $\mathbf{\Psi}_{se}$:

$$\mathbf{\Psi}_{se} = \mathbf{\Psi}_s - X_k \mathbf{I}_s \tag{5.89}$$

where
  $\mathbf{\Psi}_s$ is the vector of the stator linkage
  $\mathbf{I}_s$ is the stator current vector, $X_k = 0.5(X_d'' + X_q')$

The components of $\mathbf{\Psi}_s$ are determined by using (5.58). The first-order links are used instead of the integrators, in order to avoid the errors of the numerical integration; it means that the estimations are

$$\Psi_{s\alpha,\beta} = \frac{TU_{\alpha,\beta}}{Ts+1} \tag{5.90}$$

where $U_{\alpha,\beta}$ are the components of the vector $\mathbf{U}_s - R_s\mathbf{I}_s$. The relations (5.91) are used as compensation for phase errors that arise [6]

$$\Psi_{s\alpha}^* = \Psi_{s\alpha} + k_d\Psi_{s\beta}, \quad \Psi_{s\beta}^* = \Psi_{s\beta} - k_d\Psi_{s\alpha} \tag{5.91}$$

According to the recommendation in [16], it is taken $T = 1/\omega\lambda$, $k_d = \lambda$, $\lambda = 0.1 - 0.5$.

The described system functions properly, if the voltage of the stator windings is sufficient for the commutation of the inverter TV2 thyristors. Usually, it takes place at 5%–10% of the rated speed. In order to speed up SM to this level of speed, the intermittent control of the reactor current is used. At any instant, two thyristors conduct, one in the anode group and another in the cathode group, which draw a current. Since SM is excited, for proper selection of the conducting thyristors as a function of the rotor position, this current produces a torque, and SM picks up speed. During rotor rotation by every 60°, at first the firing pulses are removed from the conducting pair of the thyristors and the reactor current is set to zero. The earlier conducting thyristors restore their blocking ability. Afterwards, the firing pulses are applied to the next pair of the thyristors, and the reactor current is reestablished. The torque appears again, and SM continues to speed up. After reaching the said minimum speed, the control is transferred to the load commutation. In order to make current interruption faster, the reactor is shorted out with the thyristor T if there is a demand for null current.

If SM start occurs with a small load, the initial start can be carried out without the position sensor, with open-loop system. In this case, the saw-tooth signal is formed whose frequency increases smoothly from null to 5%–10% of the rated value, and this signal replaces the rotor position signal for switching-over of the thyristors in the order described. At that point, if the position of the vector Ψ_{se} is used during normal operation, the control system without the position sensor results, increasing essentially the reliability of the electrical drive.

In model5_27 the salient-pole SM with the main parameters 1850 kVA, 6 kV, 50 Hz, 1000 rpm is modeled. The base values are $I_b = 251$ A, $R_b = 19.5$ Ω. It is supposed that SM is fitted with the speed and, perhaps, with the rotor position sensors (usually, it is the same device). The subsystem **Load** models SM load as the product of the load torque by the speed, and the load torque is the product of the *Coulomb friction value* of the block **Friction** and the input *Change*. For the models of the rectifier **Rect** and the inverter **Inv**, the blocks **Universal Bridge** with *Thyristors* are used. Supply is carried out from the source 6.3 kV, 50 Hz. The circuit *R-L* at the source output (block **Three-Phase Series RLC Branch**) simulates the internal source impedance and, perhaps, the reactance of the current-limiting reactor or the supplying transformer. The reactor at the bridge output (block **Series RLC Branch**) has the inductance of 0.1 H, and its parameters provide the reasonably small pulsation of the current I_d. The thyristor is set parallel to the reactor; its function has already been explained.

There are three subsystems in the subsystem **Control**: **Id_Contr** (TV1 control), **Inv_Contr** (generation of TV2 firing pulses), and **F_Controller** (excitation current control). Besides, there is the subsystem **Flux_Est** for estimation of the vector Ψ_{se} position. Generation of the firing pulses for TV1 is carried out in the subsystem **Id_Contr** with the block SPG (**Synchronized 6- pulse Generator**). The block is synchronized with the phase-to-phase supplying voltages that are measured by the block **Three-Phase V-I Measurement**. The firing angle α_1(degree) is set by the PI current controller with the limiter at the output.

The subsystem **Id_Contr** can operate in two modes: speed control and current control. Switch-over is carried out by **Switch1** with the help of **Relay** at 10% of the rated speed. In the first mode, the switch is in the upper position, and **SR (Discrete PI Controller)** output comes to the **CR** input through the module detector and the limiter. As a feedback, the measured reactor current I_d is used. The **SR** operate with signals in SI, so the rotor rotation speed signal measured in pu is recalculated in rad/s, and I_{f_ref} reference signal is divided by the base current value. As it was said earlier, I_{f_ref} increases with increasing of I_d or is controlled in order to provide the reference value of Ψ_{se}.

In the current control mode, **Switch1** is in the lower position, **SR** is disconnected, and at **CR** input, either the constant reference start current (it is set 150 A) or null are given through **Switch**; when the latter signal is active, the blocking ability of the earlier conducting thyristors of TV2 is restored. The switch-over signal *Switch* ("Pause", "Mode") is produced in the subsystem **Inv_Contr.**

This subsystem consists of two subsystems: **PulseGen** and **Start**; the former runs the processes during normal operation, and the latter operates during the start. The unit that generates the saw-tooth signal changing from 0 to 360 (similar to the rotor position signal Teta(θ)) is placed in the upper-left part of the subsystem: When the integrator output reaches the value of 360, the signal at its reset input changes the sign that results resetting integrator output to 0. The saw-tooth signal frequency is determined by the parameters of the input circuits of the integrator that are selected so that at Step = 1 the frequency is 50 Hz. Depending on the position of **Manual Switch** and the operating level of **Switch 7**, various input signals can be chosen for start and normal work. If **Manual Switch** is in the upper position and the threshold of **Switch 7** is 0.5, the saw-tooth signal is used during start; when the speed increases to 0.1, **Switch 7** switches over to the upper position, and the signal of the vector Ψ_{se} position (that is shifted in certain way) comes to the input of **PulseGen**; thus, operation without rotor position sensor is accomplished. If **Manual Switch** is in low position and the threshold of **Switch 7** is 0.5 as said before, during start signal θ and during normal operation signal Ψ_{se} is used. If the threshold of **Switch 7** is more than 1, switching-over does not take place, and the signal θ comes to the input PulseGen always. It is determined with the help of **Switch**, whose subsystem sends the pulses to the inverter bridge.

It is possible to change easily the firing inverter angles in the rectifier and inverter modes with the help of the constant **d_alpha_motor, d_alpha_generator,** and **d_alpha1**. With the help of **Switch 2, 3, 8**, the total signals change from 0 to 360°.

The subsystem **Start** consists of the subsystem **Inv_Start** and the logical elements. The main part of **Inv_Start** is the subsystem **Fr_Pulse**. It consists of 11 blocks **Relay**; with their help, the pulses of 45° duration are produced every 60°; these pulses come to the thyristors in the natural sequence: +A, −C, +B, −A, +C, −B. The logic null signal of 15° duration is formed at the outputs of **OR** gate every 120°, and at the output of the **AND** gate, such a signal is formed every 60°. By this signal, the pulses are removed from TV2, and the output *Pause* is used for setting of the reference I_d to zero and for turning on the thyristor connected parallel to the reactor.

The subsystem **PulseGen** consists of the subsystem **Form_Pulses** and additional blocks; at the subsystem input, the signal θ_f comes that is equal either to the signal θ

or to the angle φ of the vector Ψ_{se}, perhaps with phase shift. **Form_Pulses** contains the relays and the logical gates that form the pulses in the duration of 30° every 60°. For example, it follows from the **Selector1** scheme that the output 2 of the subsystem **Form_Pulses** comes to the thyristor +A. The logical signal "1" appears at this output when **R1** is turned on, that is, when $\theta_f = 2 \times 30° = 60°$. Since the moment when $\theta_f = \theta = 0$ corresponds to the voltage phase A maximum that leads by 60° crossover point (the point with zero firing angle), the corresponding firing angle $\alpha = 120°$. It corresponds to the inverter mode of TV2. With this load, the rotor lags behind the supplying voltage and the firing angle increases. With the selected parameters, it is close to 150°.

For SM in the generator mode (TV2 is in the rectifier mode), θ_f is shifted about θ with the help of **Sum2, Switch1, Switch2** by 270° in the lag direction, so that the pulse comes to +A when $\theta = 60° + 270° = 330°$ that corresponds to $\alpha = 30°$. Switching-over is carried out depending on the sign of the **SR** output. The second pulses for the same thyristors are formed with the help of the block **Selector** and **OR** gates 60° later (double pulsing).

The subsystem **Flux_Est** estimates Ψ_{se} components by (5.89) through (5.91), and afterwards computes the angle $fi(\varphi)$, setting it in the range of 0°–360°. The three-phase signal U_{abc} is generated at the output **Sum4**, and the signal $F_{s\alpha} = F_{salpha}$ is formed as

$$F_{s\alpha} = \int \left(U_a + \frac{\lambda(U_b - U_c)}{\sqrt{3}} - \lambda\omega\, F_{s\alpha} \right) dt \tag{5.92}$$

which is equivalent to (5.90), (5.91), analogous for F_{sbeta}. Then by (5.89), the components F_{alpha}, F_{beta} of the vector Ψ_{se} are calculated. It is accepted that $\lambda = 0.3$.

Two options are provided for the reference of I_f. When **SW_If** is in the upper position, this reference is equal to I_{f_ref} from the subsystem **Id_Contr**; when **SW_If** is in the lower position, the reference is determined by the output of the flux controller **F_Controller**; at its inputs, the stator flux linkage module reference F_ref and the value of the vector Ψ_{se} module are compared. The controller begins to operate after the initial SM start.

The scope **Motor** fixes the stator currents, the speed, the reactor current, and the excitation current of SM. The scope **Invert** permits to observe the inverter voltage and current. The scope **Power** fixes the active and reactive powers of the grid (Source) and of SM (Motor) that are smoothed by the filters with the time constant of 0.02 s.

This model gives the opportunity to acquaint oneself with the operation of the electrical drive with SM and load-commutated converters and with the influence of the control system parameters on its function in different modes (speeding-up, deceleration, load changing). As an example, in Figures 5.92 through 5.94 the responses are shown for the setting of **Manual Switch** in the lower position and for the threshold of **Switch 7** equal to 0.5; it means an operation without the rotor position sensor. **SW_If** is in the lower position. The schedule is given by the blocks

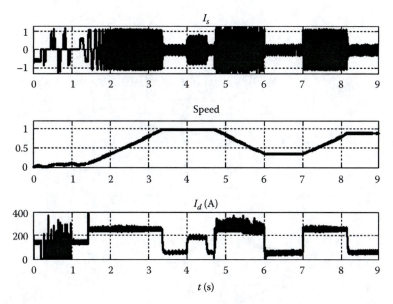

FIGURE 5.92 Transients in the electrical drive with SM and load-commutated converters.

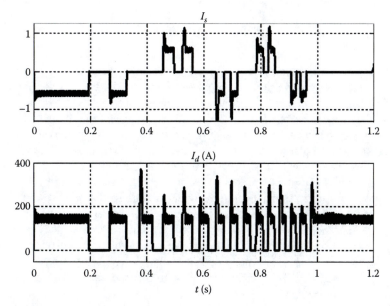

FIGURE 5.93 DC and SM currents at the start.

Wref and **Timer1**: speeding up to 100 rad/s with the load $T_m = 0.25$, the load tre-bling at $t = 4$ s, the load reducing down to $T_m = 0.25$ at $t = 4.5$ s, slowing down to the speed of 35 rad/s at $t = 4.7$ s, and speeding up to 90 rad/s at $t = 7$ s. It is seen from Figure 5.94a that when slowing down the active power becomes negative, that is, returns to the grid. It follows specifically from Figure 5.94b that during

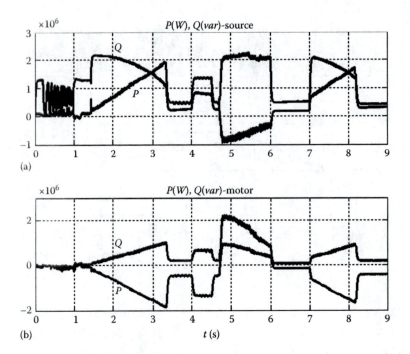

FIGURE 5.94 Power during changes in SM speed. Active and reactive power of (a) the source and (b) the motor.

the operation with the load close to the rated one (at $t = 4.3$ s) the active power is 1.35 MW, and the reactive power is 0.65 Mvar; that is, the phase displacement is $\varphi = 25.7°$. The reactive power is positive always and corresponds to its generation.

With the use of the scope **Invert**, the actual firing angles of TV2 can be found. It is not a simple task because SM voltage is distorted essentially by the commutation processes. Since the phase voltages are fixed, the firing angle of A phase is counted off from the point in that $V_a = V_c$ out of the commutation area. The steady-state area ($t = 4.3$ s) is under consideration. It is necessary for measurement, using the options of the block Scope, to zoom the picture on the screen so that not more than two periods of the voltage and current are fitted in. The period of the first harmonic is 20.8 ms. In this mode, one can see the crossover points rather distinctly. It is seen that the moment of the pulse origin (the moment when the phase current begins to rise) leads the nearest crossover point by 2.05 ms that corresponds to the angle of the inverting $\beta = 360 \cdot 2.05/20.8 = 35.5°$. As the commutation duration is 1 ms, the commutation angle is $\gamma = 360/20.8 = 17.3°$.

The reader can investigate themselves the processes involved with other variants of the start and normal work are used. While performing these investigations, changing of some controller parameters can be necessary. For example, with the use of the rotor position sensor and the flux controller (**Manual Switch** is in the upper position, threshold of Switch **7** is 2, **SW_If** is in the lower position), satisfactory responses are obtained, such as F_ref = 1.25, d_alpha_motor = 0; parameters of **F_controller** are $K_p = 0$, $K_i = 50$; and parameters of **If_controller** are $K_p = 77$, $K_i = 154$.

5.3.3 MODEL OF SIX-PHASE SM

The power SM with 2 three-phase stator windings that are mutually displaced by 30° are used for the same reason as would apply in the case of IM with two stator windings as described previously. Both stator windings are usually placed in the same slots, so that they are coupled magnetically. Modeling is carried out in the reference frame that rotates synchronously with the rotor. The equations that suit for modeling are given in [17].

Denote two windings as *abc* and *xyz*, the rotor angle θ counts off from the position of the first winding, and the magnetic axis of the second winding leads the first one by 30° (electrical). In [17] it is shown that in the rotating reference frame, the winding magnetic coupling is determined by the parameters

$$L_{sm} = L_{ax} \cos g + L_{ay} \cos\left(g + \frac{2\pi}{3}\right) + L_{az} \cos\left(g - \frac{2\pi}{3}\right)$$

$$L_{sdq} = L_{ax} \sin g + L_{ay} \sin\left(g + \frac{2\pi}{3}\right) + L_{az} \sin\left(g - \frac{2\pi}{3}\right), \quad g = \frac{\pi}{6}$$

(5.93)

where L_{ax}, L_{ay}, and L_{az} are the mutual leakage inductances of indicated phases of both windings. The values of these inductances are found by the calculation of magnetic circuits of SM; some relationships for this aim are given in [17]. Denote L_{sl}—the leakage inductance of one winding, $k = L_{sm}/L_{sl}$—the winding coupling coefficient, $L_s = (1 - k) L_{sl}$, $L_{sdq} = bL_{sl}$. The stator voltage equations are

$$s\Psi_{qs1}^r = U_{qs1} - \omega_r \Psi_{ds1}^r - R_s I_{q1}$$

(5.94)

$$s\Psi_{ds1}^r = U_{ds1} + \omega_r \Psi_{qs1}^r - R_s I_{d1}$$

(5.95)

$$s\Psi_{qs2}^r = U_{qs2} - \omega_r \Psi_{ds2}^r - R_s I_{q2}$$

(5.96)

$$s\Psi_{ds2}^r = U_{ds2} + \omega_r \Psi_{qs2}^r - R_s I_{d2}$$

(5.97)

Here index "1" is related to the windings *abc* and index "2" to *xyz*. Note that Park's transformation angle is θ for the former and $\theta + \pi/6$ for the latter. The equations for damper and excitation windings are given by (5.64)–(5.66).

Solution of the equations becomes simpler with the introduction of the flux linkages Ψ_{mq}, Ψ_{md} by the following relationships:

$$\Psi_{mq} = L_{mq}(I_{q1} + I_{q2} + I_{kq})$$

(5.98)

$$\Psi_{md} = L_{md}(I_{d1} + I_{d2} + I_{kd} + I_{fd})$$

(5.99)

where
I_{kq}, I_{kd}, I_{fd} are the currents of the damper and excitation windings
L_{mq}, L_{md} are the main SM inductances

Then, from the equations given in [17], the following expressions can be received for the stator currents:

$$I_{q1} = \frac{\Psi_{q1} - k\Psi_{q2} - (1-k)\Psi_{mq} + b(\Psi_{d2} - \Psi_{md})}{L_{sl}d} \tag{5.100}$$

$$I_{q2} = \frac{\Psi_{q2} - k\Psi_{q1} - (1-k)\Psi_{mq} - b(\Psi_{d1} - \Psi_{md})}{L_{sl}d} \tag{5.101}$$

$$I_{d1} = \frac{\Psi_{d1} - k\Psi_{d2} - (1-k)\Psi_{md} - b(\Psi_{q2} - \Psi_{mq})}{L_{sl}d} \tag{5.102}$$

$$I_{d2} = \frac{\Psi_{d2} - k\Psi_{d1} - (1-k)\Psi_{md} + b(\Psi_{q1} - \Psi_{mq})}{L_{sl}d} \tag{5.103}$$

$$d = 1 - k^2 - b^2$$

It is seen that when $k = b = 0$ these relationships come to (5.71), (5.72). The currents I_{kq}, I_{kd}, I_{fd} are determined by (5.73) through (5.75). Substituting these expressions for the currents in (5.98), (5.99), the formulas for Ψ_{mq}, Ψ_{md} are derived:

$$\Psi_{mq} = L_{aq} \left\{ \frac{(\Psi_{q1} + \Psi_{q2})(1-k) + b(\Psi_{d2} - \Psi_{d1})}{L_{sl}d} + \frac{\Psi_{kq}}{L_{lkq}} \right\} \tag{5.104}$$

$$\Psi_{md} = L_{ad} \left\{ \frac{(\Psi_{d1} + \Psi_{d2})(1-k) + b(\Psi_{q2} - \Psi_{q1})}{L_{sl}d} + \frac{\Psi_{kd}}{L_{lkd}} + \frac{\Psi_f}{L_{lf}} \right\} \tag{5.105}$$

$$L_{aq} = \left[\frac{1}{L_{mq}} + \frac{1}{L_{lkq}} + \frac{2(1-k)}{L_{sl}d} \right]^{-1} \tag{5.106}$$

$$L_{ad} = \left[\frac{1}{L_{md}} + \frac{1}{L_{lkd}} + \frac{1}{L_{lf}} + \frac{2(1-k)}{L_{sl}d} \right]^{-1} \tag{5.107}$$

SM torque is computed as

$$T_e = \left[(I_{q1} + I_{q2})\Psi_{md} - (I_{d1} + I_{d2})\Psi_{mq} \right] \tag{5.108}$$

The model of six-phase SM is given in model5_28. The subsystem SM_6Ph consists of the input part and the subsystem SM_eq that solves SM equations. The former contains the sensors of the phase-to-phase voltages supplied and the controlled sources of the phase currents drawn by SM; the resistors having rather big resistance and connected in parallel to the sources increase the accuracy of the differential equation integration in the discrete simulation mode. Their resistance is selected by experiment.

The subsystem **SM_eq** consists of the two subsystems that model electromagnetic and mechanical processes. The latter is the same as in the said IM models and is not considered here.

The subsystem **SM_Electr** consists of a number of subsystems. The subsystem **Vabc_Vqd** computes the voltages U_{qs1}, U_{ds1}, U_{qs2}, U_{ds2}, using the measured line voltages and the known angle θ. The subsystem **Stator_Rotor** has seven subsystems that solve the equations (5.94) through (5.97), (5.64) through (5.66) and the subsystem **SD_Flux**, in which calculations of the flux linkages Ψ_{mq}, Ψ_{md} and currents by (5.104), (5.105), (5.100) through (5.103), (5.73) through (5.75) are fulfilled. Torque calculation is carried out in the functional block **Te** by (5.108). The currents of both three-phase stator windings are computed in the subsystems **Iqd_abc** and **Iqd_xyz** with the help of Park's transformation.

There are circuits for the computation of the electromagnetic active and reactive SM power P_{eo}, Q_{eo}, and the load angle δ placed in the subsystem **SM_Electr**. At the output m, there are 27, partly in the vector form, signals that are arranged in the next way (the vectors are shown in the brackets): (I_a, I_b, I_c), (I_x, I_y, I_z), (I_{q1}, I_{d1}), (I_{q2}, I_{d2}), I_{fd}, (I_{kq}, I_{kd}), (F_{mq}, F_{md}), (V_{q1}, V_{d1}), (V_{q2}, V_{d2}), (P_{eo}, Q_{eo}), (δ_1, δ_2), ω, P_e—the mechanical power, θ, T_e, it means that the output demultiplexer has 15 outputs.

Before SM simulation, the program has to be fulfilled, in which SM parameters are given as follows:

Program SM_MODEL_6ph

```
Unleff=6000%V
Pn=1850000%VA
p=3%pair of poles
%Computation:
Vb=1.4142*Unleff/1.732;
Ib=2*Pn/(3*Vb)
Wb=100*pi/p%ωHOM,rad/s
wb_b=100*pi
Mb=Pn/Wb
Rb=Vb/Ib
%SM parameters in pu
Rs=0.0085*2%stator winding resistance
Rf=0.005%field winding resistance
Rkd=0.0915%d-damper winding resistance
Rkq=0.079%q-damper winding resistance
k=0.85%winding coupling coefficient
b=0.1%additional winding coupling coefficient
Xsl=0.13%stator internal leakage reactance
Xmd=1.29%main d-reactance
Xmq=0.695%main q-reactance
Xfl=0.307% field winding leakage reactance
Xkdl=0.208% d-damper winding leakage reactance
```

```
Xkql=0.185% q-damper winding leakage reactance
J=300%kgm^2% total moment of inertia of rotating parts
F=0% viscous friction coefficient
%Calculation
H=0.5*Wb*Wb*J/Pn
Xs=(1-k)*Xsl
Xsm=k*Xsl
d=1-k^2-b^2
Xaq=(1/Xmq+1/Xkql+2*(1-k)/(Xsl*d))^-1
Xad=(1/Xmd+1/Xkdl+1/Xfl+2*(1-k)/(Xsl*d))^-1
Xd=Xmd+Xsl, Xq=Xmq+Xsl
Xkd=Xmd+Xkdl,Xkq=Xmq+Xkql
Xf=Xmd+Xfl
%Given
wmo=0%initial rotor speed, rad/s
tho=0%initial rotor position, rad
phifdo=1%initial conditions of the fluxes
phikqo=0
phikdo=0
phido=0
phiqo=0
phidol=0
phiqol=0
```

In Model5_28, SM has the same main parameters as in model5_27. The model's main circuits are shown in Figure 5.95 The rectifier has a series connection of the thyristor bridges and provides 12-pulse rectification. Also, the inverter has two thyristor bridges connected in series. Two three-phase SM windings are connected to the bridge outputs.

The control systems of the rectifier and inverter are essentially the same as in model5_27. As the firing angles of the both rectifier bridges are accepted to be equal, one block **Synchronized 12-pulse Generator** is used for the generation of firing pulses. For the sake of simplicity, only one inverter control method is used: the use of rotor position signal. The additional circuits **Form_Pulses1** are added in the subsystem **Inv_Contr/PulseGen** for generation of the firing pulses *Pulses1*

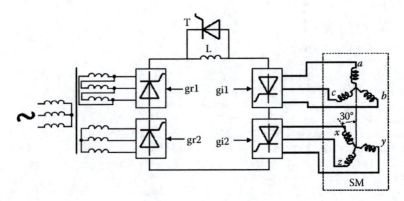

FIGURE 5.95 Main circuits of the six-phase SM with load-commutated converter.

that lead the firing pulses *Pulses* by 30°. Since the current has to flow through four thyristors, the pulse duration is increased to 60° that, together with their doubling, provides the full pulse duration of 120°. For the flux linkage estimation, the flux linkage magnitudes for each of two windings are calculated; computed values are added to the factor of 0.5.

By the scope **Currents**, one can observe the currents of both stator windings, the excitation current, and the inductor (which has the inductance of 0.2 mH) current. The scope **Motor** fixes the flux linkages F_{mq}, F_{md}, the SM active and reactive powers, and its rotation speed and torque. All the values, except for the inductor current, are shown in pu. The scope **Motor1** shows the voltage (V) and the current (A) of the stator windings *abc*, and the scope **Power** records the active (W) and reactive (var) powers of the grid and SM.

The following schedule is simulated: speeding up to 105 rad/s with the load torque $T_m = 0.25$, load increasing to $T_m = 0.75$ at $t = 2.5$ s, load reducing to $T_m = 0.25$, and slowing down to 25 rad/s. In Figure 5.96 the SM speed, its torque, and its excitation current are shown. In Figure 5.97 the plots of phase *a* and *x* currents in the motor and generator SM modes are given. The phase *x* current leads the phase *a* current by 30°. It is seen from the scope **Power** that the active SM power minus the loss is equal to the power consumed from the grid or returned to the grid (when there is a slowing-down), and the reactive SM power is always negative.

It is supposed in the described model that both rectifiers and both inverters are connected in series. The drawback of this scheme is that when there is a failure of anyone of the switches, the entire system loses serviceability. But one of the main advantages of using the six-phase SM is its ability to operate with just half the power when there is a supply voltage failure with one winding. For this purpose,

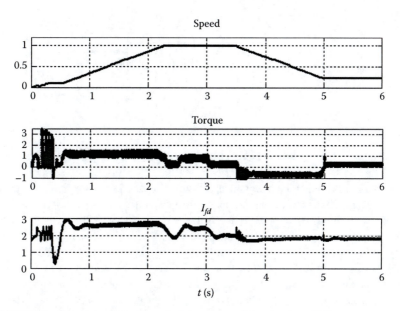

FIGURE 5.96   Speed, torque, and excitation current of the six-phase SM.

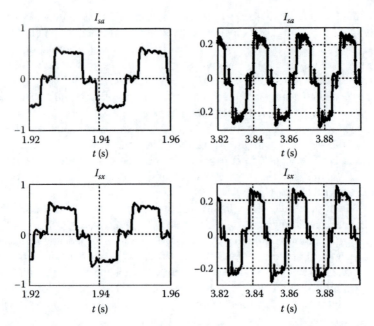

FIGURE 5.97 Phase currents in the motor and braking modes of the six-phase SM.

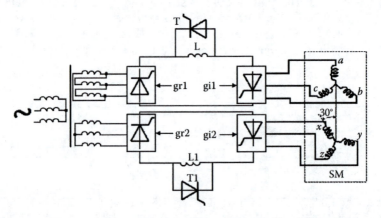

FIGURE 5.98 Main circuits of the six-phase SM with independent winding supply.

the main circuits have to be changed as shown in Figure 5.98. Note that this scheme contains more devices in the main and control circuits than the previous one.

Model5_29 simulates this system. In comparison with model5_28, the main circuit is modified according to Figure 5.98, and the second **CR** with the same reference is added in the control system. As the rectifier firing angles can be different, two SPGs are used whose reference voltages have different phasings. The inverter control system is the same.

The responses do not differ practically from the ones for model5_28. In Figure 5.99 the currents of the phases *a, x* and the phase-to-phase voltage on the first winding (lagged) are given.

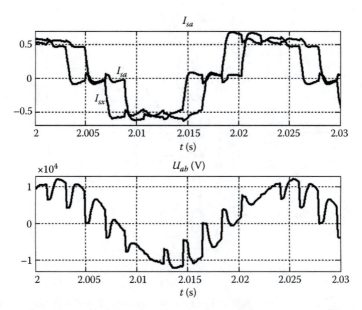

FIGURE 5.99 Currents and voltage in the six-phase SM with independent winding supply.

5.3.4 CYCLOCONVERTER SIMULATION

It has been said in Chapter 4 already that for converting the AC voltage of the constant amplitude and frequency into the AC voltage of the changing amplitude and frequency, together with VSI, the systems of the direct, without DC link, converting are used. The matrix converters whose modeling was considered in Section 4.11 are the new trend in power electronics. At the same time, the direct converters with thyristors (cycloconverters) have been in use since long ago [5]. The main circuits are shown in Figure 5.100. Cycloconverter consists of three pairs of thyristor six-pulse bridges; the bridges of each pair is connected in antiparallel direction one to another, as shown in model4_2, building the reversible rectifier. Each of the reversible rectifiers is supplied from a separate winding of the three-phase transformer; three separate transformers can be used, which gives the opportunity to use the standard two-winding transformers. Each rectifier is controlled in such a way that its output voltage averaged for a small time interval has the sinusoidal form. This average voltage is the fundamental harmonic of the output voltage. By changing the frequency and the amplitude of this harmonic, it is possible to control the SM current and flux.

As it is seen from Figure 5.100, the rectifiers and SM windings have wye connection. So, if the potential difference between the midpoints of the rectification scheme and SM that changes with the high frequency is neglected, the phase SM voltage is equal to the fundamental harmonic of the rectifier connected to this phase. This output voltage U_d is

$$U_d = E_{d0} \cos \alpha \tag{5.109}$$

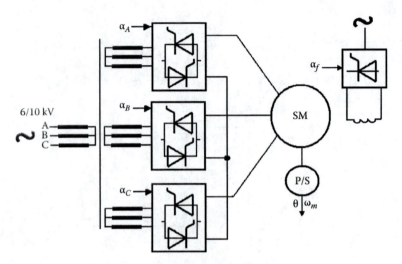

FIGURE 5.100 Main circuits of the cycloconverter.

$E_{d0} = 3/\pi U_{2max}$, U_{2max} is the amplitude of the secondary phase-to-phase transformer voltage, and α is the firing angle of the rectifier. Usually, some type of rectifier characteristic linearization is used in the control system, so its output voltage is

$$U_d = E_{d0} U_{con} \qquad (5.110)$$

Where U_{con} is the control system output signal that determines α, $|U_{con}| < 1$. Since with the rise of the frequency of the SM supplying voltage its amplitude has to increase, in order to have the constant value of the stator flux linkage for three rectifiers, one can write

$$U_{cona} = \gamma\omega_s \sin(\omega_s t), \quad U_{conb} = \gamma\omega_s \sin\left(\omega_s t - \frac{2\pi}{3}\right), \quad U_{conc} = \gamma\omega_s \sin\left(\omega_s t + \frac{2\pi}{3}\right)$$

$$(5.111)$$

γ is the proportionality factor and ω_s is the frequency of the stator voltage.

The electrical drive under consideration has the following features: SM does not usually have damper windings, and the output frequency is not more than 33%–40% of the grid frequency.

The vector control is used in the electrical drive control system. The firing angles for the rectifiers α_A, α_B, α_C are determined by the phase current controllers, whose harmonic reference signals have the form (5.111) and depend on the requested quantities I_q, I_d, I_{fd} where the parameters $I_q = I^r_{qs}$, $I_d = I^r_{ds}$ are used. These quantities depend in turn on the required torque, which is determined by the output of SR. The SM control system has to provide the stator flux linkage constancy (1 in pu) and zero displacement between stator voltage and current; it means the current space vector must be perpendicular to the flux linkage space vector Ψ_s (ignoring the active stator resistance). Under these assumptions, the SM torque in pu is equal to the module of the stator current space vector I_s. Thus, the following relationships have to be fulfilled [18]:

$$\Psi_d^2 + \Psi_q^2 = (X_d I_d + X_{md} I_{fd})^2 + (X_q I_q)^2 = 1 \tag{5.112}$$

$$I_d^2 + I_q^2 = I_s^2 \tag{5.113}$$

$$(X_d I_d + X_{md} I_{fd})I_d + X_q I_q^2 = 0 \tag{5.114}$$

I_s is the output of the SR, from which it follows in pu

$$I_q = \frac{I_s}{\sqrt{1 + I_s^2 X_q^2}} \tag{5.115}$$

$$I_d = -\frac{I_s^2 X_q}{\sqrt{1 + I_s^2 X_q^2}} \tag{5.116}$$

$$X_{md} I_{fd} = \frac{1 + I_s^2 X_q X_d}{\sqrt{1 + I_s^2 X_q^2}} \tag{5.117}$$

The control system can be built with the intermediate I_q, I_d current controllers or without them. In the second case, the quantities that are calculated by (5.115), (5.116) are transformed in the harmonic references for the phase current by relations (5.23) with the known angle θ (Figure 5.101). In the first case, these quantities are the references for the I_q, I_d current controllers, and the outputs of these regulators are

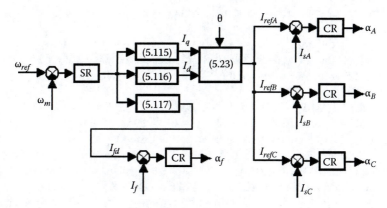

FIGURE 5.101 Cycloconverter control structure without I_q, I_d control.

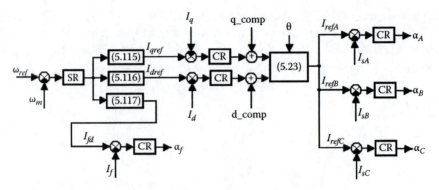

FIGURE 5.102 Cycloconverter control structure with I_q, I_d control.

transformed in the references for the phase current (Figure 5.102). From (5.62), (5.63) it follows that in the rotating reference frame the flux linkage equations for d- and q-axes are coupled, analogous, and the current equations are coupled too. For decoupling, the compensation signals are added to the regulator outputs [18]:

$$q\_comp = (X_{md}I_f + X_dI_d)\,\omega, d\_comp = -X_qI_q\omega \qquad (5.118)$$

Model5_30 is developed according to the aforementioned principles. The standard SM 2000 kVA, 400 V is used, with the difference that its damper windings are removed. For this purpose, as it follows from (5.62) to (5.68), it is sufficient to take the leakage inductances of these windings as very large values and their active resistances equal to zero. It is supposed that the rotor position and its speed are measured. Three SM phases are supplied from the similar subsystems **Power**. Each of them contains the subsystem **Power_Block** and **CR**. The former contains two thyristor bridges, transformer, SPG, and LPU. Such a scheme was described while discussing model5_4. The difference is that in model5_30 the primary transformer windings are connected in Delta11. In order to keep the right phasing of SPG, the phase-to-phase voltages coming to its inputs are lagged by 30°, with the help of the first-order links with time constant of 1.8 ms, which for 50 Hz gives the requested lag. The secondary transformer voltage is 150 V; the primary windings are connected in parallel and are supplied from the grid with voltage of 6.3 kV.

When using the transformer with three secondary windings, the transformers are excluded from the subsystems **Power_Block**; instead of this, three blocks **Multi-Winding Transformer**, each of them with three secondary windings, are used (model5_31).

The subsystem **Regulators** is built according to the block diagram shown in Figure 5.102 and contains the controllers of the speed and the currents I_q, I_d with decoupling signals and the generator of the sinusoidal references for the phase current controllers. For the sake of simplification, it is assumed that the currents I_q, I_d are measured, but in reality, only the phase currents are measured, and the block **abc_to_dq0 Transformation** has to be used for I_q, I_d calculation. A number of subsystem blocks demand the SM parameters, and when they change (e.g., for another SM), the functional expressions of the blocks must be change also, which is not convenient. Besides, the base current value is demanded because the computations are fulfilled in pu but control of the rectifier currents are run in SI. Therefore, it is reasonable to make the program analogous to the following, which is part of SM_Param_2000.

Program SM_Param_2000

```
Pn=2e6;
Un=400;
Ib=1.41*Pn/Un/1.73;
Ll=0.05;
Lmd=2.06;
Lmq=1.51;
Llfd=0.3418;
Xmd=Lmd;
Xmq=Lmq;
Xfl=Llfd;
Xd=Xmd+Ll;
Xq=Xmq+Ll;
```

The program is very simple because the SM model is selected in pu. With the selection of the SM model in SI, it would be necessary to compute the inductances in pu, as it was made for IM in the program IM_MODEL_6end. In the window *File/Model Properties/Callbacks/InitFcn* the command *run SM_Param_2000* is put in; it is possible also to write the given sequence of commands in the said window directly, similar to how it is done in the case of model5_30.

During speed control simulation, **Switch** is in the lower position. By the scope **Motor** one can observe the current, the speed, the excitation current, the torque of SM, and by the scope **Supply**, the supply current.

Figure 5.103 shows the responses arising during the speeding-up to speed 0.4 (20 Hz) with the load torque of 0.25. Modeling time is 1.6 s. Simulation takes place for a rather long period, even with the use of Accelerator. This fact makes using of the model for electromechanical process investigations that last seconds and even dozens of second and for the choice of the controller parameters, when the calculations have to be made repeatedly, embarrassing. Therefore, it seems reasonable to use this model for only the confirmation of the final results and for investigation of the steady states, but for the investigation of the electromechanical processes to have a simpler model (see further).

During an analysis of the steady states, it can be found that the modeling time can be short. Suppose we want to observe SM and supply currents for the SM current frequency of 5 Hz and of 20 Hz. For this purpose the following settings have to be made: **Switch** in the subsystem **Regulators** should be in the upper position, and in the SM dialog box, the initial slip set as $dw = -90\%$, the inertia coefficient as $H = 10^{10}$, and the modeling time as 0.8 s. As a result of simulation, the plots of the specified currents given in Figure 5.104 are obtained. The Fourier coefficients and THD are calculated with the help of **Powergui**, using the option *FFT Analysis* therein. For SM current analysis, *Structure* is *ScopeData1* and *Fundamental frequency* is 5 Hz; for supply current

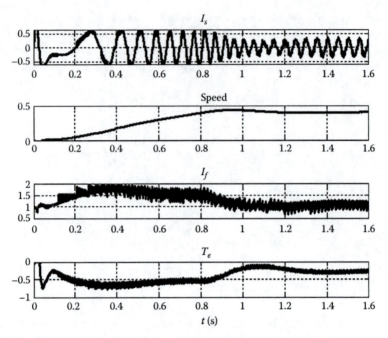

FIGURE 5.103 Transients in the electrical drive with the cycloconverter.

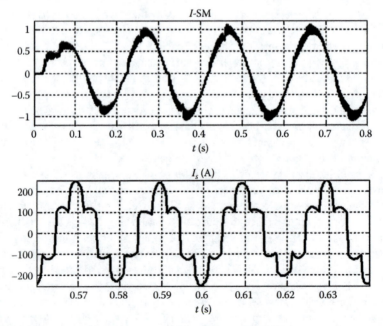

FIGURE 5.104 SM and mains currents with a frequency of 5 Hz.

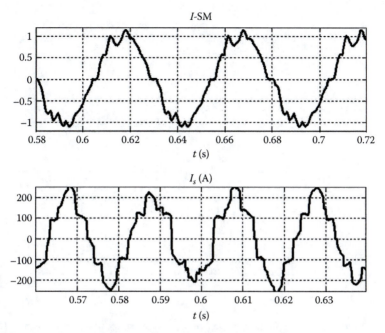

FIGURE 5.105   SM and mains currents with a frequency of 20 Hz.

analysis, *Structure* is *ScopeData* and *Fundamental frequency* is 50 Hz. As a result of simulation and computation, one can obtain the following: SM current, THD = 7.6%; harmonics pronounced have the frequency of 295 Hz and 305 Hz and essentially to a lesser extent 25 Hz and 35 Hz; for the supply current, THD = 27.2%, and the harmonics pronounced are 5th and 7th. If *dw* = −60% is set (SM current frequency is 20 Hz), the plots in Figure 5.105 are obtained after simulation. In the way described previously, it can be found that for SM current, THD = 11%, the most considerable harmonics identified have the frequency of 100 Hz and 140 Hz and, to a lesser extent, 280 Hz and 320 Hz. For the supply current, THD = 15.1% and the harmonic pronounced is the 5th.

As it has already been said, for analysis of the electromechanical processes in the electrical drives with cycloconverters, it makes sense to have a less complicated model (model5_32). Simplification is achieved by replacing the thyristor rectifier with the controlled voltage source, whose output voltage is computed by (5.110). The phase subsystem, besides **Controlled Voltage Source** and the phase current controller, contains *R-L* circuit that models the inner resistance and inductance of the transformer; what is more, the value of the resistance takes into account the voltage drop, owing to the thyristor commutation [2,5]:

$$R = \left[2(R_1 + R_2) + \frac{3}{\pi}(L_1 + L_2) \right] R_{b2} \qquad (5.119)$$

$$L = \frac{(L_1 + L_2)R_{b2}}{\pi f} \qquad (5.120)$$

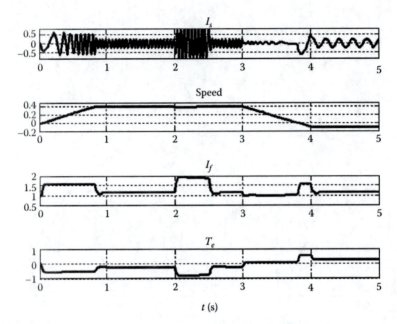

FIGURE 5.106 Transients in the electrical drive with the simplified cycloconverter model.

where R_1, R_2, L_1, L_2 are the resistances and leakage inductances of the transformer windings in pu given in its dialog box, f is the grid frequency, and R_{b2} is the base resistance of the transformer referred to the secondary winding (for its calculation by (2.16), instead of U_i, the rms value of the phase-to-phase secondary voltage must be used). For the transformer in model_5.30, $R_{b2} = 150^2/10^6 = 0.0225$ Ом, $R = (2\ 2\ 0.003 + 3/\pi\ 2\ 0.036)\ 0.0225 = 0.0018$ Ом, $L = 0.0720.0225/(\pi\ 50) = 10^{-5}$ H. The subsystem **Regulator** has **SR** and the blocks for reference fabrication of the phase currents and the excitation current and is made according to Figure 5.101.

The following process is scheduled: speeding-up to the speed 0.4 (20 Hz), with the load torque of 0.25, load torque increasing to 0.75 at $t = 2$ s, load torque decreasing to 0.25 at $t = 2.5$ s, and reverse to the speed minus 0.1 (5 Hz). One can make sure that simulation is accomplished much faster. The responses obtained are shown in Figure 5.106.

In conclusion, it is worth to draw attention to the fact that in SimPowerSystems there is DEMOS `power_cycloconverter` that simulates the cycloconverter operation with an active load.

5.3.5 SM with VSI Simulation

5.3.5.1 Standard Model

The standard model of the electrical drive with SM and VSI is included in SimPowerSystems (model AC5). The active front-end rectifier is used whose control system is similar to the one used in model5_10 (Figure 5.45). SM control system includes **SR** and the vector control system that generates the gate pulses for the

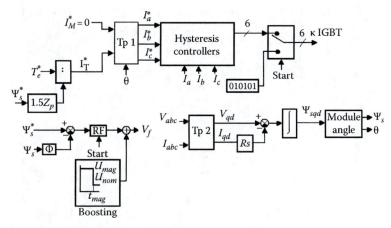

FIGURE 5.107   Vector control structure of AC5.

inverter and the excitation voltage. **SR** is the same as in the model AC3 and produces the reference signals of the torque T_e^* and the flux linkage Ψ^*. The speed reference comes to the **SR** input once the SM initial excitation is formed (signal Mag_C). The vector control structure that is aligned with the stator flux linkage is shown in Figure 5.107. With such an alignment, the torque equation (5.76) in SI becomes

$$T_e = 1.5Z_p \Psi_s I_T \tag{5.121}$$

where I_T is q-component of the stator current that produces the torque. From this relationship, for certain values T_e^* and Ψ^* that are computed in **SR**, the reference value I_T^* is calculated; the reference value of d-component $I_M^* = 0$. For a known position θ of the vector Ψ_s and using components I_T^*, I_M^*, the reference phase currents are calculated in the block **TP1**; these references come to the hysteresis controllers shown in Figure 5.42. The controller outputs affect IGBT only after activating the signal start (designated as *Mag_C*).

The excitation voltage V_f is produced by the PI controller **RF** of the stator flux linkage. The actual value of the excitation voltage is equal to the sum of the **RF** output signal and the signal at the output of the block **Mag-control1** (boosting) that is used for the acceleration of the initial excitation process: The stepped-up voltage U_{mag} is applied during a short time t_{mag} that afterwards decreases to the rated value U_{nom}.

The computation of the module Ψ_s and the angle θ of the vector Ψ_s is fulfilled by relations (5.24) when $\omega = 0$: At first, the components of Ψ_s are computed and then Ψ_s and θ.

The model AC5 picture is the same as the ones shown ones in Figure 5.34. The vector output Conv contains the signals of DC voltage (V) and the currents at the rectifier output and at the inverter input (A). The vector output Strl enables to gather information about the quantities: the rectifier input current active component, the error of the DC voltage regulation, the reference value of this voltage, the torque reference, the reference value of the stator flux linkage, the error of the speed or the

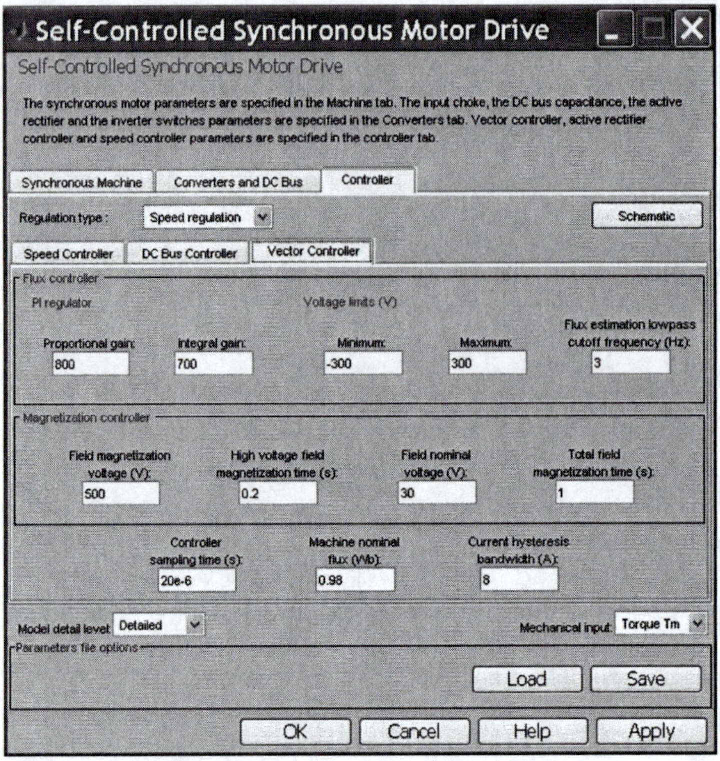

FIGURE 5.108   Vector control parameter window of AC5.

torque regulation, the speed (at the rate limiter output), or the torque reference (at the rate limiter input)—depending on the main quantity selected.

There are three windows (five pages) for the model parameter specification. The first one is the page of the SM parameters in SI. In the second window (page), the parameters of IGBT, snubbers, the capacitor in the DC link, and the input reactor (L_f in the Figure 5.45) are given. The last three pages pertain to the window *Controller*: the parameters of **SR**, DC voltage regulator, and the vector control scheme (Figure 5.108), respectively.

Pay attention to the following factors: 1. The load torque, unlike the models with DCM and IM, has the negative sign. The same sign has the electromagnetic torque in the motor mode. 2. In the dialog box, the reference DC voltage is not determined; it is given by the constant "Bus voltage set point," which is equal to 700 V; in order to see it, the AC5 model has to be opened by the command "Look Under Mask." Furthermore, in the subsystem **Active rectifier/dq2abc/dq2abc**, the grid frequency is equal to 60 Hz. In the system under simulation, these constants are set as 900 V and 50 Hz, respectively.

SM 120 kVA, 1500 rpm is used in model5_33. The capacitor capacitance is 8 mF, and the input reactor inductance is 0.5 mH. The rest of parameters can be seen in the dialog box of the model. The supply voltage is 400 V, 50 Hz. The scope **Motor**

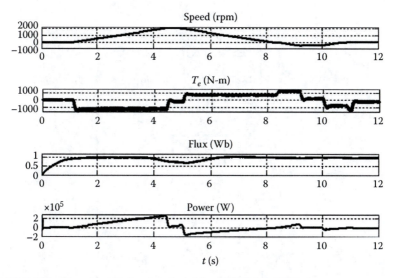

FIGURE 5.109 Transients in the electrical drive with SM and VSI.

gives the opportunity to observe the stator current, the speed, the torque, and the flux of SM and also the DC voltage. The scope **Voltages** fixes the input and output voltages of the inverter, and the scope **Power**, the active power consumed from the grid. The following schedule is simulated: initial excitation, speeding-up to the speed of 2000 rpm (with field weakening), reverse to −500 rpm, reverse to 100 rpm; the load torque is 250 N-m. The plots of speed, torque, flux linkage, and mains power are given in Figure 5.109. The processes of SM magnetization and flux weakening at speeds more the 1500 rpm can be seen distinctly.

During a slow-down, the active power is negative; it means that it returns to the supply.

5.3.5.2 Power Electrical Drive with Three-Level VSI

Today there is the large demand for the electrical drives rated for tens of MWs and rotational speeds around 3000 rpm in the oil and gas industry, especially in high-power compressor trains for liquefied natural gas [19]. SM is used in such plants usually. Since the mechanical part of the set has a number of the badly damped natural oscillations, the content of high harmonics in the SM current has to be low. It is suggested in [19] to use 4 three-level VSIs that operate in parallel for supplying of this SM. The model of such an electrical drive with a capacity of 32 MW, 3200 V is discussed in the following.

The diagram of the main circuits is shown in Figure 5.110. Four 3-level inverters that are made according to Figure 4.6 are supplied from 4 three-phase diode rectifiers. These rectifiers are powered from 4 three-phase transformers having "zigzag" connection. It gives the opportunity to have the phase displacements of the output transformer voltages that differ from one another by 15°, which ensures the low content of the high harmonics in the grid current. The three-phase reactors are set at the inverter outputs for their parallel operation. SM is supplied through these

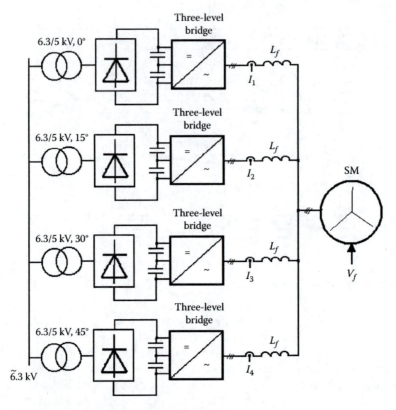

FIGURE 5.110 Supply of the power SM from 4 VSI.

reactors. Each of the reactors has the current controller **CR** whose reference is one
fourth of the total value. The inverter PWM has the triangular carrier waveforms that
are shifted by one fourth of their period about one another, that is, by 0 s, $1/F_c/4$ s,
$1/F_c/2$ s, and $3/F_c/4$ s, where F_c is the frequency of the triangular signal (it is taken
$F_c = 1000$ Hz). This ensures the mutual compensation for the majority of high harmonics in the SM current.

In the system under consideration, the energy recuperation is often needed [19].
The three-level active front-end rectifiers are set in this case, whose schemes are
analogous to those used in the inverters, and the control system is similar to the
one used in model5_10. In this case, the same simple transformers (not zigzag)
can be used.

The control system uses the principle of the vector control and is analogous to the
one used in the previous model. Supposing that the rotor angle θ is measured, the
angle α of the SM voltage space vector $\mathbf{V_s}$ is $\alpha = \theta + \delta$ and δ is the load angle that
can be computed as

$$\delta = \arctan\left(\frac{V_d}{V_q}\right) \tag{5.122}$$

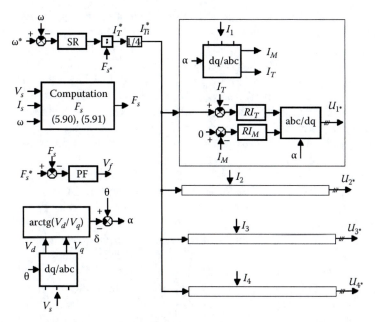

FIGURE 5.111 Control of VSI supplying the power SM.

where V_q and V_d are the projections of V_s on the rotor excitation voltage vector and perpendicular to it. The angle α is used for calculation of the components I_T and I_M from the total SM current I_s. In the motor mode, it turns out $\delta < 0$, so that in reality δ must be subtracted.

In the control system there is the PI controller **FR** of the stator flux linkage F_{smod} that affects SM excitation voltage. The F_{smod} calculation is carried out by the scheme that is analogous to the one used in model5_27.

The block diagram of the control system is shown in Figure 5.111. The three-phase output signals of the current regulators $U_1^* - U_4^*$ come to the blocks PWM.

Model5_34 simulates the described system. The resistor block is set parallel to SM; if it is absent, the program informs about the error because of series connection of SM and reactor. The rectifier bridges are in **Subsystem**. The blocks **Phase-Shifting Transformer** are used.

The blocks that realize the structure given in the Figure 5.111 are put in the subsystem **Control**. The phase-to-phase voltages V_{sabc} at SM input are measured, filtered by the filters with the cutoff frequency of 500 Hz and transformed in the phase voltages U_{ph} that are used for calculation of the stator flux linkage and the angle δ. All computations are fulfilled in pu, but the inverter current control—in SI, so the current reference value is multiplied by its base value in the block **Ki**. The components I_T and I_M of the reactor currents measured are calculated in the subsystems of the inverter current controllers **Curr_Reg1—Curr_Reg4** and are used as the feedback signals. The controller outputs are transformed in the three-phase systems $U_{ref1}—U_{ref4}$ that are inputs for PWM; it is carried out with the help of the inverse Park's transformation that uses the angle α computed earlier.

In the subsystem **Saw**, four sequences of the saw-tooth signals are produced with the frequency of F_c, and besides, each sequence, with the help of the delay blocks, is shifted about the preceding one by one fourth of the period. The gate inverter pulses are formed by comparing the U_{refi} signals with the saw-tooth signals.

A number of scopes are set to watch the electrical drive operation. The scope **Motor** records the current, the speed, the excitation current, and the torque of SM—all in pu. The rest of the scopes are in the subsystem **Measurement**. The scope **Reactor_Currents** fixes the reactor currents, that is, at the inverter outputs. The scope **Inv1** shows the phase voltage and DC voltage of the first inverter. The scope **Motor1** records the phase voltage (in pu), SM current (A), the stator flux linkage (in pu); the scope **Motor2** fixes the angle δ and the excitation voltage. Besides, the scope **Power** is set in the **Subsystem** that shows the grid-line voltage and its current. The following process is simulated: initial excitation, speeding-up to the rated speed with the load torque of 0.25 pu (speeding-up begins at $t = 0.5$ s) and the rated load surge at $t = 3.5$ s. The responses of speed, torque, excitation current, stator flux linkage, and angle δ are shown in Figure 5.112. It can be seen that the required characteristics are achieved. The phase voltage and the current of SM are given in Figure 5.113. They are in phase (remember that during SM simulation, the current flowing from SM but not flowing into SM is positive, which explains the current phase). It can be found with the help of the option *Powergui/FFT Analysis* that THD of the SM current is 0.6%. One can see by the scope **Reactor_Currents** that the reactor currents are equal practically but that THD = 4.4%. Thus, owing to interleaving, that is, to proper phase shifting of the PWM-equivalent carriers of the four converters, THD decreases by a factor of 7. The phase-to-phase voltage and current of the grid are shown in Figure 5.113. THD = 2.65% for the current that lags behind the phase voltage by 18°.

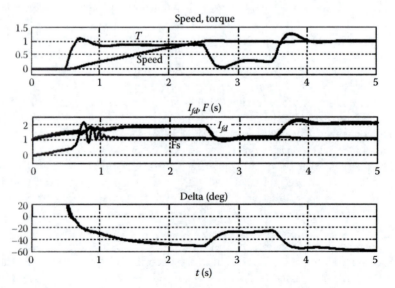

FIGURE 5.112   Transients in the electrical drive with SM and 4 VSI.

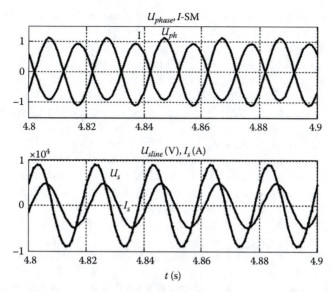

FIGURE 5.113   Currents and voltages during the supply SM from 4 VSI.

5.3.5.3   Power Electrical Drive with CHB Inverter

The rise of the electrical drive power can be obtained both by increasing the current under the moderate voltage value and increasing the voltage with the moderate current value. The latter variant has some technical and economic advantages, and the possibility of its successful use arose with the emergence of cascaded inverters. Specifically, in [20] the use of such an inverter is considered for starting and speed control of the group of 17 MW, 11 kV SMs. The gas-compressor station has three SMs like this, and one of them can operate in the speed control mode; it means it has to be connected to the inverter permanently and the others have to be speeded up softly and to be switched over afterwards to the supply of 11 kV. The single-line diagram of the plant is shown in Figure 5.114.

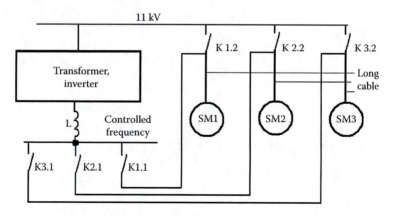

FIGURE 5.114   Supplying of the group of SM.

In the SM soft-start condition, its contactor K_.1 closes and after speeding up to the synchronous speed, this contactor opens and the contactor K_.2 closes. For the SM in the control mode, K_.1 remains closed. The synchronization of the inverter voltage powering SM and the mains voltage must be carried out before the switching-over. Since strict simultaneous commutation of the contactors is not possible, their switching is possible only with the opening of both contactors for a short time or with overlap when they are closed for a short time. Since in the first case overvoltage is possible when the SM circuit is interrupted, the second variant is used more often, but the inrush currents are possible; to avoid them, the reactor L is set.

The station construction is often such that a distance from the inverter to SM is hundred meters and more [20]. At that point, the resonance conditions are possibly caused by the inductance and capacitance of the long cable. This fact must be taken into consideration during system development.

Model5_35 simulates two-pole 16 MW, 11 kV SM with the contactors and the long cable. The CHB inverter has the same scheme as provided in model5_13, but with the difference that the transformers 2 MVA have the voltage 11/3 kV. The reactor with the inductance of 3 mH is set at the inverter output. (The parallel resistors 1100 Ω do not affect the responses but make the simulation faster). SM connects with the inverter by the cable of 500 m length, which is modeled by the subsystem **Cable**; its scheme and parameters are given in Figure 5.115 and, are essentially close to the ones given in [20]; the cable has a large capacitance because four parallel single-conductor cables are used in the each phase. With the help of the breakers **Br**, **Br1**, SM is switched over from the inverter to the mains and the breakers **Br2–Br4** are turned on or turned off for studying an influence of the cable.

There are two variants of speed reference setting. When the switch **S1** is in the lower position, the generator **V_ref** produces three-phase sinusoidal signal with the amplitude of 1, whose frequency, initially equal to 5 Hz, rises at the rate of 10 Hz/s, beginning from $t = 1$ s and up to $t = 5.5$ s and reaching by this way a value of 50 Hz. This signal is multiplied by the integrator output signal that is equal to 0.1 initially and rises with the rate of 0.2 1/s, beginning from $t = 1$ s and reaching the value of 1 at $t = 5.5$ s. When the switch **S1** is in the upper position, the control system as in the previous model is activated. There are two variants of the excitation voltage setting also—the constant voltage (the switch **S2** is in the lower position) and the stator flux linkage controller (the switch **S2** in the upper position). PWM is the same as in model5_13.

Switching-over of SM to the mains is carried out as a function of time. When the block **Relay2** turns on, the scheme of the voltage synchronization (of the mains and

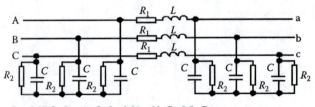

$R_1 = 0.05\ \Omega,\ R_2 = 4\ \mathrm{k}\Omega,\ L = 0.04\ \mathrm{mH},\ C = 0.5\ \mu\mathrm{F}$

FIGURE 5.115 Cable model.

the SM) is activated. At that point, the output I of the block **Integrator** increases from 0 to 1 for 1 s. When this signal is equal to 0, the output of the block **Product1** is 0 too, and the PWM input is determined by the output of the block **Product**, whose second multiplier is 1. With I rising, the output of **Product** decreases to 0, and the output of the **Product1** increases up to the value that is defined by three-phase signal coming to the first input of this block, which is in-phase with the mains voltage. The subsystem **Phase_Lead** puts some phase leading for the inverter delay compensation. After completion of the process of synchronization, **Relay1** gives the command for turning on **Br** and 0.01 s later for turning off **Br1**.

The scope **Motor** fixes the stator currents, speed, excitation current, and torque of SM. The scope **Network** records the phase voltage and current of the mains and the phase SM voltage that is smoothed out by the filter; the scope **SM_Voltage** shows the phase-to-phase SM voltage.

Let us conduct some experiments with the model. Let **S1, S2** be in the lower positions; **Br2** is closed; **Br3, Br4** are open; the block **Timer1** increases its output from 1 to 4 at $t = 8$ s, that is, the load increases from 0.25 rated to 1; and the modeling time is 10 s. The block **Relay2** turns on at $t = 6$ s, and **Relay1** at $t = 7$ s. The initial SM speed $dw(\%) = -99.9\%$. The said values of the initial frequency and the rate of its change are set in **V_ref**. As a result of the simulation, in Figure 5.116 the speed, current, and torque of SM are shown, and in Figure 5.117, the mains voltage and the inverter voltage (after the filter) are shown for before synchronization (a), after synchronization (b), and the mains current (c).

In the Figure 5.118, 5.119 the same curves are shown for simulation carried out with S1, S2 in the upper positions.

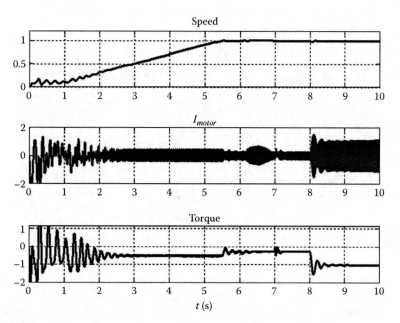

FIGURE 5.116 Transients at the start of SM with open control system.

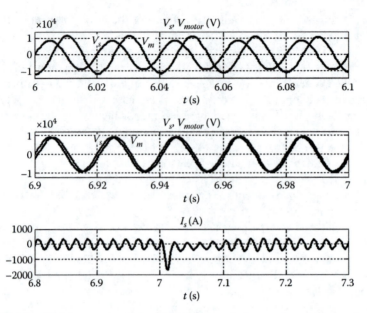

FIGURE 5.117 Mains current, mains, and SM voltages during synchronization and connection to mains with the open control system.

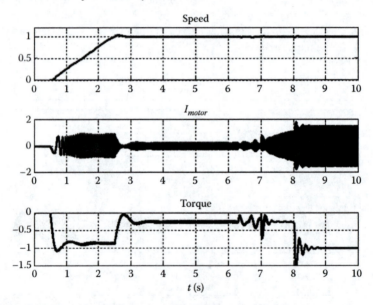

FIGURE 5.118 Transients at the start of SM with closed control system.

The influence of the long cable is investigated under the rated speed of 0.8 and, hence, under the inverter frequency of 40 Hz. The tumblers **S1, S2** are in the lower positions; the amplitude is 0.8; the frequency is 40 Hz in the block **V_ref**, the option of the frequency variation, is excluded; the initial value is 1 in the block **Integrator Limited**; in the SM model $dw(\%) = -20\%$; the saw-tooth signal has the frequency

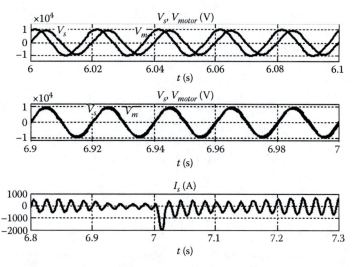

FIGURE 5.119 Mains current, mains, and SM voltages during synchronization and connection to mains with the closed control system.

$F_c = 1000\,\text{Hz}$; with the help of the block **Timer1**, the load torque is set to 1; the modeling time is 1 s. If simulation is carried out with **Br2** closed and **Br3**, **Br4** open, the SM voltage that is fixed by the scope **SM_Voltage** will be obtained as shown in Figure 5.120a. In Figure 5.121a the harmonic spectrum of this voltage is shown. This voltage shown is sinusoidal enough and the harmonics center around the triple commutation frequency of 3000 Hz. If simulation is repeated with **Br2** open

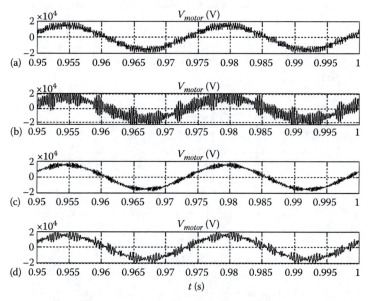

FIGURE 5.120 SM voltage with long cable connection. (a) Without cable, $F_c = 1000\,\text{Hz}$. (b) With cable, $F_c = 1000\,\text{Hz}$. (c) With cable, $F_c = 1750\,\text{Hz}$. (d) With cable, $F_c = 800\,\text{Hz}$.

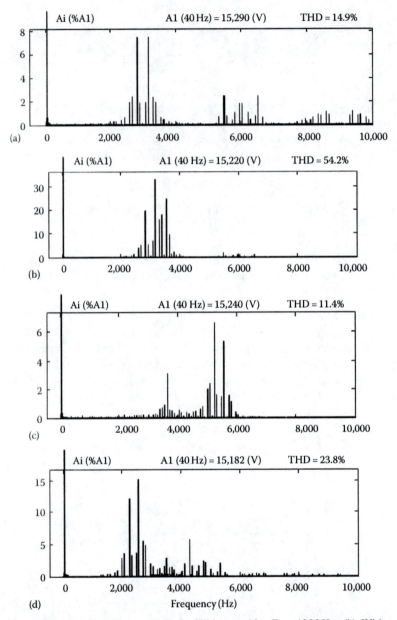

FIGURE 5.121   SM voltage spectrum. (a) Without cable, F_c = 1000 Hz. (b) With cable, F_c = 1000 Hz. (c) With cable, F_c = 1750 Hz. (d) With cable, F_c = 800 Hz.

and **Br3**, **Br4** closed, the voltage plot in Figure 5.120b and the harmonic spectrum in Figure 5.121b will result. The essential influence of the long cable causing the resonance is observed. The cable resonance frequencies are close to the equivalent inverter frequency. In Figures 5.120c, and 5.121c the voltage and its harmonic spectrum are shown when F_c = 1750 Hz, and the resonance is absent practically. To a certain extent, the resonance conditions can be diminished by a decrease in the commutation frequency. So, for example, when $F_c = 800\,\text{Hz}$, THD decreases from 54% to 24% (Figures 5.120d and 5.121d).

5.3.6 SIMPLIFIED SM MODEL

The previously described SM model is rather complex and can demand much time during simulation of the forked circuits with many SM. So a simplified SM model is included in the folder *Machines* of SimPowerSystems. In this case, a round-rotor SM with equal-axes inductances is modeled, the damper windings are excluded, and the stator circuit consists of the stator resistance and the SM inductance $L = L_d = L_q$. It can be written for phase A [4] (supposing the operation takes place in the generator mode, damper windings are absent and excitation current is independent from processes in SM)

$$U_{sa} = E_{fda} - R_s I_{sa} - L\frac{dI_{sa}}{dt} \tag{5.123}$$

$$E_{fda} = \omega_s L_m I_{fd} \sin\theta = E\sin\theta \tag{5.124}$$

$L = L_m + L_{sl}$, ω_s is the rotor electric rotation frequency, θ is the rotor angle about the axis of the phase A, and E is the voltage amplitude that induces in the stator by the rotating rotor. For B and C phases, $\sin(\theta - 2\pi/3)$ and $\sin(\theta + 2\pi/3)$ are used instead of $\sin\theta$, respectively.

Pictures of simplified SM model are given in Figure 5.122. SM parameters can be given in both SI and pu. The dialog box for the second case is given in Figure 5.123. Two variants of the supply connection can be chosen: insulated or accessible neutral. The new parameter is the damping factor K_d. It takes into account the dampening ability of the damper windings. It can be calculated by the formula [1]:

$$K_d = 4\xi\sqrt{\frac{\omega_s H P_{max}}{2}} \tag{5.125}$$

where

ξ is the level of dampening

ω_s is the synchronous frequency, rad/s (314 under 50 Hz)

H is the inertia constant s

P_{max} is the maximum power that is equal to $U_s E/X$ (all the quantity in pu)

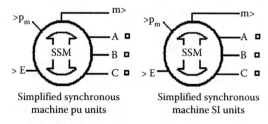

Simplified synchronous machine pu units   Simplified synchronous machine SI units

FIGURE 5.122 Pictures of the SM simplified model.

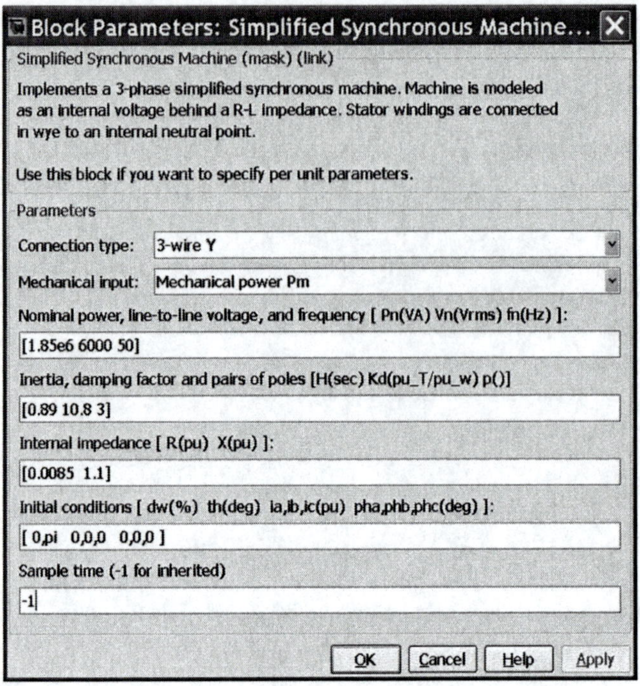

Block Parameters: Simplified Synchronous Machine... ✕

Simplified Synchronous Machine (mask) (link)

Implements a 3-phase simplified synchronous machine. Machine is modeled
as an internal voltage behind a R-L impedance. Stator windings are connected
in wye to an internal neutral point.

Use this block if you want to specify per unit parameters.

Parameters

Connection type: 3-wire Y

Mechanical input: Mechanical power Pm

Nominal power, line-to-line voltage, and frequency [Pn(VA) Vn(Vrms) fn(Hz)]:

[1.85e6 6000 50]

Inertia, damping factor and pairs of poles [H(sec) Kd(pu_T/pu_w) p()]

[0.89 10.8 3]

Internal impedance [R(pu) X(pu)]:

[0.0085 1.1]

Initial conditions [dw(%) th(deg) ia,ib,ic(pu) pha,phb,phc(deg)]:

[0,pi 0,0,0 0,0,0]

Sample time (-1 for inherited)

-1

OK   Cancel   Help   Apply

FIGURE 5.123   Dialog box of the SM simplified model.

Here it is not clear how to determine the coefficient ξ. One of the possible ways
is to use originally the full SM model, for example, to develop a simple scheme
with the full SM model that is supplied with the rated voltage, to fix the process
of load angle θ changing under the load surge, and from this curve, to find ξ with
the help of known methods of the control theory.

Three phase-to-phase voltages, the voltage E that can be constant or inconstant
and is formed by the Simulink blocks, and the load power P_m come to the model
inputs. The model transforms the phase-to-phase voltages into phase ones U_a, U_b, U_c;
computes the voltages induced in the phases $E_{fda} = E \sin \theta$, $E_{fdb} = E \sin(\theta - 2\pi/3)$,
$E_{fdc} = E \sin(\theta + 2\pi/3)$; and calculates the phase currents I_{sa}, I_{sb}, I_{sc} by formulas (5.123)
and, afterwards, the electrical power as $P_e = E_{fda}I_{sa} + E_{fdb}I_{sb} + E_{fdc}I_{sc}$.

For the SM mechanical part modeling, the dynamic torque is computed as

$$T_{din} = \frac{P_m - P_e}{\omega_m} \tag{5.126}$$

(SM is modeled as a generator), and afterwards, the speed change $\Delta\omega_m$ is calculated as

$$\frac{d\Delta\omega_m}{dt} = \frac{T_{din} - K_d\Delta\omega_m}{2H} \tag{5.127}$$

which is added to the rated speed ω_s (1 under simulation in pu). The integral of $\omega_s + \Delta\omega_m$ determines the angle θ that is used for E_{fd} computation. Thus, this simplified model can be used almost exclusively for modeling of the synchronous generators with minor rotation speed deviations from the synchronous one.

It is possible to measure 12 quantities that determine SM operation, namely: three-phase currents, three-phase voltages, three quantities of the inner emf of the phases, the rotation speed, and the values of θ and P_e. For this purpose, the same measuring block as the one used for the full model but with the selection of the option *Simplify synchronous*, or the block **Bus Selector** from the folder *Simulink/Signal Routing* can be used. The examples of use of this model will be given in the next chapter.

5.4   SYNCHRONOUS MOTOR WITH PERMANENT MAGNETS

Synchronous motor with permanent magnet (PMSM) has only one three-phase stator winding, and excitation is carried out by the permanent magnets fixed to the rotor. There are PMSMs with the surface magnet setting (SPM) and with the interior magnet setting (IPM). In the former, the magnets are glued onto the rotor, and SPM looks like salient-pole SM in section. In the IPM, the magnets are mounted inside the round rotor. Since a relative permeability of PM is very close to 1, it does not influence the stator winding inductance. SPM has the same inductances along d- and q-axes: $L_d = L_q = L_s$, whereas IPM has $L_d < L_q$, because the d-axis effective gap is larger, owing to the need to put the magnets. PMSM is described by four electromagnetic parameters: the stator resistance R_s, the inductance L_d, L_q, the rotor flux Ψ_r, and also by the number of the pole pairs Z_p (designated as p in the PMSM dialog box).

PMSM is modeled in the rotating reference frame, whose axis is aligned with the rotor-pole axis. It is possible to derive the PMSM equations from the said SM equation if we assume that the rotor current i_r is equal to the equivalent current i_m for which $L_m i_m = \Psi_r$. The voltage equations are

$$v_{sd} = R_s i_{sd} + L_d \frac{d i_{sd}}{dt} - \omega_s L_q i_{sq} \tag{5.128}$$

$$v_{sq} = R_s + L_q \frac{d i_{sq}}{dt} + \omega_s (L_d i_{sd} + \Psi_r) \tag{5.129}$$

Ψ_s components are

$$\Psi_{sd} = L_d i_{sd} + \Psi_r, \quad \Psi_{sq} = L_q i_{sq} \tag{5.130}$$

From (5.26), (5.130) the torque is

$$T_e = 1.5 Z_p \left[(L_d i_{sd} + \Psi_r) i_{sq} - L_q i_{sq} i_{sd} \right] = 1.5 Z_p \left[\Psi_r i_{sq} + (L_d - L_q) i_{sd} i_{sq} \right] \tag{5.131}$$

The picture of the PMSM model is shown in Figure 5.124. The first page of the dialog box has three lines. In the first one, one of two possible PMSM constructions

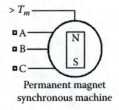

Permanent magnet
synchronous machine

FIGURE 5.124 PMSM picture.

is selected: with the sinusoidal back emf waveform and with the trapezoidal one. The latter is called also a switched PMSM, an electronic PMSM, or a brushless DC motor. Only a sinusoidal PMSM is considered in this book. The rest of the lines are the same as the ones seen for the other machines.

The second page of the dialog box is shown in Figure 5.125. Here, the aforementioned parameters are given in SI; what is more, three options for Ψ_r setting are provided: by direct Ψ_r(wb) specification, with the help of the maximum phase-to-phase voltage V that reaches the terminals of nonload PMSM rotating with the speed of 1000 rpm, or as the torque per ampere constant TC. All these quantities are connected to one

---

Block Parameters: Permanent Magnet Sy... ✕

internal neutral point.

The preset models are available only for the Sinusoidal back EMF machine type.

| Configuration | Parameters | Advanced |

Stator phase resistance Rs (ohm):

0.15

Inductances [Ld(H) Lq(H)]:

[3.5e-3, 10.5e-3]

Specify: Flux linkage established by magnets (V.s)

Flux linkage established by magnets (V.s):

0.45

Voltage Constant (V_peak L-L / krpm):

163.2419

Torque Constant (N.m / A_peak):

1.35

Inertia, friction factor, pole pairs [J(kg.m^2) F(N.m.s) p()]:

[0.05 0.01 2]

Initial conditions [wm(rad/s) thetam(deg) ia,ib(A)]:

[0,0, 0,0]

OK   Cancel   Help   Apply

FIGURE 5.125 PMSM dialog box.

another. So, for the PMSM, whose parameters are given in Figure 5.125, $\Psi_r = 0.45$ Wb, $V = 1.732\ \Psi_r p\pi\ 1000/30 = 163.15\ B$, $TC = 1.5p\Psi_r = 1.35\ H.M/A$.

The electrical part of the PMSM model consists of four blocks: the block for transformation of the input phase-to-phase voltages in the V_d, V_q voltages; the block that computes i_{sq}, i_{sd} currents by solution of Equations 5.128 and 5.129; the block that transforms i_{sq}, i_{sd} into the phase currents i_a, i_b, i_c; and the block that calculates the torque by (5.131). The mechanical part of the model is described by Equations 5.15 and 5.16. For the PMSM quantities measurement, the same measuring block can be used with the option *Permanent Magnet Synchronous* or the blocks of **Bus Selector**. It is possible to measure the stator currents i_a, i_b, i_c (A), the currents i_{sq}, i_{sd} (A), the voltages V_q, V_d (V), the rotation speed ω_m (rad/s), the rotor angle θ_m, (*tetam*) (rad), and the torque T_e (N-m). The signal "Hall effect signal" is formed in addition, which provides the logical indication of the emf space vector position. This signal changes when the phase-to-phase voltages cross zero.

The control principles are the following: For SPM, the torque does not depend on i_{sd}; it means $i_{sd} = 0$ is the best choice to decrease the heating of the motor and increase its load-carrying capacity. It follows for IPM from (5.107) that, because $L_d - L_q < 0$, keeping up $i_{sd} < 0$ increases the motor torque, but at that point, the maximum allowable current i_{sq} decreases, because it has to be

$$i_s = \sqrt{i_{sd}^2 + i_{sq}^2} \le I_0 \qquad (5.132)$$

where I_0 is the maximum allowable current of the electrical drive (of the PMSM or of the inverter). Hence, i_{sd} can change optimally.

Furthermore, it follows from (5.128), (5.129) that in the steady state, assuming $R_s = 0$, the maximum attainable speed is

$$\omega_{smax} = V_{smax}/\Psi_s \qquad (5.133)$$

where V_{smax} is the amplitude of the phase voltage that is determined by the inverter DC link voltage and the accepted modulation method and Ψ_s is the stator flux linkage. When a higher speed is required, the stator flux is to be weakened, which can be done, as it follows from (5.130), by setting $i_{sd} < 0$. At that point, however, the stator current rises substantially and the torque decreases, which limit the efficacy of this control method [5]. In [10], detailed relationships between different methods of control are given.

In model5_36, the DTC that was applied for IM already (model5_7, model5_8) is adapted for PMSM. The subsystems **Feedback** and **DTC_Reg** are the same as in model5_8. The PMSM parameters are given in Figure 5.125. DC voltage is 300 V, the resistor of 0.2 Ω models the inner source resistance, and the capacitor has the capacitance of 1 mF. The stator flux linkage components $\Psi_{s\alpha}$, $\Psi_{s\beta}$ are computed in the subsystem **Flux**: At first, Ψ_{sd}, Ψ_{sq} by (5.130) are calculated, then, as a result of Park's transformation, Ψ_{sa}, Ψ_{sb}, Ψ_{sc} are determined and afterwards $\Psi_{s\alpha} = \Psi_{sa}, \Psi_{s\beta} = (\Psi_{sb} - \Psi_{sc})/\sqrt{3}$.

The new block is the subsystem **Control_1**. It contains the speed controller **SR** with the regulated output limit that is set initially equal to T_{max} and diminishes, if the module of the current space vector i_s (5.132) exceeds I_0; the controller output gives the demanded PMSM torque. The specified flux linkage value is defined as follows [10].

Denote $I_q = L_d i_{sq}/\Psi_r$, $I_d = L_d i_{sd}/\Psi_r$, $dl = (L_q - L_d)/L_d$, $m = T_e L_d/1.5 Z_p \Psi_r^2$, $f_s = \Psi_s/\Psi_r$, then from (5.130), (5.131) it follows:

$$f_s = \sqrt{(1+I_d)^2 + (1+dl)^2 I_q^2} \qquad (5.134)$$

$$m = I_q(1 - I_d dl) \qquad (5.135)$$

The values of I_q, I_d are selected so that under a specified torque the motor current (5.132) is minimum or is equivalent under the specified current the torque is maximum. If we substitute $I_q = \sqrt{I_0^2 - I_d^2}$ in (5.135), find minimum for I_d and exclude $I_0 = \sqrt{I_q^2 + I_d^2}$ in the resulting expression, the following relationship is received:

$$I_q^2 = \frac{I_d^2 - I_d}{dl}. \qquad (5.136)$$

Furthermore, for a number of I_d values, I_q by (5.136), m by (5.135) and f_s by (5.134) are computed, and this way the requested dependence of the stator flux linkage on the torque for a specified dl value is calculated. In practice, it is preferable to have this dependence in an explicit form, as in the analytical formula. With this aim, the received curve is approximated by the relationship $f_s = am^2 + bm + c + 1$ where the coefficients a, b, c depend on dl [10]. If the approximated value of $f_s < 1$, it is accepted $f_s = 1$.

Thus, before simulation, the program MATLAB Contr_PMSM_DTC has to be executed

```
Ld=0.0035;
Lq=0.0105;
Fr=0.45;
dl=(Lq-Ld)/Ld;
I0=60;
p=2;
Km=Ld/(1.5*p*Fr*Fr);
    for j=1:30,
    Id(j)=-j*0.01;
    M1(j)=(1-dl*Id(j))*sqrt(Id(j)^2-Id(j)/dl);
    fs(j)=sqrt((1+Id(j))^2+(1+dl)^2*(Id(j)^2-Id(j)/dl));
    end;
p1=polyfit(M1,fs-1,2)
a=p1(1)
b=p1(2)
c=p1(3)
```

At first, the PMSM parameters are given, and then, as described earlier, the dependence of f_s on m is computed, which is later approximated by the second-order polynomial, using the MATLAB command p1=polyfit(M1,fs-1,2); the coefficients a, b, c of the polynomial p1 are desired quantities that, as a result of execution of this program, are put in the workspace and can be used by the model. The block **Fcn** of the subsystem **Control_1** realizes the computed dependence, forming the flux reference: At first, the torque reference T_{ref} is scaled to m by multiplication by K_m; then the value of f_s is calculated that is converted into the flux linkage reference by multiplication by Ψ_r. As shown in [10], $V_{smax} = 0.55V_{dc}$ in DTC system; thus, when $\omega_s > \omega_0 = 0.55V_{dc}/(f_s\Psi_r)$, the flux weakening has to be achieved. The block **Fcn1** computes ω_0 that is compared with the actual electrical speed $Z_p\omega_m$. If the latter value is less than the former, the same signal ω_0 comes to the second and the third inputs of Mux1, and as result, the output F_{ref} is equal to the output Fcn. When $\omega_s > \omega_0$, **Switch** turns over; at the third **Mux1** input comes $Z_p\omega_m$ and the flux reference decreases.

It is supposed that the rotor position sensor is jointed with PMSM; this sensor can be used as the speed sensor too. For the sake of simplification, it is supposed also that the currents and torque are measurable, but in reality, only the currents i_a, i_b, i_c (or two of them) are measured, and it is necessary to transform them into i_{sq}, i_{sd} by using the block **abc_to_dq0 transformation** and then to compute the torque by said formula.

In the field *InitFcn* of the option *File/Model Properties/Callbacks*, the command run Contr_PMSM_DTC is written, so when the model starts, this program is executed automatically.

This model gives the opportunity to investigate the processes that occur in various blocks of the model, an influence of the parameters of the main and the control circuits. As an example, PMSM speeding-up to the speed of 250 rad/s with flux weakening is shown in Figure 5.126. In the no-load condition, the speed of the field weakening beginning is $\omega_0 = \omega_{s0}/Z_p = 0.55\ 300/0.45/2 = 183$ rad/s. From the figure one can see that when the speed is close to the indicated one,

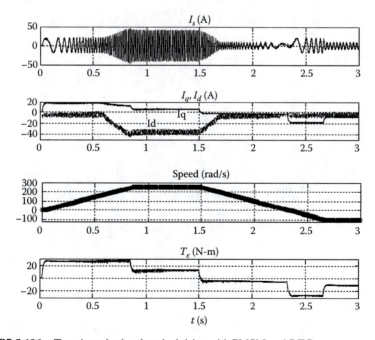

FIGURE 5.126 Transients in the electrical drive with PMSM and DTC.

a noticeable rise (with the sign minus) in the current i_{sd} can be seen, the flux decreases and speeds go above ω_0.

In SimPowerSystems a standard model of the electrical drive with PMSM called AC6 is available. The main circuits are the same as the ones used in models AC2–AC4 (Figure 5.31). The control system consists of **SR** and the vector control controller. The reference for i_{sq} is proportional to the reference torque and the reference for i_{sd} is zero. Thus, the flux weakening is not intended. Supposing the rotor angle is known, i_{sq}, i_{sd} components are transformed in the phase current references that go to the hysteresis controllers, as seen in models AC3 and AC5.

AC6 has the same picture as the other models of AC type (Figure 5.34); the output m provides access to the above-mentioned PMSM quantities and at the output $Conv$ the vector signal is formed: the capacitor voltage, the output current of the rectifier, the input current of the inverter; at the output $Ctrl$ the vector signal is formed: the torque reference, the speed error, the output of the **Rate Limiter**. In the parameter window "PMSM", the same PMSM data are given in the parameter window of PMSM itself (Figure 5.125). The parameter window "Converters and DC bus" is the same as shown in Figure 5.35. The parameter window "Controller" has the fields for specifications of the same parameters that have been shown in Figures 5.36, 5.40, 5.43 already, and their meaning is clear from the explanations provided earlier.

Model5_37 demonstrates the use of AC6. The electrical drive is supplied from the grid 220 V, 50 Hz, the circuit with resistance of 0.15 Ω, and inductance of 0.5 mH models the inner source impedance. The PMSM parameters are specified in the AC6

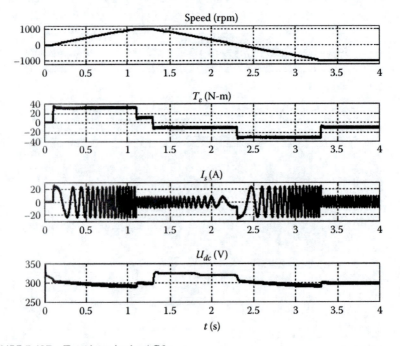

FIGURE 5.127   Transients in the AC6.

dialog box. Speeding-up to the speed of 1000 rpm and reverse to −1000 rpm is simulated (Figure 5.127). It can be seen that during deceleration, the capacitor voltage U_{dc} increases but remains limited, owing to the dynamic braking circuit.

5.5 SWITCHED RELUCTANCE MOTOR SIMULATION

From (5.131) it follows that the SM torque is not equal to zero when $L_d \neq L_q$, even when the source of excitation on the rotor is absent ($\Psi_r = 0$). The motors that use this particular feature are called reluctance motors. They are subdivided into sinusoidal and the switched reluctance motors. The former are in reality an usual SM without windings or magnets on the rotor, the design of which requires that special steps are carried out for obtaining $L_d \gg L_q$. For simulation of these motors, the previously described PMSM model can be used with $\Psi_r = 0$.

The switched reluctance motors have an entirely different construction. The motor in Figure 5.128 has six poles on the stator. Each pole carries a winding, and the opposite windings (a and a', b and b', c and c') are connected in series, so that the winding system can be regarded as three-phase one. The rotor has four poles. The current pulses apply to the stator windings in the times that depend on the rotor position, so the motor has a position sensor.

If saturation is neglected, the torque applied about the rotor can be written as

$$T_e = 0.5 I^2 \frac{dL}{d\beta} \tag{5.137}$$

where L is the inductance of the phase in which the current I flows and β is the rotor angle. If the current pulse is given in phase b under the rotor position shown in Figure 5.128, with indicated direction of the rotation, the current will produce a positive torque because the phase b inductance is increasing. If, however, the pulse is given in phase a, the torque will be negative (braking) because phase inductance is

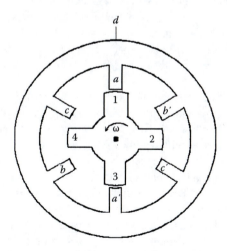

FIGURE 5.128   Switched reluctance motor (in section).

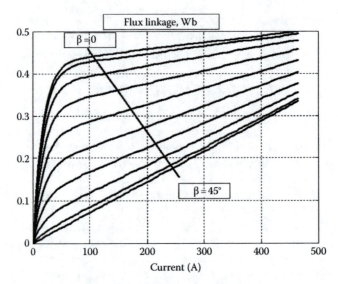

FIGURE 5.129   Characteristics of the switched reluctance motor.

decreasing (because the rotor pole 1 moves away from the stator pole). After the setting pulse generation in phase *b* and the rotation of the rotor by 30°, the rotor pole 1 takes the same position about the phase *c* that the pole 2 took about the phase *b* earlier. The pulse can be given now in phase *c*, and the rotor rotates again by 30°; afterwards the pulse can be sent in phase *a* and so on.

The angle β is counted off from the position when the axis of the rotor pole is aligned with d-axis. If the current pulse is given in phase *a* under the small angle β > 0, the torque will be negative, and for a large angle β (e.g., more as 45°), the torque will be positive; thus, by changing the angle when the pulses are sent to the phase, the mode of operation can be altered.

The switched reluctance motor (SRM) parameters are selected such that its poles are saturated essentially in the aligned position. At that point, the SRM flux depends not only on the current but also on the rotor position, that is, $\Psi = \Psi(I, \beta)$. The typical curves are shown in Figure 5.129 [1]. Analogously, the torque is a function of *I* and of β: $T_e = T(I, \beta)$. This nonlinearity makes difficult both computations and realization of the SRM control system.

SRM can have different numbers of stator phase and rotor poles. Usually the stator has 3–6 phases and the rotor 4–10 poles. The SRM type is given as a fraction whose numerator is the double-phase number and denominator is the pole number; that is, there are SRMs of 6/4, as seen in the types shown in Figure 5.87, 8/6, 10/8, 12/10.

SRM has a number of merits: higher reliability, ability to operate at high speeds, and absence of windings and other elements on the rotor, which provides low inertia moment. Such a model is used SimPowerSystems, whose picture is given in Figure 5.130a for SRM 6/4.

The points A1, A2, B1, B2, C1, C2 are the terminals of the three stator windings. For the SRM 8/6 and 10/8, the number of terminals and picture change. The output of the load model comes to the input *TL*, and the output *m* contains the vector of the

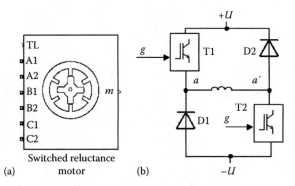

FIGURE 5.130 Picture of SRM model (a) and the diagram of a stator phase (b).

measured SRM quantities arranged as follows: *V*—the stator voltage, V; *flux*—the flux linkage, V · s; *I*—the stator current, A; *Te*—the torque, N-m; *w*—the rotor speed, rad/s; and *teta*—the rotor position, rad.

The block diagram of the SRM model is shown in Figure 5.131. The scheme consists of three (for SRM 6/4) identical circuits. The phase fluxes are calculated by (5.11), using measured winding voltages. These fluxes come to the tables ITBL for calculation of the phase currents. Besides, these tables get the values of the rotor angles about the phases that vary with different phases. The tables provide data that are analogous to those given in Figure 5.129. The outputs of tables are the values of phase currents that, together with the angles, come to the tables TTBL, whose outputs are the electromagnetic torques of different phases. The total torque is the sum

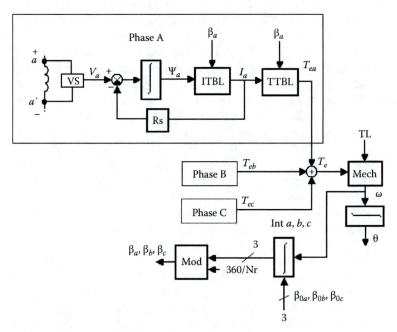

FIGURE 5.131 Structure of SRM model.

of these torques. The equation of the motion (5.15) is solved in the block **Mech,** and the speed and the rotation angle of SRM are found.

The speed value comes to three identical integrators, Inta, Intb, Intc, which are distinguished by the initial conditions: SRM Inta = 0, Intb = −30°, and Intc = −60° (see Figure 5.131). The output integrator quantities are sent to the blocks that are realized by the mathematical function of the modulo operation of $360/N_r$ (N_r is the number of the rotor poles), in the case under consideration—of 90, that is, the angles β_a, β_b, β_c change from 0 to 90°. The winding shown in Figure 5.131 is the model of variable inductance that is analogous to the one used in model5_5 (Figure 5.24 and 5.25).

On the first page of the dialog box, one of three SRM types can be selected—6/4, 8/6, and10/8—and also either the generic or specific models. The former are used for some typical curves $\Psi = \Psi(I, \beta)$, which are shown in Figure 5.129. These curves can be modified by setting the following parameters: the unaligned inductance that corresponds to the low (straight) line in Figure 5.129, the aligned inductance that is the initial section of the upper curve in Figure 5.129, the saturated aligned inductance that is the tangent line to the upper curve at the point of the maximum current, the maximum current, and the maximum flux linkage that corresponds to the maximum current for the upper curve. The curves $\Psi = \Psi(I, \beta)$ selected can be seen if the option "Plot magnetization curves" is chosen.

For selecting a specific model, curves $\Psi = \Psi(I, \beta)$ are to be given as tables, and the data for these tables can be received either by experiment (if there is a prototype) or by computation with the help of finite-element analysis. Since we do not have such data, we confine ourselves to the typical model.

The electrical drive with SRM is simulated in model5_38. The SRM parameters are given in Figure 5.132. The subsystem **Converter** forms three-phase system of the voltages for supplying the SRM windings and consists of three circuits shown in the Figure 5.130b. When the gate pulse is applied to the transistors, the DC voltage is applied to the phase winding; during the pulse removal, the circuit for current closes through the diodes and the supply source; moreover, the current direction is opposite to the applied voltage that provides for faster current extinction. The current control is accomplished by the hysteresis controllers **Relay**. The phase current reference comes to the current controller of this phase only in the case when at the corresponding output of the subsystem **Position Sensor** is the logical 1.

In this subsystem, the same scheme for angle β forming is realized as shown in the SRM model (Figure 5.131). The values of the angles are compared with the specified turn-on and turn-off angles (degree) that are different for speeding-up and slowing-down modes. The values of the specified angles are constant, but in principle it is reasonable to change them in certain way for system optimization [21]. Switching-over of the modes is executed according to the sign of the signal at the **Limiter** 200 A output. When the angle β of a certain phase is in the setting range, the logic signal 1 is generated at the corresponding output *sig* of the subsystem.

The scope **Motor** fixes the flux linkages of the phases, their currents, the SRM torque, and its rotation speed (rpm). The scope **Torque** shows the mean torque value.

If **Mswitch** is put in the low position, the speed feedback is excluded, the current value is determined by the **Limiter** 200 A, and the mode of the operation—by the

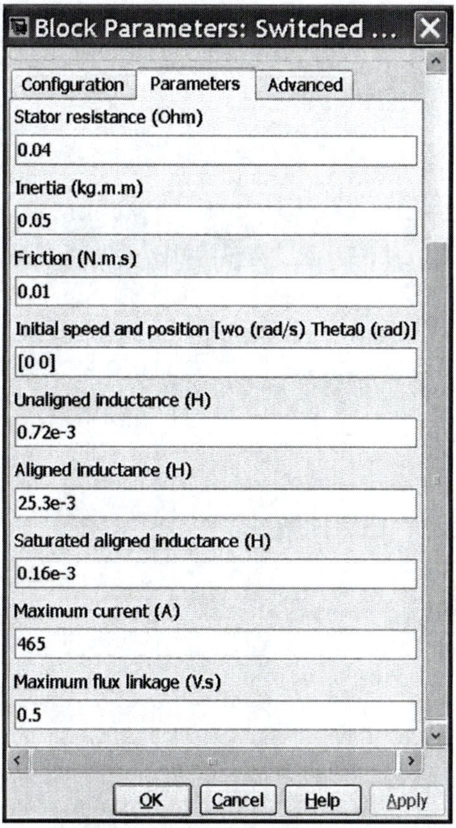

FIGURE 5.132   Dialog box of SRM.

sign of the block Step **Wref** whose initial and final values are rather large (e.g., ±2000) and the step time is 0.3 s. The load torque is set to 0 (instead of 50 N-m).

In Figure 5.133 the speed, the phase current, SRM torque, and its average value are shown as a result of simulation. During speeds about 3500 rpm, the current is equal to the reference value and the average SRM torque is constant. The big torque oscillations inherent in SRM take place. At a higher speed, the current and the torque decrease because emf essentially increases and the phase currents cannot reach their reference values. At that point, switching-over of the transistors cease; they remain in the conducting state entirely the interval of their activity, which is determined by the specified turn-on and turn-off angles; in these intervals, the constant voltages are applied to the windings. In this mode, the torque is approximately inversely proportional to the speed, and the torque oscillations diminish because the pulsations that occur due to the hysteresis current controller disappear. When the sign of **Wref** output changes, the SRM torque becomes negative and SRM slows down.

When **Mswitch** is in the upper position, the speed feedback is closed. The block **Wref** schedules speeding-up to 2000 rpm, and reverse to −2000 rpm at $t = 0.3$ s. The load torque is 50 N-m. The responses are shown in Figure 5.134.

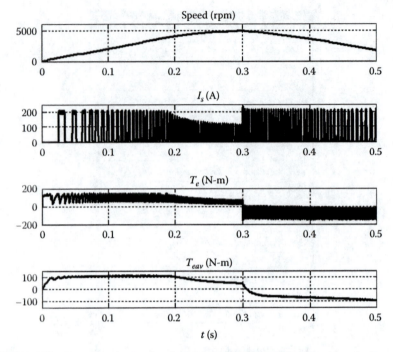

FIGURE 5.133   Transients in the electrical drive with SRM during current control.

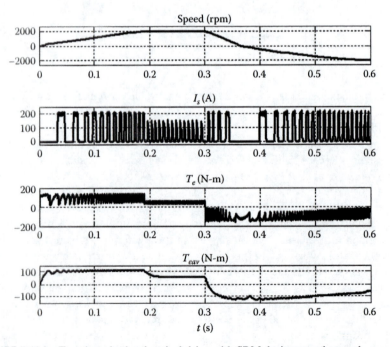

FIGURE 5.134   Transients in the electrical drive with SRM during speed control.

5.6 MECHANICAL COUPLING SIMULATION

It was supposed in the previous models that the motor and the load form the united concentrated mechanical system (so-called single-mass system). In reality, the torque is transferred from the motor to the load by means of a number mechanical elements: shafts, gears, and couplings with specific elasticities. Therefore, the torque can be transferred only by means of deformation of the mechanical parts. When there is only slight deformation, the transferred torque is proportional to the deformation value and, to a certain extent, its speed. So, the torque transferred by a shaft can be determined as

$$T_y = C \int (\omega_1 - \omega_2)\, dt + B(\omega_1 - \omega_2) \tag{5.138}$$

where
ω_1 is the speed of the input (driving) shaft end
ω_2 is the same for the output shaft end
C and B are the coefficients

These speeds are found as

$$\frac{d\omega_1}{dt} = \frac{T_e - T_y}{J_1} \tag{5.139}$$

$$\frac{d\omega_2}{dt} = \frac{T_y - T_l}{J_2} \tag{5.140}$$

where
T_e is torque applied (driving) at the shaft input
J_1 is the moment of inertia of the device that produces the driving torque (e.g., of the motor)
T_l is the torque of resistance at the output shaft end
J_2 is the moment of inertia of the rotating mass that is connected to the output shaft end

For simplification of simulations of such systems, SimPowerSystems disposes of the models **Mechanical Shaft** and **Speed Reducer**. The latter contains two models **Mechanical Shaft** and the additional model **Reduction Device**—the block that can be used independently. In Figure 5.135a the pictures of these blocks are shown.

The first block solves the equation (5.138) and has two parameters: C (*Stiffness*, N-m) and B (*Damping*, N-m-s). The block **Reduction Device** decides the equation of a type as shown (5.139), in reference to the gear unit whose block diagram is shown in Figure 5.136. Here T_h is the input torque, T_l is the output torque, i is the gear ratio, ef is the gear efficiency, J is the gear moment of inertia reduced to the input shaft, and N_{rdh} and N_{rdl} are the rotation speeds of the input and output shafts (rpm). The structure of

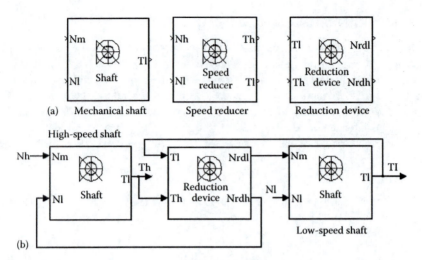

FIGURE 5.135 Pictures of the mechanical coupling models (a) and block diagram of the model Speed Reducer (b).

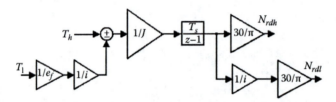

FIGURE 5.136 Structure of the model Reduction device.

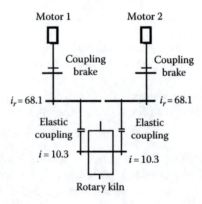

FIGURE 5.137 Rotary kiln kinematics.

the block **Speed Reducer** is shown in Figure 5.135b. The input quantities are the shaft rotation speeds; the outputs quantities are the torques of these shafts. In the model dialog box, the said parameters of the blocks included in the model are fed in.

Use of these blocks is demonstrated for a model of electrical drive of the rotary kilns used in cement industry, whose mechanical part structure is illustrated in Figure 5.137.

It consists of two similar kinematic chains, with each containing a DC motor, a multistage reducing gear with total gear ratio $i_r = 68.1$, an elastic coupling that is intended for the dynamic load suppression, and a pinion with $i = 10.3$ that is coupled with the gear wheel on the kiln body. Both the gear and the kiln with the gear wheel are multimass elastic systems; however, for the purpose of simplification they are considered single-mass ones, and only the elasticity of the elastic coupling is taken into consideration. Since its stiffness is essentially less than the other elements of the kinematic chain, its value can be determined, and the precision of calculations possible under this assumption is quite sufficient for initial estimations.

The total inertia moment of the motor, the brake and the input gear referring to the motor shaft is 21.7 kg m², the stiffness of the elastic coupling $C = 1,125,000$ N-m, $B = 0$, and the kiln moment of inertia referred to the elastic coupling shaft is 65,000 kg · m². For simulation, model5_39 is developed, in which the model DC3 of the irreversible three-phase thyristor electrical drive described earlier is used. The rated parameters of DCM are as follows: $P = 265$ kW, $U = 545$ V, $I = 510$ A, $n = 1480$ rpm. The rated torque is $T = 265,000 \cdot 30/(\pi \cdot 1480) = 1710$ N-m, $K_e = 1710/510 = 3.353$. Supposing that the excitation current $I_f = 10$ A (the excitation voltage is 150 V, and the winding resistance is 15 Ω), then $L_{af} = 3.353/10 = 0.335$. The rest of the parameters are shown in the DC3 dialog box.

The electrical drives are supplied by the voltage of 500 V; moreover, the supplying voltages are displaced by 30° one about another. In reality, the three-winding transformer is used for this purpose, as in model4_3 and model4_4.

The block **Mechanical Shaft** with the single parameter C is used for elastic couple modeling; for modeling of the second mass, the block **Reduction Device** is used, with above-specified values of the gear ratio and moment of inertia, with an expected efficiency of 0.9. The second mass is set in motion by the sum of the output torques of both elastic couplings. The kiln load torque referring to the DCM shaft of each motor is 600 N-m and increases by 25% at $t = 6$ s.

The first electrical drive operates with the speed controller whose output determines the torques for both DCMs. It means that the second DCM operates with the torque control. Since in the first DCM the current reference can be measured, this signal is multiplied by $L_{af}I_f$ in order to determine the torque of the second DCM. In addition, the signal that is proportional to the difference of the both DCM speeds comes to the input of the second DCM control system. It has two implications: limiting the second DCM speed when the mechanical connection with the kiln is severed and reducing the oscillations when the parameters of the mechanical chains are different.

The block **Wref** sets speeding-up to the speed 1500 rpm. The scopes **Motor1** and **Motor2** fix the speeds, currents, and torques of the motors, and the scope **Load** records the speed and the elastic torque of the load.

In order to understand the influence of the elastic coupling, system simulation as a single mass is carried out at first. Taking into account that the moment of inertia of the kiln (of the second mass) referred to the shaft of the motor is $0.5 \cdot 65000/68.1^2 = 7$ kg m², the moment of inertia of the first motor is set as $21.7 + 7 = 28.7$ kg m², the load is detached by setting $I_r = 0$, and the proportional gain of **SR** is 50. After simulation, the process shown in Figure 5.138 whose results are nearly ideal.

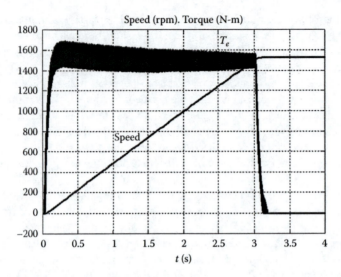

FIGURE 5.138   Transient in the single-mass system.

If the simulation is carried out for the initial two-mass system (the moment of inertia of each motor 21.7 kg-m², I_r = 68.1), one can see that the current and torque curves have pronounced oscillatory character (the swing to 40%). The various actions that can be taken to reduce oscillations are not considered here. Let us decrease the proportional **SR** gain to 10. The processes are shown in Figure 5.139. For clarity, the speed and the torque are referred to the motor shaft. Although in a steady state

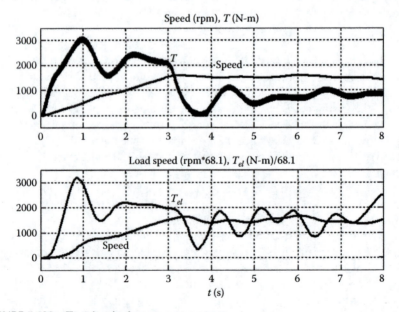

FIGURE 5.139   Transient in the two-mass system.

the torque of each motor is equal to the half of the load torque, during the transition condition the relative motor torque peak is twice as much. The processes in both motors are identical.

It is possible in practical situations that the values of C are different in the kinematic chains, for example, when some springs in one of the elastic couplings are broken. Suppose that the value C decreases by 20% in the second chain. If the simulation is carried out with parameter *Stiffness* = 0.9×10^6 in the block **Mechanical Shift1**, one can see that the oscillations of both the motors and the load torques increase essentially under **Gain4** = 0. If the simulation is repeated with **Gain4** = 50, it can be seen that oscillations reduce drastically. So, it proves that aforesaid purpose of the signal that is proportional to the motor speed difference is correct.

The mechanical system models described do not give an opportunity to simulate a backlash in transmissions that exists almost always, particularly in the electrical drives of medium and big power capacities. The subsystem **Shaft_Luft** takes a backlash into consideration. It differs from the block **Mechanical Shaft** by the block **Dead Zone** at the integrator output, where the signal corresponds to the angle of the shaft deformation (model5_40). If the values of dead zones are set equal to ±7°, the result of simulation demonstrate an increase in the torque, which peaks at about 25% and the rise of oscillations. The influence of the backlashes is especially large when their values are different in the kinematic chains. If, for example, the backlash = 0 in the kinematic chain of the first motor, the oscillations increase essentially under **Gain4** = 0. But under **Gain4** = 50 one can observe oscillation reducing considerably. Therefore, in this case, when there is a large difference in backlashes in the kinematic chains, the use of additional action of the speed difference to the second (slave) motor turns out reasonable. Such a system was placed in service during the reconstruction of one cement plant in Hungary and produced a positive effect.

REFERENCES

1. Mathworks, *SimPowerSystems™, User's Guide*, Natick, MA.
2. Perelmuter, V.M. and Sidorenco, V.A., *DC Thyristor Electrical Drives Control Systems* (in Russian), Energoatomizdat, Moscow, 1988.
3. Kovacs, K.P. and Racz I., *Transiente Vorgänge in Wechselstrommaschinen*, Verlag der Ungarischen Akademie der Wissenschaften, Budapest, Hungary, 1959.
4. Krause, P.C., Wasynczuk, O., and Sudhoff, S.D., *Analysis of Electric Machinery*, IEEE Press, Piscataway, NJ, 2002.
5. Bose, B.K., *Modern Power Electronics and AC Drives*, Prentice Hall PTR, Upper Saddle River, NJ, 2002.
6. Perelmuter, V.M., Direct torque control improvement, *10th EDPE Conference*, pp. 134–139, Dubrovnik, Croatia, 1998.
7. Perelmuter, V.M., Three-level inverters with direct torque control, *IEEE-IAS-2000 Conference Record*, vol. 3, pp. 1368–1374, Rome, Italy, 2000.
8. Salomäki, J., Hinkkanen, M., and Luomi, J., Sensorless control of induction motor drives equipped with inverter output filter, *IEEE Transactions on Industrial Electronics*, 53(4), 1188–97, June 2006.
9. Patent USA 6058031, Five level high power motor drive converter and control system, Issued on May 2, 2000.

10. Perelmuter, V.M., *Direct Control of the AC Motor Torque and Current* (in Russian), Osnova, Kharkov, 2004.

11. Wu, X.Q. and Steimel, A., Direct self control of induction machines, fed by a double three-level inverter, *IEEE Transactions on Industrial Electronics*, 44(4), 519–527, August 1997.

12. Zhao, Y. and Lipo, T., Space vector PWM control of dual three-phase induction machine using vector space decomposition, *IEEE Transactions on Industry Applications*, 31(5), 1100–1109, September/October 1995.

13. Hadiouche, D., Razik, H., and Rezzoug, A., Study and simulation of space vector PWM control of double-star induction motor, *Conf. Rec. CIEP 2000*, Acapulco, Mexico, pp. 42–47.

14. Singh, G., Nam, K., and Lim, S., A simple indirect field-oriented control scheme for multiphase induction machine, *IEEE Transaction on Industrial Electronics*, 52(4), 1177–1184, August 2005.

15. Bose, B.K., *Advances and Trends in Power Electronics and Motor Drives*, Elsevier, Amsterdam, the Netherlands, 2006.

16. Hinkkanen, M. and Luomi, J., Modified integrator for voltage model flux estimation of induction motors, *IEEE Transaction on Industrial Electronics*, 50(4), 818–820, August 2003.

17. Schiferl, R.F. and Ong, C.M., Six phase synchronous machine with AC and DC stator connections, part I: Equivalent circuit representation and steady-state analysis, *IEEE Transactions on Power Apparatus and Systems*, 102(8), 2685–2693, August 1983.

18. Slezhanovski, O.V., Dazcovski, L. et al., *The Cascaded Control Systems for AC Drives with Power Converters* (in Russian), Energoatomizdat, Moscow, 1983.

19. Schröder, S., Tenca, P., Geyer, T., Soldi, P., Garcés, L.J., Zhang, R., Toma, T., and Bordignon, P., Modular high-power shunt-interleaved drive system: A realization up to 35 MW for oil and gas applications, *IEEE Transactions on Industry Applications*, 46(2), 821–830, March/April 2010.

20. Endrejat, F. and Pillay, P., Resonance overvoltages in medium-voltage multilevel drive systems, *IEEE Transactions on Industry Applications*, 45(4), 1199–1209, July/August 2009.

21. Krishnan, R., *Switched Reluctance Motor Drives*, CRC Press, Boca Raton, FL, 2001.

6 Electric Power Production and Transmission Simulation

6.1 COMPUTATION OF TRANSMISSION LINE PARAMETERS

The information about three models of the transmission lines that are included in SimPowerSystems™ is given in Chapter 2. The parameters of these models depend on many factors: Is the line overhead or underground? What is the construction of the overhead line (height, distance between the phase conductors)? How are the phase conductors arranged one after the other? How many lines the power transmission line support (tower) carries? How many neutral wires are used and how are they arranged? The others include material and construction of the phase conductors (the outside diameter, the thickness of conducting material, the number of the wires in the bundle) and so on. There are a number of parameter computation methods that can be found in the technical literature and handbooks. SimPowerSystems assists with calculation of the parameters are necessary for functioning of the said transmission line models. The matrixes of resistances, inductances, and capacitances are computed for overhead line conductors arranged freely, and the symmetrical components are computed also for the three-phase line. For the N-wire system, three N*N matrixes are calculated: matrix $\mathbf{R}$ of series resistances, matrix $\mathbf{L}$ of series inductances, matrix $\mathbf{C}$ of parallel capacitances. It is interesting to note that, owing to influence of the ground, matrix $\mathbf{R}$ is not diagonal.

It has been said in Chapter 3 already that any system of the three-phase quantities, for example, voltages V_a, V_b, V_c, can be represented by the so-called the symmetrical components V_0, V_1, V_2 by relations (3.17) through (3.19) that are written in the matrix form as

$$\mathbf{V}_{012} = \mathbf{T}^{-1}\mathbf{V}_{abc} \tag{6.1}$$

where $\mathbf{V}_{012}$, $\mathbf{V}_{abc}$ are the vectors with components V_0, V_1, V_2 and V_a, V_b, V_c, respectively; the reverse conversion ((3.14) through (3.16)) is

$$\mathbf{V}_{abc} = \mathbf{T}\mathbf{V}_{012} \tag{6.2}$$

$$\mathbf{T} = \begin{bmatrix} 1 & 1 & 1 \\ 1 & a^2 & a \\ 1 & a & a^2 \end{bmatrix} \quad a = e^{j2\pi/3} \tag{6.3}$$

$$\mathbf{T}^{-1} = \frac{1}{3}\begin{bmatrix} 1 & 1 & 1 \\ 1 & a & a^2 \\ 1 & a^2 & a \end{bmatrix} \tag{6.4}$$

If in the three-phase system the voltages and the currents are coupled with the impedance matrix $\mathbf{Z}_{abc}$

$$\mathbf{V}_{abc} = \mathbf{Z}_{abc}\mathbf{I}_{abc} \tag{6.5}$$

then for the symmetrical components

$$\mathbf{V}_{012} = \mathbf{Z}_{012}\mathbf{I}_{012} \tag{6.6}$$

$$\mathbf{Z}_{012} = \mathbf{T}^{-1}\mathbf{Z}_{abc}\mathbf{T} \tag{6.7}$$

For the total symmetrical system $\mathbf{Z}_{abc}$ (all the diagonal elements are the same and all other elements are equal), the matrix $\mathbf{Z}_{012}$ is the diagonal one and has only two different elements that are called the impedances of zero and positive sequences. This matrix is close to diagonal in practical cases [1,2]. The parameters of the block **Distributed Parameter Line** can be specified both by matrixes **R, L, C** and by impedances of positive and zero sequences; at that point, for $N = 6$, the quantity of mutual impedance of zero sequence appears with components R_{0m}, L_{0m}, C_{0m}. Only the impedances of positive and zero sequences are specified for the block **Three-Phase PI Section Line**. In order to compute the line parameters, it is necessary either to fulfill the command power_lineparam in the MATLAB® command window or to use the option *Powergui/Compute RLC Line Parameters*. The window shown in Figure 6.1 opens in the both cases. On the upper left, one can choose between the metric system of units and the English system of units. We shall use the former. Furthermore, the line frequency and the ground resistance, $\Omega \cdot m$, are specified; the latter value can be taken as zero. In the field *Comments* any text can be fed in that relates to this line.

In the top right field, information about line geometry is given. The number of the phase conductors (bundles), usually three or six, and the number of the ground wires, usually one or two, are defined. The information about position of the phases is given in the following table. In Figure 6.2a the arrangement of conductors corresponding to the data given in Figure 6.1 is shown. On the left, names of the conductors are given (P1–P3, g1, g2), and in the first column, the phase numbers. It is seen from Figure 6.2a that all the phase conductors are on the same height above the earth, 20.7 m, and the neutral wires are on the height of 32.9 m. In the columns X and Y_{tower} the data is fed in about the conductor horizontal position relative to the middle (null) tower line and about the conductor vertical position relative to the earth. The conductors between towers sag usually, so the minimum distance to the earth is less than the specified distance in the column Y_{tower}; this distance (mid sag) is defined as the column Y_{min}, and this circumstance is neglected in Figure 6.1. The number of the conductor type is to be added in the last column: 1 for the phase conductors, and 2 for the neutral ones.

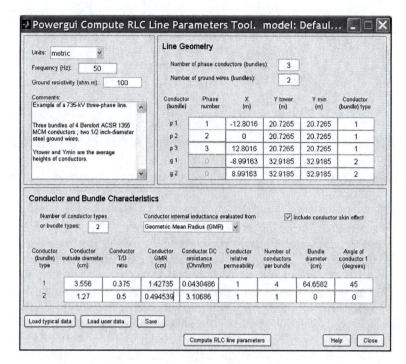

FIGURE 6.1   Powergui window "Compute RLC Line Parameters."

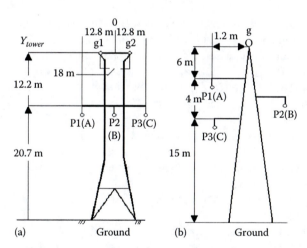

FIGURE 6.2   Constructions of the transmission lines. Conductors arrangement for (a) the line 735 kV and (b) the line 25 kV.

The data of the separated conductors are given in the lower part of the window. In the field *Number of conductor types or bundle types*, the number of the conductor types is given that corresponds to the data of the last column of the previous table and defines the number of rows in the following table. The field named *Conductor internal inductance evaluated from…* has the variants *T/D ratio, Geometric Mean Radius (GMR)*, and

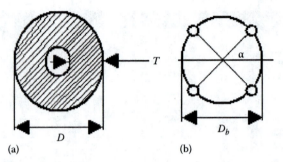

(a)   (b)

FIGURE 6.3 Conductor construction. (a) Conductor geometry. (b) Wire arrangement in the bundle.

Reactance X_a of 1 m *spacing.* The fact is that the conductor total flux linkage determining its inductance consists of the flux within the conductor whose lines of force embrace only a part of the current flowing in the conductor and the external flux that is determined by the total current. The inside flux linkage is defined by the conductor geometry—by its diameter and the thickness of the conducting material, as shown in Figure 6.3a. *GMR* is equal to the radius of the hollow conductor with zero thickness that has the same self-inductance as the conductor under consideration. The value X_a is determined by the value *GMR* and the frequency. The choice of the variant depends on available information about the conductor. In the field to the right, one can mark the necessity to take into account a skin effect (Figure 6.1).

Now it is possible to add conductor data to the table. The outside diameters of the separate wires in cm, the ratio *T/D* (it is seen, specialty, from the table that the ground wire is solid) or the value *GMR*, the DC resistance of the conductor, and the conductor material's relative permeability (1 for the copper, aluminum) are added in the columns successively. Furthermore, the number of the wires with the previously mentioned data in the cable assembly (bundle) of one phase is added. Usually, several wires in a phase are used during transmission of the high voltages—220 kV and more—for reducing the losses and electromagnetic interference owing to the corona effect. In the case under consideration, the number of the wires in a phase is 4. It is supposed that the separated wires are arranged even on the circumference of the diameter D_b. The value D_b and the angle α (degree) between the horizontal axis and the direction to the first wire in the bundle (Figure 6.3b) are specified in the following two columns. If the number of the wires in the bundle is 1, these two parameters are zero.

After the table is filled in, the computation program can be started by clicking on "Compute RLC line parameters"; the results appear in the opening window as three 3*3 matrixes, **R_matrix** (Ω/km), **L_matrix** (H/km), and **C_matrix** (F/km), and three vectors, $\mathbf{R}_{10} = [R_1, R_0]$ (Ω/km), $\mathbf{L}_{10} = [L_1, L_0]$ (H/km), and $\mathbf{C}_{10} = [C_1, C_0]$ (F/km), where the vector first components are the positive sequence parameters and the second ones are the zero sequence parameters. By clicking on "Send RLC parameters to workspace," these matrixes and vectors are sent to the workspace, which can be stored by the command Save "name." In the dialog box of the block **Distributed Parameter Line**, the aforesaid designations of the matrixes can be indicated, and then, at the start of simulation the model with this block, the line parameters will be taken into account automatically. This applies to the blocks **Three-Phase PI Section Line** (only as for the vectors $\mathbf{R}_{10}$, $\mathbf{L}_{10}$, $\mathbf{C}_{10}$) too. It is possible also to specify the parameter numerical values in the previously discussed models. For this purpose, the command "Send RLC parameters to block" is used in the window of the computation results. At first, the requested block on the model scheme is selected; after clicking on the button "Selected block," the block name appears. There are two opportunities with the use of the block **Distributed Parameter Line**: RLC matrixes or positive and zero sequences, and only the second variant for the block **Three-Phase PI Section Line**. After activating the selected option, the numerical values of the matrixes or vectors enter into the block dialog box. The block **PI Section Line** is used often for a simplified simulation of the symmetrical three-phase line as a single-phase one. At that point, the values R_1, L_1, C_1 are put in the block dialog box. The computation results, together with initial data, as a text file, can be received with the help of the command "Create a report" in the window of the calculated results.

The transmission line 25 kV of another construction is shown in Figure 6.2b; the phase conductors have only one wire and are placed about both tower sides, and there is only one neutral wire. The conductors sag by 2 m. The window for line parameter calculation is shown in Figure 6.5. The reader can compare it with the window in Figure 6.1. Action with the former is the same as with the latter. The window data are stored in the file LineParameters1.mat.

The following example considers more complex line consisting of 2 three-phase circuits, whose construction corresponds to the one shown in Figure 6.4.

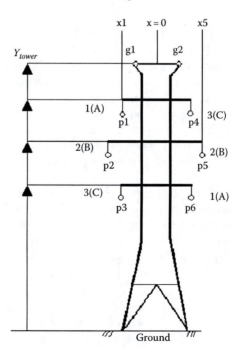

FIGURE 6.4 Variant of the transmission line construction.

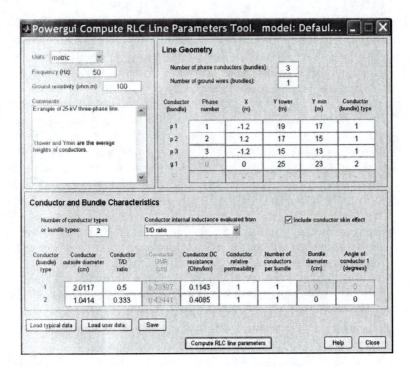

FIGURE 6.5   Window of parameter calculation for the line in Figure 6.2b.

The window for parameter calculation is shown in Figure 6.6. Since all necessary rows about line geometry do not have enough room in the corresponding part of the window, the scrollbar appears on the right; the lower, invisible in Figure 6.6, that is, the table part, is shown in Figure 6.7. It is suggested at first that one split three-phase line be considered; it means that the conductors of the same phases are connected in parallel, supposing even current distribution between them. Then the conductors of the same phases receive the same numbers (1, 2, 3), and the line is modeled as the three-phase one. The calculation results are created in the same form as discussed earlier for the said lines. The results are stored in the file LineParameters2.mat.

If the line is in the form of 2 three-phase lines, then the numeration of the phases with the same names on the left and the right is different, as shown in Figure 6.8. Only the middle part of the table is shown, in that the distinctions are seen. The same situation appears when for some reason it is desirable to consider the conductors of the same phase individually. In this case, the 6*6 matrixes of the resistances, inductances, and capacitances are found, and also five-dimensional vectors $\mathbf{R}_{10} = [R1\_1, Ro\_1, Rom, R1\_2, Ro\_2]$ (Ω/km), $\mathbf{L}_{10} = [L1\_1, Lo\_1, Lom, L1\_2, Lo\_2]$ (H/km), $\mathbf{C}_{10} = [C1\_1, Co\_1, Com, C1\_2, Co\_2]$ (F/km) where the first two components are the positive and zero sequence parameters for circuit 1, the forth and the fifth components are the same for

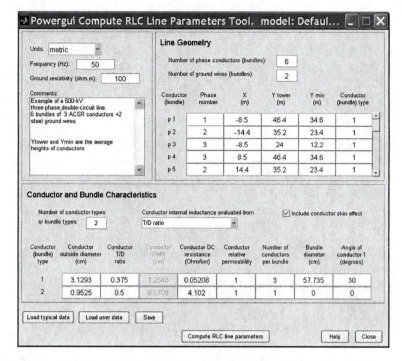

FIGURE 6.6 Window of parameter calculation for the line in Figure 6.4.

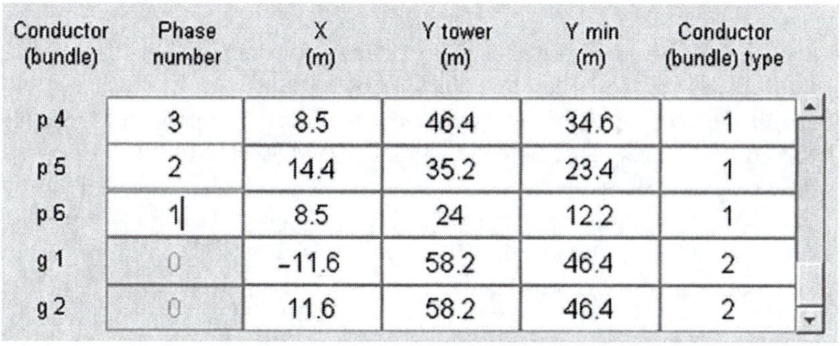

Conductor (bundle)	Phase number	X (m)	Y tower (m)	Y min (m)	Conductor (bundle) type
p 4	3	8.5	46.4	34.6	1
p 5	2	14.4	35.2	23.4	1
p 6	1	8.5	24	12.2	1
g 1	0	−11.6	58.2	46.4	2
g 2	0	11.6	58.2	46.4	2

FIGURE 6.7 Part of the parameter calculation window for the line in Figure 6.4.

circuit 2, and the third component is the mutual zero sequence component. In the case under consideration

$$\mathbf{R}_{10} = [0.018374, 0.18599, 0.17116, 0.025837, 0.20652] \ (\Omega/\text{km}),$$

$$\mathbf{L}_{10} = [0.92952, 3.3630, 2.1556, 0.92119, 3.3399]10^{-3} \ (\text{H/km}),$$

$$\mathbf{C}_{10} = [1.2571, 0.78555, -0.27957, 1.4829, 0.92398]10^{-8} \ (\text{F/km}).$$

The matrix values for $N = 6$ can be transferred to the dialog box of the block.

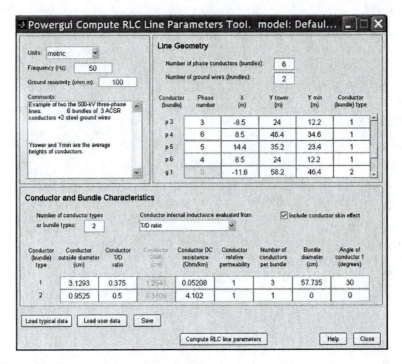

FIGURE 6.8 Window of parameter calculation for the second variant of the line construction in Figure 6.4.

Distributed Parameter Line by one of the previously mentioned ways. As for the parameters of the sequences, this block accepts only three-dimensional vectors; therefore, if this method of the line parameter specification is selected, the average values of positive and zero sequences computed for circuit 1 and 2 can be accepted. The line data are stored in the file LineParameters3.mat. Note in the conclusion that SimPowerSystems contains parameters of several standard transmission lines. For access to these data, the command Load/Typical line data in the line data window (Figure 6.1) has to be carried out.

6.2   USE OF THE SIMPLIFIED SM MODEL

The simplified SM model was described in the previous chapter; it was noted there that the use of this model is reasonable for simulation of the systems with several generators. As an example of the use of this model, the model of the isolated set with two generators (SG) of the same power that feed the common load (model6_1) is given. One can see in the subsystem **System** that both SG feed the three-phase resistive load 3 MW; moreover, the SG1 feeds through the line with parameters 1 Ω, 6 mH, and the SG2—through the line with parameters 1 Ω, 12 mH. Besides, there is the permanent load of 40 Ω for each phase. The line elements are shunted by the large resistances of 100 Ω, in order to prevent the series connection of the inductances of SM and the line, which is inadmissible in SimPowerSystems. The SG1 parameters are given in Figure 5.123. The factor K_d will be estimated later. The SG2 parameters are the same, with the difference that the rotor initial angle is π. The load is connected at $t = 5$ s. The output voltages and currents, the rotation speed, and the electrical power are measured for each SG. Besides, the angle δ is calculated for SG1 as a difference between the phase angles of phase A inner emf and phase A voltage; these quantities are measured with the help of the blocks **Fourier Analyser** that are adjusted for the frequency of 50 Hz and for the first harmonic.

Let us consider one possibility for computation of the factor K_d. In the example under consideration, the same SM is used as in model5_27, with the difference that $L_{md} = L_{mq} = 1$. Model6_2 is used, in which the full and the simplified models of the same SM in the motor mode are supplied with the same voltage and experience the same actions: the load increasing from 0 to 0.75 at $t = 2$ s with simultaneous excitation increasing from 1 to 1.5. Set $K_d = 9.9$ and carry out simulation, recording the angle δ. It is seen that both curves are sufficiently similar (Figure 6.9). Explain how the value K_d was determined.

The response δ_1 has the overshoot $\Delta_{max} = (52 - 33)/33 = 0.57$. By known relationship [3]

$$\Delta_{max} = \exp\left(-\frac{\pi\xi}{\sqrt{1-\xi^2}}\right) \tag{6.8}$$

we find $\xi = 0.18$. By (5.125), keeping in mind $E = 1.5$, $P_{max} = 1 \times 1.5/1.1 = 1.36$, one receives $K_d = 4 \times 0.18\sqrt{314 \times 0.89 \times 1.36/2} = 9.9$.

Return to model6_1. In order to keep constant the requested frequency of the SG output voltage, the SGs have speed controllers that affect SGs input power with the help

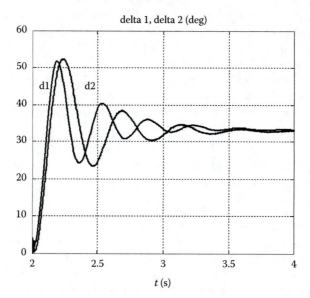

FIGURE 6.9 Angle δ changing during full and simplified SM simulation.

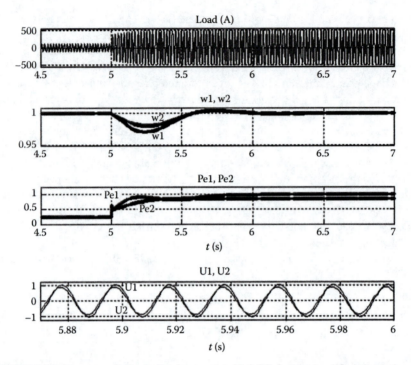

FIGURE 6.10 Processes in the isolated network with 2 SG during simplified simulation.

of the prime mover. It is seen in the subsystem **F_Reg** that these controllers are of PI type with the output limitation of 1.5; the first-order links at the outputs model the prime mover inertia. SG1 is provided by the PI voltage controller that affects the excitation voltage. Its output limitation is equal to 4; the first-order link at the output models the time lag of exciter. The block **Rectifier** realizes the relationship $U_{out} = \text{Max}(U_{sa}, U_{sb}, U_{sc})$.

Working with this model, one can watch the different variables of SG1 and SG2, estimate the influence of grid and load parameters, values of power inputs and excitation voltage, and so on. It is accepted in the model that SG2 has the constant excitation; it can cause uneven SG load (as shown in Figure 6.10). If it is impermissible, the SG2 has to have the voltage controller too. The responses of load current, rotation speeds, powers, and voltages of SG1, SG2 are shown in Figure 6.10.

6.3 SIMULATION OF SYSTEMS WITH HYDRAULIC-TURBINE GENERATORS

A large amount of electric energy is produced by converting the water-flow energy. A simplified diagram of the hydraulic-turbine—SG is shown in Figure 6.11. The water stream with the velocity U is incident upon the blades of the turbine; this incidence is caused by the height difference H of the water level and turbine position; L is the length of the penstock. The volume of the water flowing in is controlled by the position of the gate that is driven by a servomotor. SG is coupled

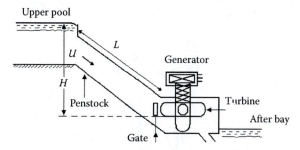

FIGURE 6.11 Simplified construction of the block **hydraulic turbine-SG**.

with the turbine shaft. With a number of simplifications, the turbine behavior is described by the following equations [1]. The water velocity in the penstock is

$$U = K_u G \sqrt{H} \tag{6.9}$$

where
K_u is the proportionality factor
G is the position of the gate
H is the hydrostatic pressure

The turbine mechanical power is

$$P_m = K_p H U, \tag{6.10}$$

where K_p is the proportionality factor.

The water column acceleration due to a change in the hydrostatic pressure is

$$T_w \frac{d\Delta U^*}{dt} = -\Delta H^* \tag{6.11}$$

$$T_w = \frac{L U_0}{a_g H_0} \tag{6.12}$$

where the lower index 0 designates the initial values, the symbol * designates the relative incremental change, and a_g is the acceleration due to gravity. After linearization of (6.9), (6.10), taking into account (6.11), the transfer function of an ideal hydraulic turbine is obtained as

$$\frac{\Delta P_m^*}{\Delta G^*} = \frac{(1 - T_w s)}{(1 + 0.5 T_w s)} \tag{6.13}$$

$$\Delta G^* = \frac{\Delta G}{G_0}$$

For nonideal turbine, one can write its equation as

$$\frac{\Delta P_m^*}{\Delta G^*} = \frac{a_{23}(1+bT_w s)}{(1+a_{11}T_w s)}$$

(6.14)

$$b = a_{11} - \frac{a_{13}a_{21}}{a_{23}}.$$

The coefficients a_{ij} depend on the turbine type. For example, for ideal lossless Francis-type turbine $a_{11} = 0.5$, $a_{13} = 1$, $a_{21} = 1.5$, $a_{23} = 1$.

This model describes turbine operation with small increments. It is sufficient for an analysis of the steady states but is inadequate for studying large increments in power and frequency. Therefore, the more complex nonlinear model is accepted in SimPowerSystems, which for the relative quantities is depicted in Figure 6.12.

Here the input g^* is the so-called real gate opening that is related to the ideal gate opening G^* as

$$G^* = A_t g^*$$

(6.15)

$$A_t = \frac{1}{(g_{max}^* - g_{min}^*)}.$$

(6.16)

Here g_{min}^* is an initial gate opening that corresponds to the losses in the no-load condition when $G^* = 0$, g_{max}^* is the final gate opening corresponding to $G^* = 1$, for example, $g_{min}^* = 0.16$, $g_{max}^* = 0.96$. The factor β characterizes the damping occurring due to the turbine's rotational speed oscillations. Its value is about 1–2 (pu) usually; the last value means, for instance, that speed (frequency) change by 1% results in load change by 2%.

The turbine model depicted in Figure 6.12 is inside the block **Hydraulic Turbine and Governor** in the folder *Machines*. In order to use this model for simulation in other models, one can do the following: copy this block in a specific Simulink® model, select the option *Look under Mask*, and in the window that appears subsequently, pick out the block **Source** and select the option *Look under Mask* again. The block (model) **Hydraulic Turbine** appears in the obtained window and can be

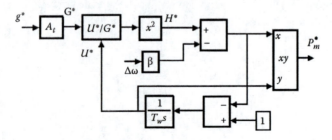

FIGURE 6.12   Nonlinear turbine model.

copied in the user's model scheme. The model parameters g_{min}, g_{max}, T_w, β can be fed in the dialog boxes of the blocks composing this model either directly or added using the MATLAB command window or specified in the option *File/Model Properties/ Callbacks*.

The turbine has a governor, whose function is to control the turbine rotational speed by affecting its mechanical power by means of inducing change in gate opening with the help of a servomotor that provides a flat (without droop) characteristic speed— power. Such a characteristic is admissible for the isolated block turbine generator, but in most cases, several generators operate in parallel, and for ensuring even load sharing between generators a droop characteristic is used. This characteristic is realized by the servomotor position feedback with the coefficient R_p. It is recommended often to choose R_p so that a speed change by 5% causes full gate opening ($R_p = 0.05$).

The controller diagram accepted in SimPowerSystems as shown in Figure 6.13a. The servomotor controls the gate position with the aid of the PID controller whose input is the difference between the reference ω_{ref} and actual ω_e turbine rotational speed. Two methods for droop characteristic realization are provided, depending on the binary signal d_{ref}: In the upper switch position, the gate position negative feedback is used; in the lower switch position, the droop depends on the differ- ence between the SG electric power P_e and the reference value P_{ref}: If, for instance, $P_e > P_{ref}$, the summer 3 output signal >0 that appears as the negative one at the output of the summer 2 causes a decrease in the water flow and, consequently, a decrease in turbine power.

Servomotor model diagram is shown in Figure 6.13b. The limitations of both gate position changing speed and gate boundary positions are provided.

The picture of the block **Hydraulic Turbine and Governor** is shown in Figure 6.14; the input signals are the reference speed (usually 1), the actual speed (formed by the block SG), the deviation of the SG rotational speed from the

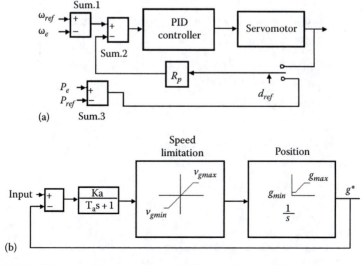

FIGURE 6.13   Diagram of the controller unit: (a) controller and (b) servomotor.

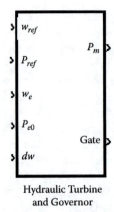

Hydraulic Turbine
and Governor

FIGURE 6.14   Picture of the block **Hydraulic Turbine and Governor**.

rated one, as well the reference power and the electric power that is formed in block SG too. The last two can be not used, if $d_{ref} = 1$ is selected. The block output is the mechanical power applied to SG.

The block dialog box is shown in Figure 6.15. The meaning of box lines are: first—the gain and the time constant of the link that models the servomotor; second—the gate

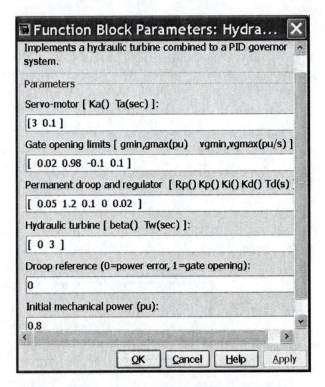

FIGURE 6.15   Dialog box of the block **Hydraulic Turbine and Governor**.

position boundary and the maximum speed of gate travel; third—the accepted droop and four PID controller parameters—the last one is the time constant of a first-order low-pass filter that limits the high-frequency gain of the PID; fourth—the turbine parameters β and T_w; fifth—the method for droop characteristic realization: 0—by power difference, 1—by gate position; and sixth—the power initial value; this value changes automatically by using of Powergui.

The use of the block **Hydraulic Turbine and Governor** is shown with some examples.

Model6_3 shows a unit turbine-SG feeding the isolated load. The load's initial power is 50 MW, and at $t = 5$ s it rises to 100 MW stepwise. SG has a power of 200 MVA and a voltage of 13.8 kV. Its inertia coefficient takes into account the inertia of the turbine rotational details.

The controller parameters are rather typical; some of them are chosen by experiment. Pay attention that the droop is set to zero ($R_p = 0$) because the unit operates island.

SG is provided by a typical voltage control system, **Excitation System**, which is placed in the folder *Machines*. Its block diagram is shown in Figure 6.16. The SG reference voltage V_{ref} and the value of the voltage space vector amplitude are compared at the block inputs. The latter value is computed by using its components V_d, V_q; just these quantities are formed by the block for measurement of SG variables. The signal of the amplitude is smoothed by the filter with the time constant T_r. The error signal is corrected in phase by the phase lead link with the time constants T_b, T_c and goes to the main controller. At its output (exciter input) the limiter is set that can operate in two modes. If $K_p = 0$ in the dialog box, the output is limited by the values E_{fmin}, E_{fmax}. If $K_p > 0$, the upper limitation is equal to the amplitude of the output voltage space vector V_{tf} is multiplied by K_p. E_{fmin}, and E_{fmax} are specified in pu.

The exciter model with the transfer function shown in Figure 6.16 is connected to the limiter output. The exciter model output is connected to the input of the SG model with the same name (input of the excitation voltage V_f). The vanishing negative feedback from the exciter output to the system input is provided to improve system stability. The model has an additional input V_{stab} for the external stabilizing signal.

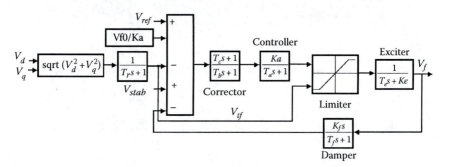

FIGURE 6.16   Block diagram of the excitation system.

The opportunity to start simulation, beginning with a certain steady state, is provided in the model. For that purpose, the output voltage V_{t0} and the excitation voltage V_{f0} have to be specified in pu. At that point, the signal V_{f0}/Ka comes to the system input, and all the system blocks receive the corresponding initial values.

The step-up transformer 210 MVA, 13.8 kV/230 kV, delta/wye is set at the SG output; this transformer feeds the said load.

Powergui permits commencing of simulation with steady state that corresponds to the initial load 50 MW. For this aim, the option *Load Flow and Machine Initialization/ Update Load Flow* is used. At that point, the vector of the SG initial values changes, and also the constant P_{ref}. **Scope** shows the load phase *A* currents (initial and connected loads), the SG voltage, and its excitation voltage; **Scope_g,Pm,w** records the turbine output power, the gain position, and the turbine rotational speed.

The processes related to the second load are shown in Figure 6.17. It is seen that right after an increase in load, the power P_m decreases to some extent, although the water feed increases (*g* increases). This is because during the gate opening, owing to the water compressibility, the water stream (water amount) does not rise instantly and pressure inside turbine decreases [1]. The change in turbine rotational speed is

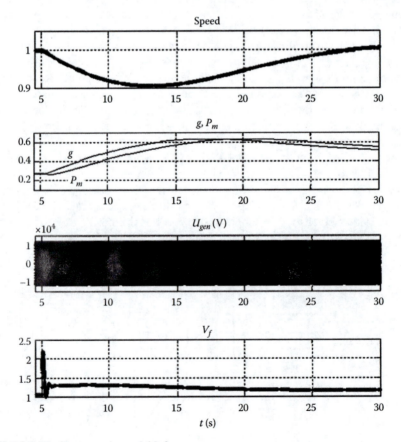

FIGURE 6.17 Processes in model6_3.

rather large; thus, one can observe the short-time frequency decreasing by about 5 Hz, but SG voltage, with the help of the voltage controller, is kept constant.

Model6_4 shows two sets having similar characteristics and feeding the common load 300 MW. The first SG is the same as in the previous model; the second SG has about 10% less power compared to the first. The transformers in the model are the same as in the previous model, and their secondary windings are connected in parallel. The turbine models are the same as in the previous model too; the droop method is set to 0 (by the difference between the reference and actual powers). The block **Three-Phase Fault** models three-phase ground short-circuit at $t = 0.1$ s; at $t = 0.2$ s the fault is eliminated.

Before simulation, the window *Powergui/Load Flow and Machine Initialisation* should be opened.

In the top right field, the list of the SG appears. For SG 200 MVA, a *Swing Bus* as the *Bus Type* is selected, and for the voltage of 13,800 V and the power of 150 MW, for SG 180 MVA, *P&V Generator* as the *Bus Type* is selected, with the same parameters. After the command *Update Load Flow* is carried out, the reference power of the first SG changes to 145 MW, the report about the calculation results appears in the left part of the window (Figure 6.18), and the power references and the initial values change in the model appropriately. The phase currents of the second transformer and the speed change with the first SG are plotted in Figure 6.19.

Model6_5 shows a parallel operation of the grid and the SG driven by hydraulic turbine with a common load of 150 MW. The grid 500 kV is modeled by the block **Three-Phase Source** that has three-phase short-circuit power for the rated voltage

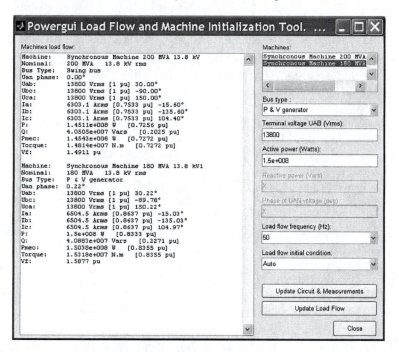

FIGURE 6.18 Window of the initialization computation results.

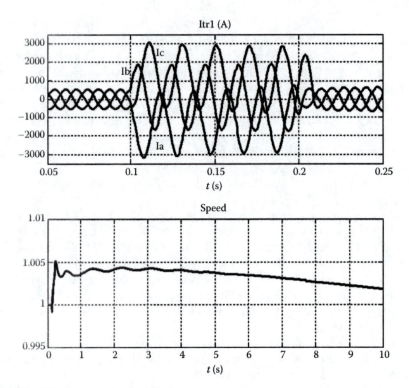

FIGURE 6.19 Processes in model6_4.

of 15,000 MVA. The parameters of the turbine and SG are the same as in the previous models; the controller parameters change a little.

SG connects to the load with the help of **Breaker** that is closed initially and which opens at $t = 0.4$ s.

At the same time, the turbine reference power is set to 0. The scope **U,I** fixes the load voltage and currents of the load, transformer, and the source, and the scope **W,P,g** records rotational speed, turbine power, and gate position.

Before simulation, in the window *Powergui/Load Flow and Machine Initialisation* as the *Bus Type: P&V Generator*, 13,800 V, 100 MW power is set for SG. After simulation one can see by **U,I** that before the **Breaker** opening, the load current is caused by the sum of the transformer current and the source current, but after opening— only by the source current. Turbine behavior during the disconnection of load from SG is shown in Figure 6.20. It is seen that a temporary rise in speed occurs, owing to the system inertia, so that the unit's mechanical strength has to be computed taking into this factor into account.

The following, more complicated model is the simpler version of the demonstrational model `power_svc_pss` from [4]. It uses the so-called Power System Stabilizers (PSS). Their main function is to increase system damping by additional action on SG excitation. In order to achieve this, the PSS has to provide the additional component of the electrical torque in antiphase to the speed deviations. PSS responds to the speed change usually, so it contains a differentiator. Besides, since the

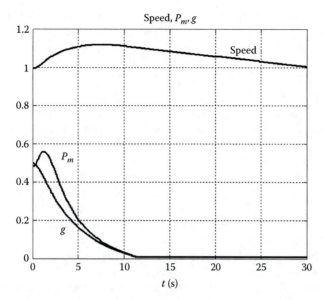

FIGURE 6.20 Processes in model6_5.

circuit for signal transfer from the exciter control input to the SG excitation current has a perceptible lag, PSS contains the phase-lead links for frequencies of 0.1–2 Hz.

The PSS with speed measurement has some drawbacks that arise due to the shaft's elastic oscillation influence. An alternative is to use the difference between the signals of the mechanical turbine power and the electric SG power because

$$\Delta\omega_e = \frac{1}{M}\int(\Delta P_m - \Delta P_e)dt \qquad (6.17)$$

where
M is the inertia constant
ΔP_m is the increment of the input mechanical power
ΔP_e is the increment of the output electric power

The block diagram of the stabilizer's main variant, **Generic Power System Stabilizer**, is shown in Figure 6.21. It is placed in the folder *SimPowerSystems/Machines* and consists of the low-pass filter of the first order, the amplifier, the high-pass filter, the two lead/lag links (depending on the chosen time constants), and the output limiter.

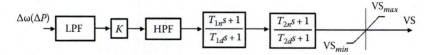

FIGURE 6.21 Block diagram of the **Generic Power System Stabilizer**.

Together with this block, a more complicated unit, the so-called **Multi-Band Power System Stabilizer**, is included in SimPowerSystems. Its necessity is motivated by the fact that diverse types of electromechanical oscillations take place in electric power systems whose frequencies range from 0.2 to 4 Hz [1,4]. This device is intended for damping of these oscillations.

The device has three parallel circuits, with each of them tuned for a certain bandwidth. Each of the circuits contains the differential band-pass filter, the amplifier, and the limiter. The outputs of three circuits are summed, and the total output is limited too.

Two different approaches are available to tune the device: simplified setting and detailed setting. In the first option, the central frequency of the band-pass filter and its gain, total gain, and limitations are defined. For the second option, the adjustment is more intricate and is not considered here.

The block diagram of the device with its simplified tuning is shown in Figure 6.22.

Take for consideration model6_6. SG M2, 6000 MVA, 13.8 kV driven by the hydraulic turbine feeds the load 6000 MW through the transformer with the secondary voltage of 750 kV. This load is fed additionally from SG M1, 2000 MVA that is also driven by the hydraulic turbine through the step-up transformer and the transmission line of 1000 km length. By (2.19) the characteristic line impedance is

$$Z_c = \sqrt{\frac{L}{C}} = \sqrt{\frac{1.05 \times 10^{-3}}{15.8 \times 10^{-9}}} = 258 \ \Omega$$

and by (2.24) surge line load is

$$N_p = \frac{V^2}{Z_c} = \frac{750^2 \times 10^6}{258} = 2180 \ \text{MW}.$$

The block **Three-Phase Fault** is set for fault modeling.

The subsystem **Turbine & Regulators** is connected to each SG. This subsystem contains the same blocks **Hydraulic Turbine and Governor** and **Excitation** as in the previous models, but unlike them the input vstab of the latter block is used and is connected to the output of the block **Multiport Switch**. The zero signal (PSS is absent)

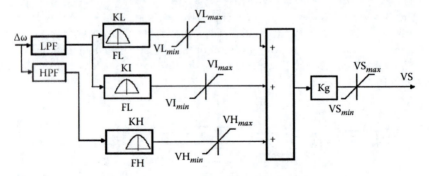

FIGURE 6.22   Block diagram of the block **Multi-Band Power System Stabilizer**.

comes to the first switch signal input, the output of PSS (as shown in Figure 6.21) with the input signal as a difference of power (6.17) comes to the second switch signal input, and the output of PSS according to Figure 6.22 comes to the third switch input. The value of the constant 0, 1, 2 that comes to the switch control input has to be specified.

A number of scopes are set to observe the processes in the system. The scope **Machines** fixes the deviations in rotor positions and their rotational speeds from the synchronous paths and also the voltage amplitudes. The scope **System** records the voltages measured by **B1** and **B2** and the power on the section where **B1** is set; it is seen from the subsystem **PV Measurements** that the positive sequence amplitudes of the said voltages come to the output V, and the active power meter output comes to the output P; this power is computed by the voltages and currents that are measured by **B1**. Note that this meter can operate only in the phasors mode.

It is worth mentioning that an initialization has been made already in model6_6: *Bus Type—P&V Generator* with the voltage of 13,800 V and the power of 1500 MW is selected for M1, and for M2—*Swing Bus* with the same voltage and the angle of 0°. One can see more detailed initialization results in the window *Load Flow and Machine Initialization*. The three-phase short circuit mode is set in the block **Three-Phase Fault**, with start set at $t = 0.1$ s and end at $t = 0.2$ s.

If the simulation is started with PSS = 0, it can be seen, observing the mutual position of the both SG rotors, that the process is unstable. With PSS = 1, when PSS acts as shown in Figure 6.21, the process is stable, but small speed oscillations in the frequency of 0.028 Hz take place (Figure 6.23). If we set PSS = 2, when PSS shown in Figure 6.22 is active, the oscillations disappear completely.

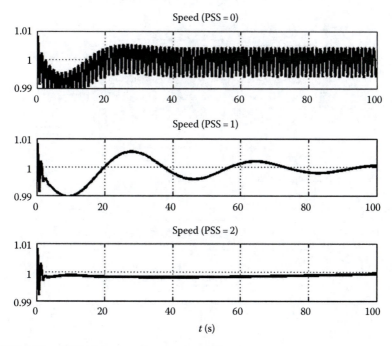

FIGURE 6.23 PSS influence on the system stability.

6.4 SIMULATION OF SYSTEMS WITH STEAM TURBINE-SYNCHRONOUS GENERATOR

A steam turbine converts an energy of high temperature and pressure steam into rotating energy and, afterward, with the help of SG, in the electric energy. The steam turbine consists of several sections placed at the same shaft. Their joint functioning create the turbine torque. The typical turbine contains the high-pressure section HP, the steam stream in which comes from a boiler and is controlled by the main control valve CV, the reheater RH, the intermediate-pressure section IP with the reheat intercept valve IV set in front of it, the crossover piping, and one or more low-pressure sections LP. Control of rotational speed (power) of the turbine is carried out mainly by CV and IV is used if there is a need to lessen the turbine torque quickly and when a turbine load is removed. The typical configuration is depicted in Figure 6.24.

The steam turbine, as a control plant, is defined by the following time constants: T_{CN} is the time constant due to charging time of the steam chest and the inlet piping that is equal to 0.2–0.3 s usually, T_{RN} is the time constant of the reheater that is equal to 5–10 s usually, and T_{CO} is the time constant associated with a crossover piping and equal to 0.5 s about.

The block diagram of the steam turbine is shown in Figure 6.25.

The nonlinear blocks NL1 and NL2 serve for the consideration of nonlinear relationship between displacement of the valve servomotor and the amount of the steam stream. Per unit system is used for parameters and variables. The turbine's maximum power for the rated value of the steam main pressure P_T and for fully open valves (position $g = g_i = 1$) is accepted as the power base quantity. The turbine factors F_{HP},

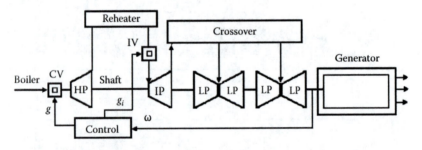

FIGURE 6.24 Typical configuration of the steam turbine.

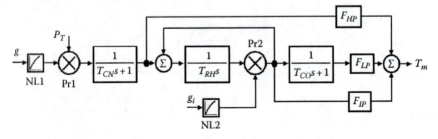

FIGURE 6.25 Block diagram of the steam turbine as a controlled plant.

F_{IP}, F_{LP} define what part of the total power is produced by each section; the factor sum has to be equal to 1. As the typical values, one can accept $F_{HP} = 0.3$, $F_{IP} = 0.3$, $F_{LP} = 0.4$. The relative torque is calculated as

$$T_m^* = \left(\frac{\omega_{base}}{P_{base}}\right) T_m \tag{6.18}$$

that is, the mechanical torque in pu is equal to the mechanical power in pu. To present the turbine in the power system, the power as calculated previously is multiplied by the ratio of the maximum turbine power to the total base power of the power system.

If we take $T_{CO} = 0$, the turbine's simplified transfer function can be written as

$$\frac{\Delta T_m}{\Delta g} = \frac{(1 + sF_{HP}T_{RH})}{(1 + sT_{CH})(1 + sT_{RH})} \tag{6.19}$$

It is seen from Figure 6.25 that if we admit that the valve IV is open fully, the turbine model can be presented as a series connection of a number of the first-order links; their outputs are summed with the weights whose sum is equal to 1.

The main function of the steam turbine control system is the rotational speed control by affecting the mechanical power with the help of CV. At that point, the droop characteristic is to be provided for parallel operation of several units. Functional block diagram of the turbine control is shown in Figure 6.26. It is seen that, when the actual speed ω, exceeds the reference speed $\omega_{ref} = 1$ essentially, for instance by 17%, the amplifier in the valve IV circuit operates, which in turn causes a rapid decrease in the turbine's torque. When the speed deviations are small, only CV operates. It is supposed that the nonlinearity of the valve is compensated. The block diagram of the control system model is shown in Figure 6.27.

The rotor of the steam turbine is a complex mechanical system having a length upto 50 m and weighing a few hundred tons. In such a system the elastic vibrations exist that should be taken into account in many cases during system analysis.

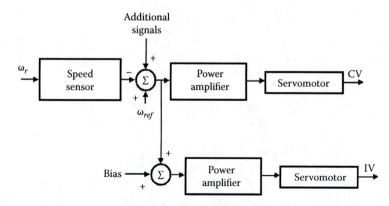

FIGURE 6.26 Functional block diagram of the steam turbine control.

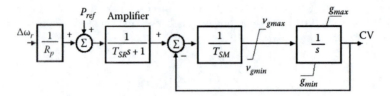

FIGURE 6.27   Block diagram of the model of the steam turbine control system.

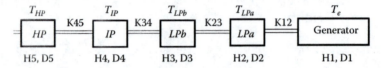

FIGURE 6.28   Structure of the five-mass model of the steam turbine.

The rotor model is presented as several localized masses connected by elastic couplings. The structure of the five-mass model is shown in Figure 6.28 where T_i are the torques produced by each section or SG, H_i, D_i are the inertia and damping coefficients, and K_{ij} are the stiffness coefficients of the corresponding shaft sections. All parameters are given in pu and are referred to the same speed and usually to the electrical speed. Denote δ_i—position of mass i in electrical radian relative to the synchronous rotating reference frame, $\delta_i = \omega_i t - \omega_0 t + \delta_{i0}$; ω_i is the electrical rotational speed of the mass i and ω_0 is the nominal rotational speed (314 rad/s with $f = 50\,\text{Hz}$). Then the motion equation of the second mass, for instance, is written as

$$\frac{2H_2 d\Delta\omega_2}{dt} = T_{LPa} + K_{23}(\delta_3 - \delta_2) - K_{12}(\delta_2 - \delta_1) - D_2\Delta\omega_2 \qquad (6.20)$$

The equation becomes simpler for the first and the fifth masses because an elastic torque acts only from one side. For example, for SG

$$\frac{2H_1 d\Delta\omega_1}{dt} = -T_e + K_{12}(\delta_2 - \delta_1) - D_1\Delta\omega_1 \qquad (6.21)$$

SimPowerSystems (folder *Machines*) contains the block **Steam Machine and Governor** that models the turbine, the rotor elastic couplings, and the control system. Its picture is shown in Figure 6.29.

The block input signals are as follows: ω_{ref} is the speed reference, usually $\omega_{ref} = 1$; ω_m is the SG rotational speed formed by SG model; P_{ref} is the SG power reference; and *d_theta* is the angle of the SG rotor relatively the synchronous rotating reference frame formed by SG model. The block output signals: *dw_5-2* is the vector containing the speed deviations of the masses 5, 4, 3, 2 from the synchronous speed; *Tr5-2* is the vector containing the torques transmitted by the masses 5, 4, 3, 2; *gate* is the level of CV opening; *Pm* is the output mechanical power, and the signal is connected to the SG model input.

The scheme of the block consists of three main blocks. The first one is the model of **Steam Turbine**. It consists of four first-order links with the gain of 1. It was

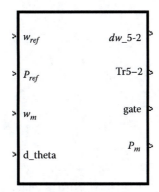

FIGURE 6.29 Picture of the block model **Steam Machine and Governor**.

mentioned earlier about the possibility of such a model. The time constants are numbered from the right to the left; that is, the next to SG block is the time constant $T_a(1)$. If some of the time constants are equal to 0, the corresponding links can be removed. The product of the CV position and the value of the maximum steam pressure (accepted 1) comes to the input of the left block (the first turbine section). The outputs of the first-order links are sent to the block outputs with the weight factors $F2$–$F5$ (the first is the nearest to SG). Thus, these outputs model the components of the total turbine torque that are produced by separate sections. Besides, the signal of the full output of the first left block is measured (turbine input section) denoted *flowHP*.

The model can operate in two modes: as single-mass model or as multimass model. If the first mode is selected, the SG constant of inertia has to take into account the unit's total moment inertia. In the model scheme, all the weighted outputs of the separated links are summed, resulting in the turbine's full torque. This sum, through the upper switch contact, comes to the input of the multiplication block where it is multiplied by SG speed, forming as a result the turbine's mechanical power P_m.

If the multimass mode is chosen, the block called *4-Mass shaft* is active. This block contains four subsystems that are called **1 mass, 2 masses, 3 masses, 4 masses**. The latter subsystem models the full four-masses system, and the others—with less number of the localized masses. To model the system with the reduced mass number, the corresponding inertia constants are set to zero. At that point, only the system that has the wanted mass number remains in the work, and just its output passes to the block output. These outputs are *dw_5-2*, *Tr5-2*, as mentioned earlier, and the mechanical torque.

If some constant of inertia is set to 0, the remaining system compresses in the direction of SG (if, for example, only two masses are considered, it will be the masses 2 and 3). The data of the considered masses are shifted respectively. If some constant of inertia is set to 0, the share of the total torque corresponding to it has to be set to 0 too. However, the torque share of the section can be zero even if there is nonzero constant of inertia. The subsystems **1 mass, 2 masses, 3 masses, 4 masses** solve the corresponding number of the equation of the type (6.20) and (6.21).

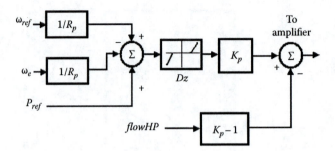

FIGURE 6.30   Input circuits of the speed controller.

The third block in the turbine model is the speed controller. If the model scheme is compared with that shown in Figure 6.27, one can see that the former has difference in the input part (as far as the amplifier), which is shown in Figure 6.30. The distinguishing features are an introduction of the dead zone Dz and addition of the feedback by the steam stream in the first section with the gain $K_p - 1$. Note that when $Dz = 0$, the gain from the amplifier input to the signal *flowHP* is equal to 1, but there are a number of lags in the circuit, whose total equivalent time constant is denoted as T. Designating as x the output Dz and $y = flowHP$, we write:

$$\frac{[K_p x - (K_p - 1)y]}{(Ts + 1)} = y \tag{6.22}$$

or

$$\frac{y}{x} = \frac{1}{(T/K_p)s + 1} \tag{6.23}$$

Hence, when $K_p > 1$, the controller speed of response increases without influence on the gain.

The block dialog box with its adjustment for multimass simulation is shown in Figure 6.31. Note that if we choose the *single-mass* simulation, the second, the third, and the fourth lines from the bottom disappear.

The controller parameters K_p, R_p, Dz (Figure 6.30) are specified in the second line. In the third line, the time constants of the amplifier (called here *Speed Relay*) and the servomotor are given. In the following line, as a vector, the maximum speed of the CV travel and its boundary (Figure 6.27) are indicated. Afterwards, the SG rated speed is designated. In the following two lines, the time constants of the turbine sections and corresponding shares of the total torque are defined, and the sections are numerated from the right to the left, supposing that the device located at the extreme right is SG, as shown in Figure 6.24. Then in three lines the coefficients of inertia for each section, stiffness coefficients between them, and the damping factors are determined. Keep in mind that for a multimass model, SG coefficient of inertia has to take into account only the moment of inertia of the rotor. In the last line, the initial value of the mechanical power in pu P_{m0} and the initial SG angle θ_0 (degree)

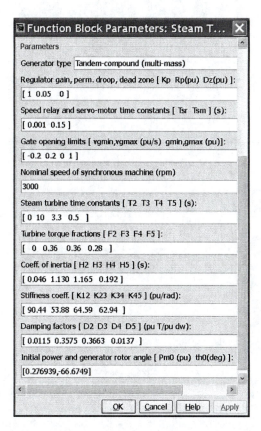

FIGURE 6.31   Parameter window of the block **Steam Machine and Governor** after adjusting for multimass simulation.

are given. For a single-mass model, only the first value is needed. P_{m0} and θ_0 are updated automatically with the use of Powergui.

Furthermore, some examples will be considered. Model6_7 simulates the same system, as model6_3, with the difference that instead of the block **Hydraulic Turbine and Governor**, the block **Steam Machine and Governor** is used. SG has a power of 200 MVA, voltage of 13.8 kV, and rotational speed of 3000 rpm. The load that is equal to 50 MW initially, increases twofold at $t = 5$ s. The block parameters are shown in Figure 6.31, $P_{m0} = 0.28$. Select the single-mass mode. The SG constant of inertia is equal to 3.2 s and takes into account the moment of inertia of the all units. Before the start of simulation, with the help of the option *Powergui/ Load Flow and Machine Initialization*, the initial values corresponding to steady-state conditions are formed. If simulation is carried out with **Switch** and **Switch1** in the upper position (input *vstab* is disconnected from the block **Multi-Band Power System Stabilizer** considered in the previous paragraph), it can be seen that the system is unstable with the oscillation frequency at about 0.056 Hz. If the simulation is repeated with **Switch1** in the lower position (the block **Multi-Band Power System Stabilizer** is connected), and **Switch2** in the right position, the system turns out

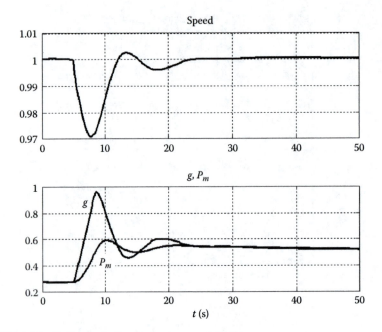

FIGURE 6.32 Processes in the single-mass model6_7.

stable, but the speed static error of 1.3% exists. To eliminate it, the additional slow-response integral speed controller is set. For its activation, **Switch** is to be set in the lower position. The obtained curves related to changes in speed, CV position and mechanical power are plotted in Figure 6.32. The frequency error is 0.

Furthermore, this system is simulated as the multimass; the demanded option is selected in the dialog box, Figure 6.31. The SG constant of inertia is taken to be equal to 0.8 s (instead of 3.2 s) because it takes into account only the inertia of SG. One can see slow oscillations of the speed with a frequency of 0.066 Hz and amplitude of 1.75% and rapid oscillations with a frequency of 13.3 Hz and amplitude of 2.5%. The oscillations with a frequency of 13.5 Hz and amplitude of 4–4.5 pu are seen in the curves of the torque and the power. For reducing these oscillations, the rejection filter of the second order with the central frequency of 13.3 Hz is set at the input of the block **Multi-Band Power System Stabilizer**. The block **Second–Order Filter** from the folder *Extra Library/Control Blocks* is used for this purpose (the filter type *Band stop (Notch)*), the cutoff frequency is 13.3 Hz, the damping factor Z = 0.5). If the option *Plot Filter Response* is marked, it will be plotted in the new window, as shown in Figure 6.33. If the simulation is repeated again with **Switch2** in the left position, one can see that the oscillations reduce essentially (they are absent practically in the speed; the power oscillation amplitude is about 4% in the steady state). However, the problem of the complete elimination of oscillations remains.

Model6_8 simulates the same system as Model6_5 (load throwing-off), with the difference that the block **Steam Machine and Governor** is used instead of the block **Hydraulic Turbine and Governor**. SG load whose initial value is 100 MW (0.5 pu)

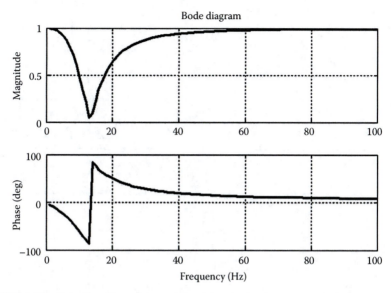

FIGURE 6.33   Rejection filter characteristics for reduction in speed oscillation.

at $t = 0.4$ s is thrown off, and only the small load 5 MW remains. The block parameters are shown in Figure 6.31, $P_{m0} = 0.5$. The turbine is modeled as multimass. Before the start of simulation, with the help of the option *Powergui/Load Flow and Machine Initialization*, the initial values corresponding to steady-state conditions are formed. The SG speed and CV travel responses are shown in Figure 6.34. It is seen that there is a rather large increase in speed, and the use of IV necessary for reducing such an increase that not, however, provided by the model in SimPowerSystems.

The reader can observe the power and torque oscillations caused by the elastic couplings.

Model6_9 demonstrates the so-called subsynchronous resonance (SSR) that appear in the system with the series capacitor-compensated transmission lines [1]. When the transmission line does not have the capacitors that are connected in series with other line elements, the faults cause the direct component of the current. These components induce rotor currents of 50 (60) Hz and, therefore, produce an additional torque of the same frequency. The natural frequencies of the multimass rotor are much less usually. If the series capacitors are set in the transmission line, in order to compensate for the inductive voltage drop, the direct component of the current cannot flow, and the current of the natural frequency circulates (Figure 6.35):

$$\omega_n = \frac{1}{\sqrt{LC}} = \omega_0 \sqrt{\frac{X_c}{X_L}} \tag{6.24}$$

or

$$f_n = f_0 \sqrt{\gamma} \tag{6.25}$$

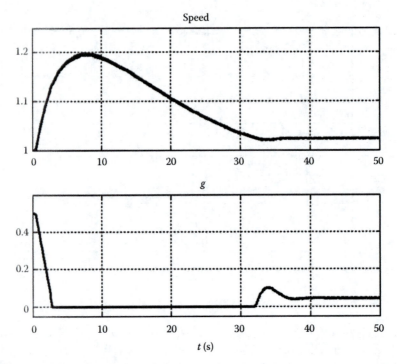

FIGURE 6.34   Processes in model6_8.

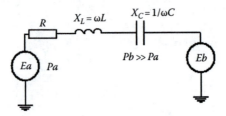

FIGURE 6.35   Series compensation scheme.

where γ is the level of the inductance compensation. The current with the slip frequency $f_c = f_0 - f_n$ induced in the rotor can be close to the mechanical natural frequencies. If, for example, the level of the inductance compensation is 55%, that is, $X_c = 0.55X_L$, then the slip frequency is 12.9 Hz.

The demonstrational model power_termal [4] is included in SimPowerSystems that is a benchmark for SSR demonstration developed by IEEE. Model6_9 is more simple one. Two-pole SG 600 MVA, 13.8 kV is connected to the source of the large power through the transformer 500 kV/13.8 kV and the transmission line. The line is modeled as the circuit with the lumped parameters that has the resistance of 18 Ω and the inductance of 0.53 mH. The series capacitor bank is set in the line that provides a compensation level of 50%. The block **Three-phase Fault** models ground short circuit through inductance of 0.1 mH and resistance of 0.001 Ω at $t = 0.03$ s, and at $t = 0.05$ s this fault is cleared. SG excitation voltage is constant. The turbine is modeled as two-mass (plus SG as the third mass). The scope **Turbine** gives the opportunity to observe the speed deviations of turbine masses and SG from the synchronous speed and the elastic torques between masses 1–2 and 2–3, and the scope **Grid** shows the voltages and currents of various elements in the main circuit.

Before the start of simulation, with the help of Powergui, the command *Load Flow and Machine Initialization/Update Load Flow* is fulfilled. For SG, in the field *Bus type*, the option *P&V generator* is chosen for active power equal to 0. It is made to start simulation from no-load conditions. At that point, the needed values of P_{ref} and V_f and the initial values for SG are set.

After simulation one can see that the elastic torque between SG and LP section can reach a value of 5 pu, which is dangerous for the rotor. The current peak in the short circuit is about 7 kA. If simulation is repeated without the capacitor bank, the said elastic torque is not more than 2 pu.

The rather complicated task of the multimachine system stabilization is considered in the demonstrational model power_PSS [4]. The diagram and the parameters of the main circuits are adopted from [1]. The system to go through simulation consists of two areas (Area1, Area2) connected to two parallel transmission lines. The lines have the same parameters: $R = 0.0529\ \Omega/\text{km}$, $L = 0.0014\ \text{H/km}$, $C = 8.77 \times 10^{-9}\ \text{F/km}$ and are modeled by the blocks **Distributed Parameter Line**. The distance between areas is 220 km. The block **Three-Phase Fault** is set in the middle of one line. The breakers **Brk1**, **Brk2** disconnect damaged section eight periods later of the ground short circuit. The system diagram is shown in Figure 6.36.

The power supply structure of each area is the same as shown in Figure 6.36b. It consists of two SG 900 MVA, 20 kV and step-up transformers 20 kV/230 kV. The SG constants of inertia in Area1 and Area2 are some different. The transformers have the impedance voltage $e_k = 15\%$. The first transformer is connected in parallel to the

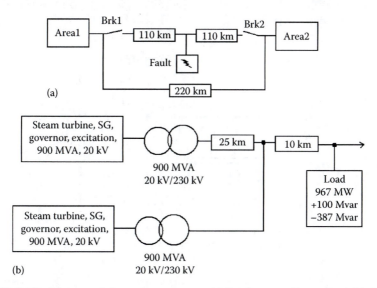

FIGURE 6.36 Diagram of the two-area system: (a) Total system diagram and (b) the diagram of the first area.

second one through the line 25 km in length and then both are connected to the load and to the inter-area line through the line 10 km in length. The short line parameters are the same as the parameters of the long lines, but they are modeled by the blocks **Three-Phase PI Section Line**.

The active load of Area1 is 967 MW and the active load of Area2 is 1767 MW, and each SG is adjusted for the power of 700 MW; therefore, a power of about 400 MW has to be transmitted from Area1 to Area2 that, for the line natural power of 140 MW (Table 2.1), stresses the line.

Each of the four subsystems **Turbine & Regulators** contains the blocks modeling the steam turbine and SG considered earlier. The turbine together with SG is modeled as a single-mass system. The subsystem contains the models of three PSS blocks that were also considered during the description of hydraulic turbine models. The output of any block can be connected to the input *vstab* of the SG excitation block (Figure 6.37). The first PSS model is realized according to the scheme in Figure 6.22. The deviation of the SG speed from the synchronous one ($\Delta\omega$) comes to its input. The second PSS model is made according to the scheme in Figure 6.21, and the same signal $\Delta\omega$ comes to its input. The third PSS model is made according to the scheme in Figure 6.21 also, but the signal, the so-called acceleration power (the difference between the mechanical and electrical powers), comes to its input. PSS choice is made based on the specification of the constant *PSS Model* the values 0–3: 0—without PSS, 1, 2, 3—the PSS model is selected, the same for all subsystems, according to the description provided previously. Simulation is fulfilled in the *Phasors* mode.

The aim of simulation is to estimate the influence of various PSS schemes on the system stability. Both small-signal stability (changing of the M1 reference voltage by 5% for 0.2 s) and large disturbance stability (the three-phase ground fault in the central point of one line) are investigated. It was found that PSS according to Figure 6.22 provides better damping.

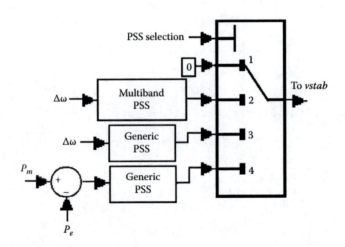

FIGURE 6.37   The diagram of PSS selection.

6.5 SIMULATION OF WIND GENERATION SYSTEMS (WG)

6.5.1 WG with an Induction Generator (IG)

At present, much attention is paid to development of the renewable energy sources, to which wind generators (WG) belong. WG of small power in areas with stable winds can be used isolated, but more often they are included in the power system, operating in parallel with other energy sources. The power of WG wind turbine is written as

$$P_m = K_1 V_w^3 C_p(\lambda, \beta) \qquad (6.26)$$

where
K_1 is the proportionality factor
V_w is the wind speed, m/s
C_p is the performance coefficient of the turbine that depends on the blade pitch angle β and on tip speed ratio λ that is equal to the ratio of the rotor blade tip speed to wind speed, $\lambda = R\omega_r/V_w$, R is the rotor radius, and ω_r is the rotational turbine speed, $2 < \lambda < 13$ usually

Dependence of C_p on λ is maximum when $\lambda = \lambda_m$, so, in order to receive the maximum power, ω_r is to change with a change in V_w. As for the angle β, it is set equal to the minimum feasible value and increases only during very large wind speed, in order to limit WG power.

There are various analytic relationships for $C_p(\lambda)$. In SimPowerSystems it is accepted as

$$C_p(\lambda) = c_1\left(\frac{c_2}{z} - c_3\beta - c_4\right)e^{-c_5/z} + c_6\lambda \qquad (6.27)$$

$$\frac{1}{z} = \frac{1}{\lambda + 0.08\beta} - \frac{0.035}{1+\beta^3} \qquad (6.28)$$

$c_1 = 0.5176$, $c_2 = 116$, $c_3 = 0.4$, $c_4 = 5$, $c_5 = 21$, $c_6 = 0.0068$.

The dependences C_p from λ for different β values calculated by (6.27) and (6.28) are shown in Figure 6.38.

It is supposed in the turbine model that P_m is the mechanical power in pu, C_p is the performance coefficient in pu in relation to C_{pmax} (see Figure 6.38), and V_w is the wind speed in pu in relation to the base speed. As such, some average value of the expected speed is accepted. Then K_1 is equal numerically to the power in pu when $C_p = 1$ and $V_w = 1$ and usually ≤ 1.

The wind turbine model in SimPowerSystems is specified by WG static characteristics. The picture of the model is given in Figure 6.39a, and the dialog box in Figure 6.40.

The model has three inputs: for the generator speed reference in pu, for the angle β (pitch angle) in degree, and the wind speed in m/s. The output is the generator torque in pu. It is specified consecutively in the dialog box: the turbine rated power

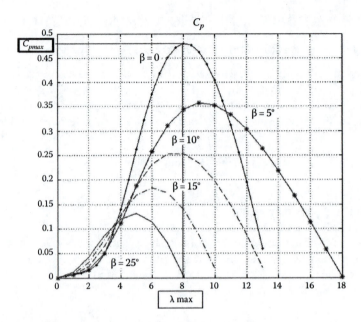

FIGURE 6.38 Dependencies C_p of λ with different values β.

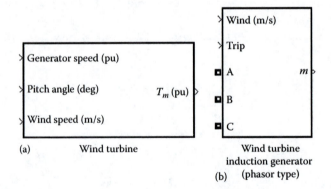

FIGURE 6.39 Pictures of WG models: (a) WG and (b) phasor model of WG with IG.

(W), the generator rated power (VA), the base wind speed (m/s), the maximum power for the base speed (in relation to the rated power, see Figure 6.41), and the turbine rotational speed corresponding to the power maximum for the base wind speed (in relation to the rated generator speed, Figure 6.41). If the option *Display wind turbine power characteristics* is marked, it has to be indicated in the field *Pitch angle beta to display wind turbine power Characteristics*, for what value β these characteristics have to be plotted. In Figure 6.41, such characteristics are given for the turbine with parameters shown in Figure 6.40 when $\beta = 0$.

In WG, the squirrel-cage induction generators (SCIG), the wound-rotor induction generators (WRIG), the synchronous generators with the excitation winding (SG), and SG with the permanent magnets (SGPM) are used. Two models of the

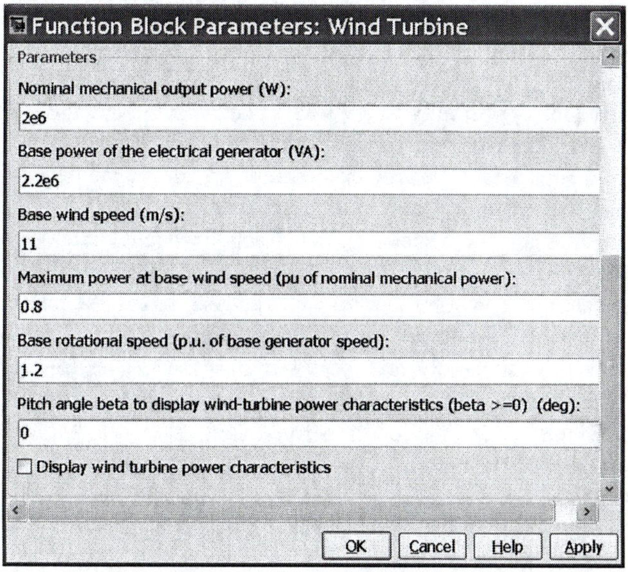

Function Block Parameters: Wind Turbine
Parameters
Nominal mechanical output power (W):
2e6
Base power of the electrical generator (VA):
2.2e6
Base wind speed (m/s):
11
Maximum power at base wind speed (pu of nominal mechanical power):
0.8
Base rotational speed (p.u. of base generator speed):
1.2
Pitch angle beta to display wind-turbine power characteristics (beta >=0) (deg):
0
Display wind turbine power characteristics
OK
Cancel
Help
Apply

FIGURE 6.40 WG dialog box.

Turbine power characteristics (pitch angle β = 0°)
Turbine output power (pu of nominal mechanical power)
Turbine speed (pu of nominal generator speed)
0
0.2
0.4
0.6
0.8
1
1.2
1.4
13 m/s
12 m/s
11 m/s
10 m/s
9 m/s
8 m/s
7 m/s
6 m/s
1.2 pu
Max. power at base wind speed (11 m/s)
and β = 0°
Tracking characteristic

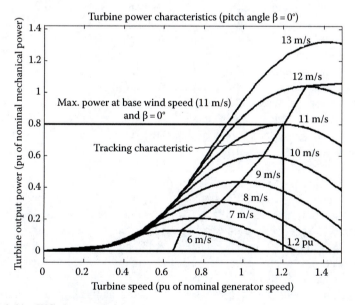

FIGURE 6.41 WG characteristics.

full WG consisting of the turbine, the IG, and the control system are included in SimPowerSystems. One of the models uses SCIG and another uses WRIG, the so-called Double-Fed Induction Generator. The black-white pictures of these models are the same (Figure 6.39b), but the drawing of WG appears in color. The points *A*, *B*, *C* are the WG power output terminals, the signal of the wind speed comes at

input *Wind*, and at the input Trip the logical signal can be sent that, being equal to 1, disconnects WG from the grid. The option "External mechanical torque" can be selected in the dialog box; at that point, the input *Wind* is replaced by the input T_m. This option gives the opportunity to use WG model with the other, developed by the User model of WG. Both models operate only in the Phasor mode.

The first model consists of the previously described turbine model and the squirrel-cage IM model, whose stator connects to the grid with the help of breaker. An only controller in the model is the β controller that is active when IG output power exceeds nominal value.

The output vector *m* contains eight internal signals: V_{abc} are the phase IG voltages (multiplexed signal), I_{abc} are the IG currents (multiplexed signal), P is the IG output active power, Q is the same for the reactive power, ω_r is the IG speed, T_m is the mechanical torque applied to IG, T_e is the IG electromechanical torque, and *Pitch_angle* β, degree (all the quantities, except the last one are in pu).

The dialog box has two pages: IG parameters and turbine parameters. On the first page, the same IG parameters are specified as was done for IM proper (Chapter 5); on the second page, in addition to the parameters shown in Figure 6.40, the gains of β controller, the maximum value of β, and the maximum speeds of β changing (deg/s) are defined.

This model is used in model6_10. The WG model has the power of 2 MW; IG voltage is 575 V. WG is connected to the power grid by means of the step-up transformer 575 V/25 kV and the transmission line 25 km in length. The grid 25 kV is modeled by the source with the short-circuit power of 40 MVA. The scope **Wind Turbine** fixes the active and reactive powers, the IG speed, and the angle β. The scope **B25 Bus** records the voltage vector amplitude, the active and reactive powers, and the current vector amplitude in the point where the measurer B25 is set.

It follows from the data in the dialog box that for a wind speed of 11 m/s and for rated IG speed, the unit power is equal to the rated one; hence, for a larger wind speed the turbine power has to be limited by action on β. The block **Wind** specifies the initial wind speed of 8 m/s with its increasing to 12 m/s at $t = 8$ s. The block **Trip** turns off the unit at $t = 18$ s.

Run simulation. The plots of the speed, the power, β, and the voltage on the bus **B25** are shown in Figure 6.42. It is seen that with the rise of wind speed, IG power increases too, and at $t = 11$ s it reaches the nominal value. The angle β begins to increase, reducing the turbine power and, in turn, limiting the IG output IG. Owing to absorbing of reactive power, the voltage on the bus **B25** remains lower than the nominal value. The input 3 of the scope **Wind Turbine** shows that the variations in IG speed are small: from 1.001 to 1.005.

The model of WG with WRIG is implemented according to the scheme in Figure 6.43. It does not distinguish from the Scherbius cascade scheme shown in Figure 5.71.

There are various ways of rectifier control for WG optimization. If the wind speed V_w is measured, the turbine rotational speed ω can be controlled such that it would be proportional to V_w with the factor of proportionality λ_m (in corresponding scale). It is possible also to control IG power P_e, depending on V_w; at that point, the reference power is equal to its maximum value in existing wind speed (Figure 6.41).

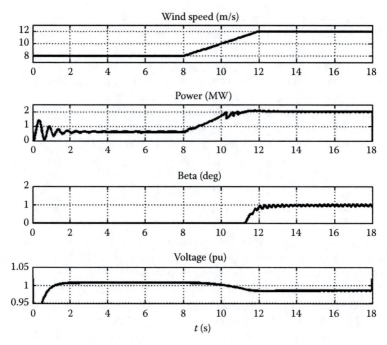

FIGURE 6.42 Transients in WG with the squirrel-cage IG.

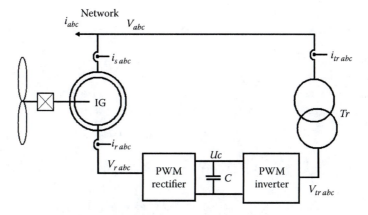

FIGURE 6.43 Structure of WG with the wounded rotor.

It is possible also to construct the control system without V_w measurement; at that point, the reference P_e^* is determined by the turbine rotational speed in conformity with the so-called tracking characteristic (Figure 6.41). The tracking characteristic is the locus of the maximum power of the turbine (maxima of the turbine power vs. turbine speed curve); the characteristic is limited by the rated power value, and it is zero in low speeds. Let, for instance, $\omega = \omega_b$, then $P_e^* = P_{max\_\omega b}$. Let V_w increase to 12 m/s, then the turbine power exceeds the IG power, and the turbine with IG increases its rotational speed. At that point, the IG power rises, following the tracking characteristic, until the IG power is getting equal to the turbine's maximum power when $V_w = 12$ m/s; analogous, but in the opposite direction, the process runs when $V_w < 11$ m/s. Just such a control method is accepted in the model under consideration.

Therefore, the rectifier control system controls the IG power and the reactive power or the stator voltage, whereas the inverter control system keeps the reference capacitor voltage. Besides, there is also the controller of angle β.

The demonstrational models *power_wind_dfig.mdl* and *power_wind_dfig_det.mdl* are included in SimPowerSystems that simulate WRIG in the Phasor mode and in the discrete mode, respectively. In the latter model, the optimal IG rotational speed is computed by using the tracking curve, based on the power measured at the IG output. The speed controller compares this speed with the actual one, and the difference, through the PI controller, determines the reference for the IG torque.

The analogous WG is simulated in model6_11, but with a different control system.

In this model, as before, the rectifier controls IG speed and the inverter controls the power transferred in the grid. DTC is used for IG speed control in model5_19. The parameters of IG and other elements are the same as in model5_19, but IG has three pole pairs. The grid is modeled by the source 150 MVA (short-circuit level), 120 kV, WG connects to it with the help of the transformer 120 kV/1200 V, power of 4 MVA. The inverter control system is the same as in model5_19, and the rectifier control system differs by the reference signal: It is not the rotor speed, but the IG torque. From (6.26) it follows that if $\lambda = \lambda_m$ and, consequently, $C_p = C_{p\,max}$ and $P_m = P_{m\_max}$, then

$$P_{m\_max} = K_1 \left(\frac{\omega}{\omega_b} \right)^3 \tag{6.29}$$

and

$$T_{m\_max} = \frac{K_1 \omega^2}{\omega_b^3} \tag{6.30}$$

With the chosen turbine parameters, $K_1 = 0.7$, $\omega_b = 1$. Taking into account the mechanical losses as $T_{los} = D + F\omega$, the demanded IG torque is

$$T_e^* = 0.7\eta_r \omega^2 - D - F\omega \tag{6.31}$$

where η_r is the turbine gear efficiency.

This quantity is computed in the block **Rect_Contr** as the torque reference for DTC. In other respects, this block is the same as in model5_19. If the turbine mechanical torque exceeds T_e^* with IG running speed (for instance, owing to an increase in the wind speed), IG rotational speed will increase to achieve a balance, and this way, the IG rotational speed will correspond to the optimal power transfer. The torque reference maximum is limited by 1.5. When IM speed exceeds 1.3, the angle β controller (**beta_contr**) comes into effect, which increases the pitch angle and, thus, limits the power received by the turbine from the wind.

The scope **Generator** fixes the speed, the torque, the stator current, and the rotor flux linkage of IG. The rest of the scopes are gathered in the subsystem **Measurements**. The four-axes scope **Supply** records the IG output voltage and the currents of the grid, the primary transformer winding, and the IG stator.

The scope **Power1** shows the active powers in the point of common coupling (PCC): grid—yellow, IG—magenta (red-purple), transformer—cyan (blue-green); the scope **Power** fixes the active and reactive powers at the IG output. The scopes **Tm, Uc, Pitch** record the turbine mechanical torque, the capacitor voltage, and the angle β, respectively.

The following schedule is simulated: turning on and operation when $V_w = 5$ m/s, increase in V_w to 10 m/s at $t = 2$ s, increase in V_w to 14.5 m/s at $t = 9$ s, and decrease in V_w down to 7 m/s at $t = 13$ s. Run simulation with the switch **Sw** in the upper position. The response of IG speed is shown in Figure 6.44a.

At first, the section where $V_w = 10$ m/s (for instance, at $t = 8.9$ s) is considered. It follows from the foregoing that the reference IG speed is equal to 1 in this case; that is, the rotor rotates with the synchronous speed. Looking at the last axis of the scope **Generator**, it can be seen that, the rotor flux frequency is zero practically. The scope **Power** shows that the power transferred in the grid is 1.08 MW, that is, 0.7 of the rated power minus losses. The reactive power is equal to zero in all the conditions because the inverter controls the grid current when $I_q = 0$. One can see by the scope **Supply** that the grid voltage and current are in antiphase (the active power transfers in the grid and the reactive power is zero), the transformer current is shifted by 90°; that is, the reactive power demanded by IG is provided by the internal processes in the unit. The plots of the active powers of the

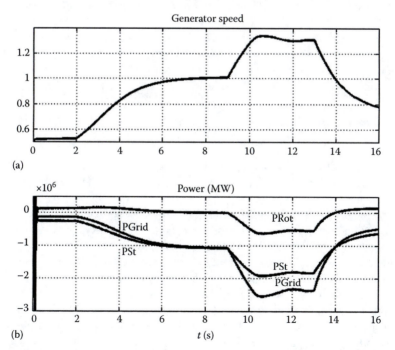

(a)

(b)

t (s)

FIGURE 6.44 Transients in WG with the wounded-rotor IG. (a) Generator speed. (b) Stator, rotor, and grid power.

grid (PGrid), the IG stator (PSt), and the IG rotor (PRot) are given in the Figure 6.44b. In the scenario under consideration, PGrid = PSt, PRot = 0.

With rising wind speed, IG tries to accelerate to the speed 1.5, but the pitch-angle controller prevents it: It keeps the limiting value of 1.3 with some overshooting. The scope **Pitch** shows that β reaches value of 5.5°; by the scope **Tm** one can see that the pitch-angle controller limits the turbine's mechanical torque.

The section where $V_w = 14.5$ m/s (for instance, at $t = 12.5$ s) is being considered. The rotational speed is 1.3. It is seen by the scope **Generator** that the rotor flux frequency is 15 Hz. By the scope **Power,** the power transferred in the grid is 2.35 MW. By the scope **Supply**, the transformer current shift is less than 90°; that is, the rotor power adds to the stator power. It is seen from Figure 6.44b that the power transferred in the grid results from the stator power of 1.79 MW and the rotor power of 0.56 MW; that is, approximately (because of losses) $P\_Rot = sP\_St$, where s is the slip ratio.

Consider the section corresponding to $V_w = 7$ m/s (for instance, at $t = 16$ s). The reference IG speed is 0.75. One can see by the scope **Power** that the power transferred in the grid is 0.43 MW. It is seen by the scope **Supply** that the shift of the transformer current is more than 90°; that is, the rotor power is subtracted from the stator power. Figure 6.44 shows that the power transferred in the grid is equal to the stator power of 0.57 MW minus the rotor power of 0.14 MW. It is worth noting that the DC link voltage U_c is equal to the reference value of 1500 V in all conditions.

If the simulation is repeated with the switch **Sw** in the lower position, the speed and the active power responses will not change, but it is seen by the scope **Power** that the reactive power is consumed from the grid. The scope **Supply** shows that the phase displacement appears between the voltage and the current of the grid, but it is excluded between the voltage and the transformer (rotor) current. At that point, the grid condition is seen getting worse, but the current stress in the rotor circuit is less; specifically, at $t = 10.9$ s, the first harmonic amplitude of the transformer current is 436 A, whereas it was equal to 894 A during the previous simulation running.

6.5.2 WG with a Synchronous Generator with Permanent Magnets (SGPM)

The demonstrational model power_wind_type_4_det.mdl is included in SimPowerSystems in which WG with wound field SG operating parallel with the grid is simulated.

At present, more often SGPM is used in WG. Its advantage is more reliable SG construction, without windings and slip rings on the rotor, but the needed converters have to be used for the full WG power, not for a part of it, as it takes place in WG with IG.

The main circuits of the unit are shown in Figure 6.45. The voltage induced in the SG stator is rectified by the rectifier with force-commutated semiconductor switching devices, for instance, with IGBT. The rectifier's purpose is to control SG speed. With the use of wind speed, optimal SG rotational speed is calculated, which is the reference for the speed controller; the signal of the actual speed ω_{sg} is generated too. According to the discussions in Chapter 5, the stator current component $I_{gd} = 0$ and the component I_{gq} is determined by the speed controller output where I_{gd}, I_{gq} are aligned alongside the rotor position θ_{sg}.

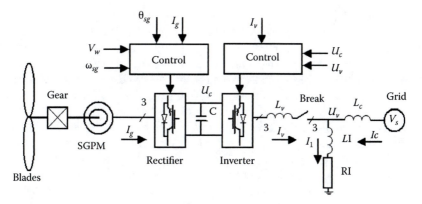

FIGURE 6.45   Structure of WG with SGPM that operates in parallel to the power grid.

The inverter has the same circuits as the rectifier, with the capacitor C set between them. The inverter's main purpose is to transfer the maximum power to the grid. An indicator of the relationship between the produced power and the power transferred in the grid is a constancy of the capacitor voltage. This voltage stays equal to the reference value by affecting the current component I_{vd}, with $I_{vq} = 0$. Here I_{vd}, I_{vq} are aligned in the direction of the load voltage space vector $\mathbf{U}_v$. Instead of names "Rectifier," "Inverter," others names are used often, for instance, "Generator converter," and "Grid converter."

PMSG in model6_12 has parameters: P_n = 280 kW, ω_n = 22 rad/s, $2p$ = 36, Ψ_r = 1.052 Wb, R_s = 0.04 Ω, $L_d = L_q = L_s$ = 1.5 mH, J = 100 kg m^2 (with turbine moment of inertia). The capacitor C capacitance is 7.5 mF. In parallel to the capacitor the limiting circuit is set, in order to protect DC link from overvoltages that can appear during wind gust. This circuit is analogous to the dynamic braking circuit in model5_8. The inverter is connected to the grid with the help of the reactor with the inductance of 0.8 mH. The grid 400 V, 50 Hz is modeled by the block **Three-phase Source** with internal R-L impedance. The load has parameters 400 V, 50 Hz, P = 300 kW, QL = 300 kvar. The high-pass filter with a tuning frequency of 350 Hz is used to improve the current form (not shown in Figure 6.45).

The subsystem **Turbine** contains the model of the wind turbine described earlier with the rated power of 250 kW. When SG rotates with the rated speed, the maximum power that is equal to the rated power is achieved in the wind speed of 11 m/s; this value is accepted as the base value. Since the turbine model operates in pu, and the SG model in SI, the blocks **Gain, Product, Gain1, and Divide** are included in the model for the direct and inverse conversion. The speed reference is limited by the value of 30 rad/s. Besides, the pitch angle controller is added that begins to act when the turbine torque exceeds 1.3 of the rated one. **Rate Limiter** limits the possible rate of wind speed change by the value of 4 m/s^2 and takes into account also that, because of the system flexibility, the turbine shaft torque does not change instantly.

The subsystem **Gen_Reg** controls the rectifier. It contains PI speed controller, whose output is the reference for I_{gq} (the reference for I_{gd} is zero), the converter of the reference quantities I^*_{gq}, I^*_{gd} in the three-phase signal of the reference quantities I^*_a,

$I_b^*, I_c^*,$ PI controllers of the phase currents, and PWM with the modulation frequency of 3240 Hz. These devices have been described already. Pay attention that the rate of decrease of the generator speed (when the wind speed decreases) is set in the block **Rate Limiter** much lesser than the rate that exists during an increase in the generator speed, so that the processes in the system could settle with the wind speed decreasing; otherwise, the short-time alterations of the generator rotational direction can also be seen; at that point, the oscillations appear in the unit.

The subsystem **Inv_Reg** controls the inverter. It contains PI capacitor voltage controller, whose output is the reference for I_{vd} (the reference for I_{vq} is zero). By measuring the two phase-to-phase voltages, the position of the grid voltage space vector is computed, with which the inverter current is aligned. The rest of the blocks are the same as in the previous subsystem. Pay attention on the signs of the outputs of the phase current controllers **Current_Reg**.

In order to observe operation of this rather complex system, a number of scopes are set in the model that are gathered in the subsystem **Measurements**. The scope **Generator** shows the currents, the speed, and the torque of SG; the scope **U_I** records the load voltage (output of the unit) and the currents of WG, the grid, and the load. The scopes **Power, Uc, Pitch** record the WG power, the capacitor voltage, and the pitch angle, respectively.

The following conditions are simulated. The system begins to function with the wind speed of 11 m/s. WG is speeding up to the speed of 22 rad/s under a turbine torque of 11.3 kN m (Figure 6.46). The capacitor voltage is set to 1200 V. The active WG power is 237 kW, and the reactive power is 8 kvar. The load power is provided by WG, whereas the grid active power is about null. The inverter current is in-phase

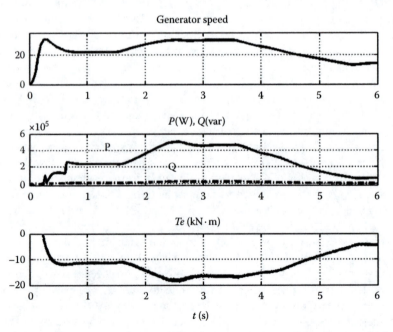

FIGURE 6.46 Transients in WG with SGPM that operates in parallel to the power grid.

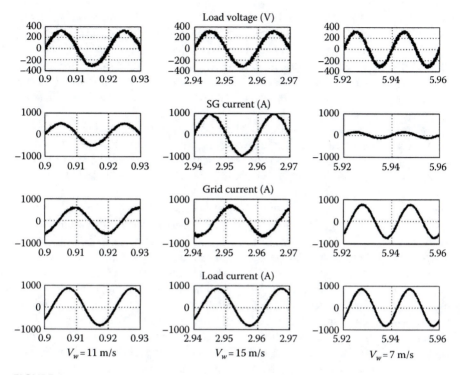

FIGURE 6.47   Load voltage and WG grid and load currents.

with the voltage; the load current lags 45°, which is possible when the active and reactive load powers are of equal values. The grid current lags 90°, so that the grid delivers only the reactive power (Figure 6.47).

At $t = 1.5$ s the wind speed rises to 15 m/s. The SG speed increases to 30 rad/s, WG power grows to 462 kW, and the turbine torque tries to exceed the specified maximum of 1.3, so that the pitch angle controller starts operating, which is set as $\beta = 3.2°$. The SG current increases also and supplies the grid partly (the angle between the voltage and the current of the grid is more than 90°).

At $t = 3.5$ s the wind speed decreases to 7 m/s, the WG speed and power are reduced to 14 rad/s and 63 kW, respectively; the current injected by WG decreases drastically, and the grid current almost completely copies the load current, operating with the phase angle of 45° and sending to the load both active and reactive energies (Figure 6.47).

6.5.3   WG with SGPM and Diesel-Generator

A more complicated model, model6_13, differs from the previous one in terms of using, instead of a grid, a source of the comparable power, specifically, the unit diesel SG. SGPM of WG has the following parameters: $P_n = 28$ kW, $\omega_n = 22$ rad/s, $2p = 36$, $\Psi_r = 0.526$ Wb, $R_s = 0.17\ \Omega$, $L_d = L_q = L_s = 4.5$ mH, $J = 15$ kg m² (including turbine). The block **Synchronous Machine pu Fundamental**, 42.5 kVA, 230 V, 50 Hz, 1500 rpm is used for the SG model. For speeding up simulation, the command *Powergui/Load Flow and Machine Initialization* is carried out before start.

The subsystem **Wind_Syst** models WG together with its electrical part that were described during the discussion on model6_12. There are the following differences: The output filter is excluded; the DC voltage is 500 V and the capacitance of the capacitor is 3 mF. WG connects to the load of 30 kW, 30 kvar with the aid of the reactor of 4 mH, 0.02 Ω.

Unlike the previous model, in which the load voltage and its frequency are equal to the parameters of the grid, and whose power is much more than the WG power, in the considered model situation is more intricate and depends on the relationship between the WG power with the present wind speed and the load power. While WG power is less than the load power, SG remains in work, providing the missing power, the voltage, and the frequency with the help of SG controllers. When the wind speed rises, it can be seen that WG power exceeds the load power; at that point SG offloads, and WG can, in principle, operate in isolation. The problem of load voltage and frequency control occurs at that point. For the voltage control, the reactive power source can be used, and in our case SG, which remains connected to the load. Excess power leads to a rise in frequency; for its limitation, the following methods can be used: a load increase either by turning on a ballast (Bal) or by connecting to useful loads, for instance, pumps; a connection parallel to the load of the energy storage elements, for instance, of the batteries or the flywheels; and increase in the pitch angle. The first method is used in the model under consideration because the WG of a small power does not usually have a device for the automatic pitch control. Thus, for providing the spinning reserve, the diesel remains in work with a small power in the rated rotational speed.

The subsystem **Bal** uses the controlled current sources that are connected in the circuits of the controlled conductance. The source current is proportional to the applied voltage and to the conductance factor. The maximum conductance is equal to 0.52 S; it means the maximum ballast power is $220^2 \times 0.52 = 25.2$ kW. The main circuit diagram is shown in Figure 6.48.

It is worth noting that there is another ballast model in the demonstrational model *power_windgen.mdl* [4]: This model contains eight groups of the three-phase resistors, whose powers during nominal voltage are equal to $P_{min} 2^{i-1}$, $i = 1..8$. With the help of the static switches that are controlled by eight-digit binary code the demanded ballast power is selected.

The subsystem **Diesel and SG Control** that is a changed model from *power_machines.mdl* is used for SG control. It contains two controllers: the rotational speed (of the SG output frequency) that affects the diesel power (**Diesel**) and the SG output voltage that acts on the excitation voltage (**Excitation**).

The diagram of the speed controller is shown in Figure 6.49. The diesel control circuits contain a number of links whose total transfer function has a rather big order; this transfer function is replaced by the transfer function of the second order that approximates well enough the original transfer function. It is taken $T_0 = 0.02$ s, $\zeta = 0.3$, $T_3 = 5$ s. A time-delay element models the delay, with which the diesel generate the output torque. The block output is the product of the torque and the speed. The voltage control is standard (Figure 6.16).

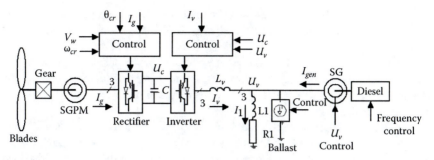

FIGURE 6.48 Structure of WG with SGPM that operates in parallel with the diesel generator.

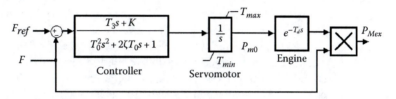

FIGURE 6.49 Speed controller of the diesel generator.

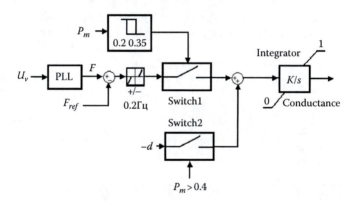

FIGURE 6.50 Controller of the ballast load.

The diagram of the Bal-controller is depicted in Figure 6.50. In fact, it is a controller of the load voltage frequency. The frequency is measured with the help of PLL. The measured frequency is compared with the reference one, and the difference comes to the integrator through the dead-zone block (is set of ±0.2 Hz); the integrator output is limited by 0–1. This output determines the ballast conductance that increases with an increase in frequency, at which point the ballast power grows.

If the block **Ballast_Power** is open, one can see that it consists of three controlled current sources; they draw the currents from the point of common coupling. The currents are in-phase with the phase voltages and their values are defined by the conductance factor that changes from 0 to 0.52. The filter and the inductance are intended for decoupling of the algebraic loop between the voltage and the current.

In the diagram in Figure 6.50 the switches 1, 2 are shown to operate in the following way: **Switch1** activates control of the ballast load, when the SG power decreases down $0.2P_{mnom}$; during its increase to $0.35P_{mnom}$, the value of the ballast that is formed at this moment is preserved because the integrator output switches off. With further increase in power to $0.4P_{mnom}$, Switch2 closes, the negative signal is applied to the integrator input, and the integrator output decreases smoothly to zero, setting the ballast power equal to zero too.

A number of scopes are set to observe system functioning. The scope **PMSM** fixes the current, the speed, and the torque of WG, and the scope **Scope_Uc**—the capacitor voltage. The five-axes scope **Currents** shows on the first axis the load voltage, and on the others—the currents in PCC in the following order: WG, SG, load, and Bal. The scope **SM** records the input power of SG, its excitation voltage, the amplitude of the output voltage, and the SG speed. The scope **Power** shows the powers of the previously mentioned four circuits in the same order (WG—yellow, SG—magenta [red-purple], load—cyan, Bal—red). The power is calculated as a product of the instantaneous voltage and current values of one phase averaged for 20 ms and multiplied by 3. The following schedule is simulated: System turns on with zero wind speed ($V_w = 0$, the mechanical system braked), WG starts with $V_w = 5$ m/s at $t = 2$ s, V_w increases to 13 m/s at $t = 5$ s, and V_w decreases to 5 m/s at $t = 9$ s. Some of the results are shown in Figures 6.51 and 6.52. It follows from studying these figures and the scope data that the load voltage during all conditions remains equal to 226 V (rms of the phase-to-phase voltage) and the frequency is limited by the permissible boundaries (the short time deviation in dynamics do not exceed 3%). The deviation

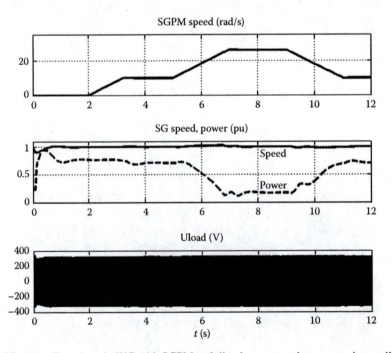

FIGURE 6.51   Transients in WG with SGPM and diesel generator that operate in parallel.

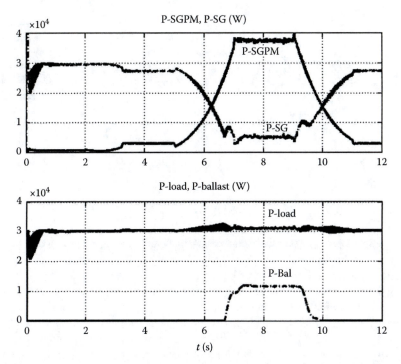

FIGURE 6.52 Power changes in WG with SGPM and diesel generator that operate in parallel.

of the capacitance voltage from the value of 500 V does not exceed 1%. It is seen in Figure 6.52 that the load power is constant. With rising of wind speed and the WG rotational speed, the SG and diesel powers decrease. At $t \approx 6.8$ s, the mechanical power decreases to $0.17 \times 42.5 = 7.2$ kW, and the ballast power begins to increase; the sum of the latter and the load power is (at $t = 8$ s) 11.6 kW + 31.1 kW = 42.7 kW, and the sum of the WG and SG powers is 37.3 kW + 5.4 kW = 42.7 kW too. When V_w decreases, the power of the Ballast is zero, WG power decreases to 3 kW, the load power is provided by SG, and the output power is 27 kW.

6.5.4 SIMULATION OF A STAND-ALONE WG

In the previous models, WG operation in parallel to a grid of large or comparative power was considered. The voltage in PCC and the frequency are defined by the grid. The main demand to WG in this condition is generation of maximum power.

The diagram of WG with SGPM is shown in Figure 6.53. The SG voltage is rectified by the diode rectifier. The chopper boost controller is used in order to have a constant DC voltage U_{ci} applied to the inverter input (to the capacitor C_i). The chopper consists of the IGBT K, the inductor L and the diode D. The three-phase sinusoidal signal that is amplitude- modulated by the output of the voltage controller comes to PWM of the inverter.

The maximum power that WG can produce depends on the present wind speed V_w. Therefore, the load has to be divided in the sections that are turned on or

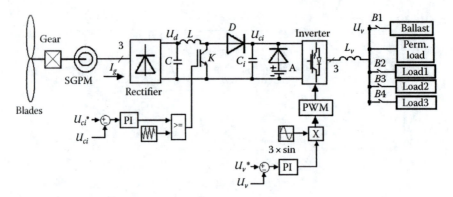

FIGURE 6.53 Diagram of the stand-alone WG.

off, depending on V_w. The nondisconnected section (or the permanent load), whose supply has to be fulfilled in any case (the systems of the emergency lighting and communication and the systems of life-support and survival), can be distinguished among these sections. The battery A is set for keeping up this section's supply, whose voltage is somewhat lesser than the reference value U_{ci}^*; this battery powers the inverter when U_{ci} is decreasing. The battery capacity has to be sufficient to supply the nondisconnected load during the time that is necessary for connection of unit of the emergency power supply. A boosting charge can be realized from inverter DC link during normal wind speed. The battery holds the output voltage when the wind speed decreases quickly—till the corresponding section (or sections) is turned off.

WG operates usually on the right, drooping part of its characteristic (Figure 6.41). The rotational speed corresponds to the load power (plus losses) under the present wind speed. With the load increasing, the rotational speed decreases and the WG power increases, till the new equilibrium position is reached. Such a process takes place while the WG power is less than the maximum value in the present wind speed. Hence, in the system under consideration, there is no need to control WG, in order to reach the maximum power, as this condition will be obtained automatically when required.

During a load decrease or wind speed increase, the WG speed increases and can reach an impermissible value. To prevent this from happening, either pitch control or ballast control can be used.

In order to augment an interval between the wind speed when the definite section is turned on and the wind speed when this section is turned off, it is possible under lessening V_w, to decrease the load voltage that still remain in the boundary of the existing standards. Result of this measure depends on the level of the load power dependence on the applied voltage.

Since the anemometers are not set always, it is desirable to have a version of WG without them. It follows from WG characteristics that, for the given rotational speed, the definite power can be received only for the definite wind speed. Thus, to estimate the wind speed, the inverse problem has to be solved: to compute the wind speed, knowing the rotational speed and the power, and afterwards, to determine the maximum admissible load. Since the number of available load steps is limited, one can dispense with solving of the inverse problem, namely to calculate several dependences of the power from the rotational speed during definite wind speeds and to compare the power's present value with the calculation results, determining in turn the power range that, at the present moment, with the existing wind speed, can be obtained.

Model6_14 simulates the described system with the WG rated power of 60 kW and with the rated rotational speed of 22 rad/s. The chopper switching frequency is 1 kHz.

The capacitor voltage control is fulfilled in the subsystem **Uc_contr**: The reference (accepted 500 V) and the actual values of U_c are compared at the input of PI controller, whose output is compared with the saw-tooth signal; as a result of comparison, the switch gate pulses are generated. The circuits for the switch current protection are provided: When the current through the switch exceeds 350 A, the switch is blocked; the following pulse generation is possible after current dropping to 250 A with delay of 0.1 s.

The inverter modulating signal has a frequency of 50 Hz. The third harmonic is added. The amplitude of the modulating signal is defined by the output of the load voltage controller (subsystem **Control**). The reference corresponds to the rms phase-to-phase voltage of 230 V. When WG is turned off, the battery Ak with the rated voltage of 400 V supplies the nondisconnected load. LC filter is set at the inverter output.

The load (the subsystem **Load**) consists of the active-inductive sections having the powers: the nondisconnected section—10 kW and the connected consecutive sections, 13, 19, and 14 kW. Two options for their connection are provided: by signal of the wind speed with the help of the blocks **RELAY** having set points (turn on/turn off, m/s): 8.4/8.2, 10/9.7, 11/10.8 and by the indirect estimation. The option is selected by the switch **Mode_Switch**: In the upper position the first option is chosen and in the lower position the second option is chosen.

The signals for load section selection are formed in the subsystem **Load_contr** that contains the subsystem **Compute**, in which, by the same formula that is used in the turbine model, the power values P1, P2, P3 are computed for the wind speeds of 8.4, 10, and 11 m/s and in the present turbine rotational speed. These values are compared with the actual power P. If P < P1, only the permanent 10 kW load remains; if P1 < P < P2, the 13 kW load is connected in addition; if P2 < P < P3, the 19 kW load is added; and if P3 < P, the 14 kW load is added. The power value is calculated as the product of the SG torque T_e (supposing that $T_m \approx T_e$) by the rotational speed ω.

It has been mentioned already that the wind speed, in which WG works with the certain load, can be lessened, if the load voltage is decreased during a decrease in wind speed. The corresponding circuits are in the subsystem **V-comp**. When a certain load is connected, the load voltage is set equal to 230/133 V; during a decrease in wind speed, the voltage is decreasing also by 15%, to 195/113 V.

For the WG rotational speed limitation, either the pitch control or the ballast control can be used (the second option is used usually for the small WG). For utilizing the first option, **Switch** in the subsystem **Turbine** is to be set in the right position, and for the second option, **Breaker** in the subsystem **Ballast** is to be closed. The diagram of the last subsystem is considered in model 6_13.

The main scopes that fix the WG quantities are gathered in the subsystem **Indication**. The shown signals are pointed on the tags **From** and in the texts above the axes.

The processes during a gradual decrease in wind speed V_w from 11.1 m/s down to 5 m/s are shown in Figures 6.54 and 6.55. It is seen that before load connection (at $t < 1$ s),

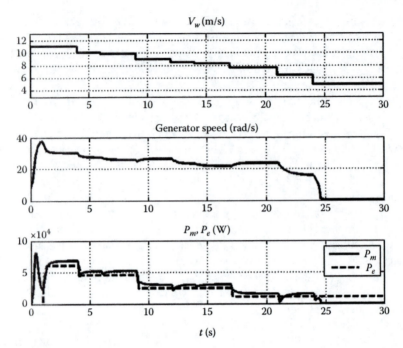

FIGURE 6.54 Processes in the stand-alone WG during a decrease in the wind speed measured.

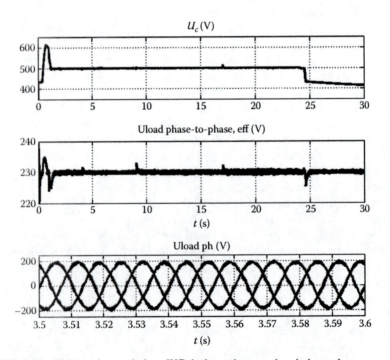

FIGURE 6.55 Voltages in stand-alone WG during a decrease in wind speed.

the ballast load controller limits the ω reference value by 35 rad/s. Furthermore, with V_w lessening, the WG power P_m and the load power P_e decrease in four steps: P_m/P_e (kW) = 75/67, 58/50, 34/28, 18/12 (the load power is more than values given previously, because the latter are specified for the voltage of 220V; in fact, the voltage of 230V is held). At $t = 24$s V_w decreases to 5m/s, WG stops production of electric energy, and the output power is provided by the battery.

The DC link voltage remains at the level of 500V rather precisely till $V_w = 5$m/s, the rms of the phase-to-phase voltage is equal to 230V in any condition, and the shape of the phase voltage U_{ph} is nearly sinusoidal.

Run simulation with the switch **Mode_switch** in the subsystem **Load** in the lower position; this way the wind speed estimation method, instead of the V_w measurement, is used. When the simulation is complete, plots are formed as shown in Figure 6.56; they do not differ noticeably from those depicted in Figure 6.54.

In order to check the method for section operation boundary broadening during a decrease in V_w, by a decrease in voltage, one can do the following: in the subsystem **Load**, in the blocks **Relay, Relay2, Relay3**, change the switch-off points that are equal to 8.2, 9.7, 10.8m/s for 7.8, 8.9, and 10m/s, respectively. Run simulation. It is seen that at $t > 4.7$s normal WG functioning stops (U_c defined by the battery). Repeat simulation with **Switch1** in the subsystem **Control** in the upper position, by turning on the circuit of the voltage control. The plots obtained are shown in Figures 6.57 and 6.58. It is seen that WG operates in a regular fashion, and the load voltage deviations are kept in the permissible boundaries.

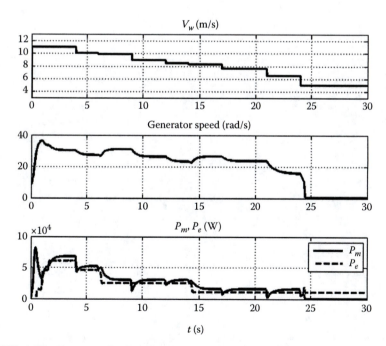

FIGURE 6.56 Processes in the stand-alone WG during a decrease in the wind speed estimated.

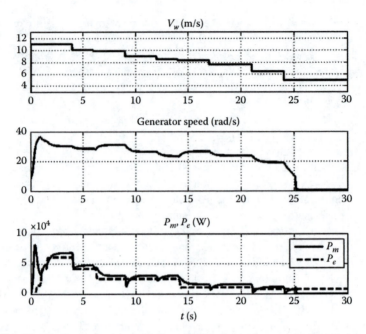

FIGURE 6.57 Processes in the stand-alone WG during a decrease in the wind speed and changes in the output voltage caused by the variations in the wind speed (speeds, powers).

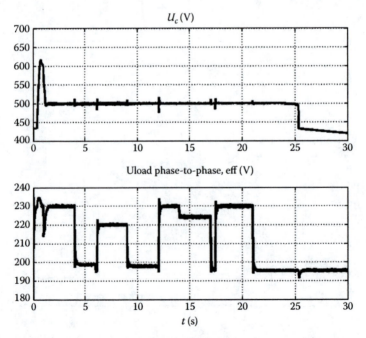

FIGURE 6.58 Processes in the stand-alone WG during a decrease in the wind speed and changes in the output voltage as a result of changes in wind speed (voltages).

6.6 SIMULATION OF THE UNIT: DIESEL—SQUIRREL-CAGE IG

The units with a diesel as a prime mover (D) and SG are the most widespread self-contained power sources. At that point, the output voltage frequency is defined by the speed of the diesel, and the voltage control is fulfilled by action on the SG excitation voltage. The disadvantage of such a unit is the complexity and insufficient reliability of SG. Lately, in connection with development of power electronics, the alternative units have emerged that use the latest power electronics—with SGPM and with the squirrel-cage IG (SCIG). The voltage source inverters (VSI) are used in these units usually. When SGPM is utilized, the circuits similar to that shown in Figure 6.53 can be used. When SCIG is used, the following problems emerge: the initial IG excitation; frequency control by action on the D torque, keeping in mind that for IG the rotational speed and the stator voltage frequency are different, owing to a slip; and the output voltage control. These problems can be solved with the use of diagram shown in Figure 6.59 [5].

SCIG is driven by the diesel D fitted out with the speed sensor ω_d that is useful for the unit protection, apart from everything else. The diesel has the speed controller; in order to obtain the requested frequency at the IG output, the feedback signal is adjusted by the integral difference between the reference and the actual frequencies. The latter signal is formed by PLL. The capacitors C_{st} form the circuit for the initial self-excitation.

The voltage control is carried out by VSI connected parallel to IG output. Control of the active and reactive components of the current I_s is fulfilled by the control system, with the help of the fast-acting three-phase current controller. Active current control is realized, depending on the VSI capacitor voltage U_c, and the reactive current control depends on the output IG voltage.

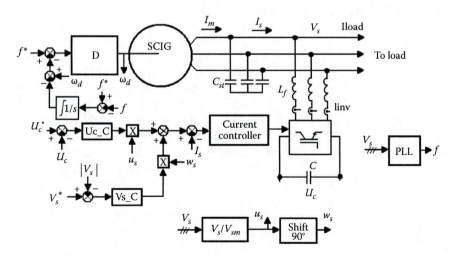

FIGURE 6.59 Diesel-IG.

The aforesaid is realized as follows. The three-phase system of the quantities u_s, whose amplitude is equal to 1, and which are in-phase with V_s, are computed in the block **Vs/Vsm** as

$$V_m = \left[\left(\frac{2}{3}\right)\left(V_{sa}^2 + V_{sb}^2 + V_{sc}^2\right)\right]^{0.5} \tag{6.32}$$

$$u_{sa,b,c} = \frac{v_{sa,b,c}}{V_m} \tag{6.33}$$

The three-phase system w_s leads u_s by $\pi/2$ that is reached by calculations:

$$w_a = \frac{(u_{sc} - u_{sb})}{\sqrt{3}}, \quad w_b = -\frac{w_a}{2} + \sqrt{3}\frac{u_a}{2}, \quad w_c = -\frac{w_a}{2} - \sqrt{3}\frac{u_a}{2} \tag{6.34}$$

The signals u_s are multiplied by the output signal of the capacitor voltage U_c controller (of PI type), forming the active current reference for I_s. In a similar way, the signals w_s are multiplied by the output signal of the output voltage V_s controller (also of PI type), forming the reactive current reference for I_s. The products are summed, forming the total current I_s reference that is reproduced with sufficient accuracy by the previously mentioned current controller affecting VSI gate pulses.

Model6_15 simulates such a system. SCIG, 160 kVA, 380 V is driven by diesel. The diesel output variable is a torque, not power as in Model6_13. The three-phase capacitor bank delta-connected with the capacitance of 1.3 mF per phase is set for IG self-excitation. VSI is connected to IG with the help of the reactor with an inductance of 2 mH. The capacitance of the VSI capacitor is 10 mF. VSI control system is fulfilled according to the diagram shown in Figure 6.59; the VSI capacitor voltage is set to 800 V, and the rms of the phase-to-phase voltage is 380 V. The hysteresis controller of current I_s is used. Various types of loads can be connected with the help of the breakers **Br, Br1, Br2**. With the help of **Switch** in the main model diagram and similar ones in the subsystem **Control**, the gate pulse generation and turning-on of the controllers take place after the VSI capacitor is charged. **Scope1** records the frequency, the load phase voltages, the VSI, and the load currents; **Scope** shows the IG speed, the rms of the phase-to-phase voltage, the VSI capacitor voltage, and the IG stator current. It is necessary to note that this model does not simulate the self-excitation process correctly because the saturation characteristic does not allow for specifying the residual flux; therefore, the initial value is specified not for the flux but for the stator current, with use of which the initial flux is computed; that does not conform to the process physics. It is accepted during simulation that the initial value of the current in one of the phases is equal to 1% of the rated IG current.

Set the switch-on times for the breakers as Br—7 s, Br1—4 s, Br2—60 s; the single-phase loads **Load1** are equal to 2 Ω each one. Thus, the operation with the

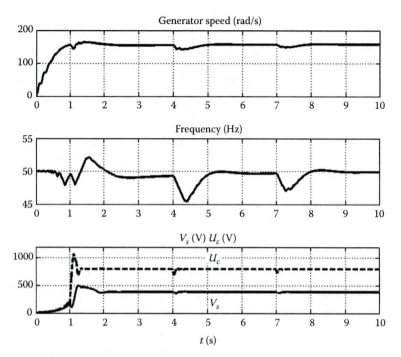

FIGURE 6.60 Processes in the unit diesel-IG.

balanced three-phase load is simulated and takes place abruptly. In Figure 6.60 IG speed ω_m, the voltage frequency that is measured by PLL, rms value of the phase-to-phase voltage, and the capacitor voltage are shown. It is seen that the system variables are restored in a rather short period after load application. The scopes **Scope** and **Scope1** show that the load voltage and the stator current are very close to the sinusoids; at $t = 9$ s the current THD is not more than 0.3%.

If the simulation is carried out with different phase resistances of **Load1**, for instance, 2 Ω, 4 Ω, 1.33 Ω, one can see that at $t = 9$ s the load currents are essentially different (amplitudes are 253.5 A, 199.4 A, 274.2 A), but the load voltage and IG currents remain balanced (the phase voltage amplitudes are 310 V, 309.6 V, 309.5 V; the phase current amplitudes are 246.7 A, 245.2 A, 244,8 A); that is, the system considered carries out symmetrization for an unbalanced load.

Now the balanced nonlinear load is simulated. For that purpose, the switch-on times are set: **Br1** = 40 s and **Br2** = 4 s, and the switching of all phases are marked in **Br2**. When the simulation is complete, we see that with the load capacitance of 1 μF, the load current THD is 17.4% at $t = 9$ s, whereas the load voltage THD is 1.3% and the IG current THD is 2.8%; that is, the effective filtration of the harmonics is achieved. If the mark about switching-on of one phase in **Br2** (it means that nonlinear non-balanced load is simulated) is removed and the simulation repeated using the load capacitance of 500 μF, one can find that the load phase current THD are 17.2%, 14.1%, 1%; THD of the phase voltages are 2.1%, 1%, 1.4%, and THD of the IG phase currents are 11.9%, 4.4%, 9.2% and the main "contribution" gives the third harmonic.

6.7 FACTS SIMULATION

The problems with improving the quality of the transmitted electric energy are always the center of attention for the electrical power engineering. As the main indicators of the electrical energy quality, the following are considered: the voltage deviation from the nominal value at the receiving line end or/and along line, the voltage frequency deviation in the line from the accepted value, the content of the higher harmonics in the line voltage and current, and the relative value of the transmitted reactive power. Most part of the steps have to decrease the values of these indicators. The wanted characteristics should be held not only in the steady states but also in transients (abrupt load changing or a fault with its subsequent clearing) as well.

Over the past decades, new power electronic devices (high-voltage thyristors, GTO- and IGCT-thyristors, IGBT transistors, etc.) are used intensively for the improvement of electric energy quality. The power systems used in such devices are called FACTs—Flexible AC Transmission Systems. Beginning from the fourth version, four blocks for the simulation of FACT-systems are included in SimPowerSystems; besides, a number of demonstrational models have been developed.

6.7.1 STATIC SYNCHRONOUS COMPENSATOR SIMULATION

Static synchronous compensator of reactive power (Static Var Compensator—SVC) is a shunt-connected unit that controls the output voltage, changing the flow of the reactive power that is injected in or absorbed from the electric power system. When the voltage is decreasing, SVC generates the reactive power, operating in the capacitance mode; when the voltage is increasing, SVC absorbs the reactive power, operating as an inductance. The reactive power flow control is carried out by switching on and switching off, with the help of the thyristor switches, the capacitors and reactors in the secondary winding of the matching transformer, whose primary winding is connected to three-phase network, or by a smooth change of the equivalent inductance value with the help of the thyristors that are connected in series with the reactor and whose firing angle is controlled (Figure 6.61). When the inductance is switched off, the capacitance reactive power is $P_c = U^2 B_{cmax}$, $B_{cmax} = 2\pi f C$ is the capacitive susceptance, and U is the phase-to-phase voltage. When the inductance is activated (when the firing angle α decreases), the SVC power becomes equal to $P = P_c - U^2 B_{l\alpha}$, $B_{l\alpha} = 1/2\pi f L_\alpha$ where L_α is an equivalent inductance for the given angle α, and for $\alpha \approx 0$ $P = P_c - P_l$, $P_l = U^2 B_l$. The value P can be both positive and negative, depending on L. The SVC control system consists of the unit for the measurement of the voltage vector amplitude in PCC, voltage controller, and **Synchronized 6-pulse Generator** that forms the firing pulses for the thyristors.

The SVC model in SimPowerSystems utilizes, in fact, the static characteristic SVC for the complex vectors and can be used only in the Phasor simulation mode. This model does not include the details of the unit realization but takes into account the delays in the control system. The current vector **I**

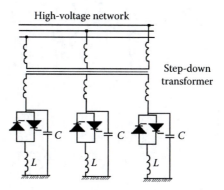

FIGURE 6.61 SVC diagram.

that is formed by SVC is determined as $\mathbf{I} = jB\mathbf{U}$ where $\mathbf{U}$ is the voltage vector in PCC and B is the SVC resulting susceptance. The model can operate in two modes: var control and voltage regulation. The block parameters are specified in several windows. In the first window, either *Power data* or *Control parameters* can be selected. In the first case, the rated phase-to-phase voltage and frequency, the base three-phase power P_{base} that is used for computation of the B changing boundary in pu and for base current calculation, the boundary for reactive power changing, and the delay time of the thyristor control system T_d are specified. Besides, the option "SVC modeled using positive-sequence" can be marked, wherein the negative sequence current is excluded from consideration; the zero sequence current is absent always.

When the window *Control parameters* is chosen further options appear that help with selecting the required block operation mode: *Var control* or *Voltage regulation*. In the first case, the reactive susceptance B_{ref} in pu is determined; in the second case, the SVC voltage reference V_{ref}, the voltage control characteristic droop (1%–4% usually) and gains of the voltage controller are specified. If the option "External control of reference voltage" is marked, the line "Reference voltage" disappears, and the additional input appears on the block picture for assignment of this voltage by the Simulink signal.

The block diagram of the SVC model is shown in Figure 6.62. Its base constitutes two controlled current sources of A and B phases (phase C current is equal to the sum of these currents with the opposite sign) that determine the reactive current of the unit. The phase current reference is defined as follows. The input unit $V_{abc} \to V_1$, V_2 computes the positive V_1 and negative V_2 voltage sequences in pu. The following relationships are used:

$$V_1 = \frac{1}{3}(V_{ab} - V_{bc}a^2), \quad V_2 = \frac{1}{3}(V_{ab} - V_{bc}a) \qquad (6.35)$$

which are obtained from (6.1) and (6.4). These quantities are multiplied by jB, forming the positive and negative current sequences (all operations in the model are

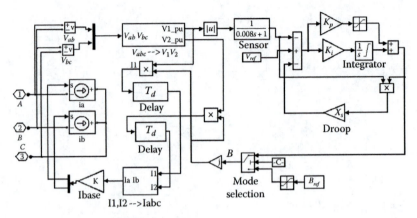

FIGURE 6.62   Diagram of the SVC model.

fulfilled with the complex numbers), so that when $B > 0$ the current leads the voltage. The complex products obtained, which are shifted in time for the delay time T_d, are multiplied by the base current value and are converted in the reference currents I_a and I_b by (6.2), (6.3).

If the constant B mode is selected, the switch for the mode selection is in the lower position, and $B = B_{ref}$. During voltage control, this switch is in the upper position, and value B is defined by the voltage controller output. This is PI controller, and the signal $-X_s BV_1$ is added to the error signal at its input, so that in a steady-state

$$V_1 = \frac{V_{ref}}{(1 + X_s B)} \tag{6.36}$$

that is, the droop appears in the external characteristic.

In model6_16 the block **SM 2000** is the equivalent model of the group of SG that are operated in parallel and are driven by hydraulic turbines. SG supplies the active load through the step-up transformer power of 2000 MVA and transmission line of 100 km length. The load can change from 1000 to 1500 MW. The line natural load is 900 MW. The load step takes place at $t = 1$ s. The SVC reactive power is 300 Mvar. Run simulation in the constant susceptance mode with $B_{ref} = 0$. Before the start of simulation, the initial values have to be set with the help of Powergui. After the simulation is complete, the curves of the voltage vector magnitudes in the points $B1$ and $B2$ are obtained as plotted in Figure 6.63a. It is seen that the load voltage changes essentially. Repeat simulation when SVC operates in the voltage control mode, with $X_s = 0.01$. One can see from the plot in Figure 6.63b that load voltage deviation (in point $B2$) decreases essentially. By the scope **SVC** one can watch the changing of the controller output signal B during the transient process.

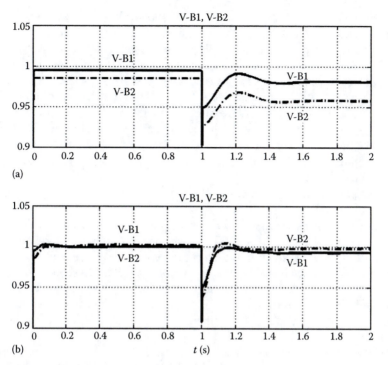

FIGURE 6.63 Transients in the transmission line: (a) without SVC and (b) with SVC.

The use of this SVC model gives only a general idea of SCV influence on the power system; the problems of SVC realization and influence of the higher harmonics generated by SVC remain uncertain, so it is reasonable to have a more detailed model. Such a model is developed in model6_17. The electric power station with parameters 500 kV, 2000 MVA, through the transmission line of 350 km length with the natural load of 976 MW, feeds the load that changes from 500 to 900 MW, 100 Mvar; load increasing takes place in two equal steps with the help of the breakers **Br1, Br2**. SVC is executed according to the diagram in Figure 6.61 and connected by the breaker **Br3**. SVC capacitive power is 350 Mvar and the inductive power is 600 Mvar. The reactors are connected through the subsystem **L_SVC** that consists of the subsystems **SVC_Power** and **Control**. The former contains the thyristors connected in series with the reactors, the latter—**Synchronized 6-pulse Generator** (SPG) and the integral voltage controller. The block **Timer** delays controller activation until process of SVC turning-on comes to a conclusion. In the case under consideration, $X_s = 0$.

The inductive part of SVC generates a nonsinusoidal current with prevalence of the fifth harmonic; therefore, the **Three-Phase Harmonic Filter** is set in parallel to the capacitors. The filter is adjusted as a single-tuned filter with a cutoff frequency of 250 Hz; the filter power is 50 Mvar.

A special feature of SPG synchronization is that the current in the reactor lags 90° behind the voltage across the reactor; hence, when the phase voltage reaches maximum, the current crosses zero level and the conducting thyristor is cut off; that is, this instance has to be taken as a natural firing point. Take into consideration the thyristor +A. The block SPG responds to zero-level crossing by the synchronizing voltages. So, if we begin counting off from that instant, when the phase voltage V_a crosses the zero level up in V_a, the firing angle α has to change from 90° to 180°, and at the input VCA that is connected with the thyristor +A the voltage $-V_a$ has to be sent (according to the SPG circuits). Instead of that, the voltage V_c is sent to the input VCA that crosses the zero level down in V_c by 60° behind the instance when V_a crosses the zero level up in V_a; the additional shift in 30° is provided by the first-order link with the time constant of 1.84 ms, so that α changes from 0° to 90°.

Switch allows investigating the model without feedback and the voltage feedback closes in the upper switch position.

Three scopes can be used to observe model behavior. The scope **V1,V2** records the voltages in the points **B1** and **B2**, that is, at the beginning and at the end of the line, at which point the first axis shows the voltage vector magnitudes and the second axis shows the voltage in real form in the point **B2**. The scope **Scope_SVC** shows the processes in SVC: the output current, the reactive power, the angle α; the scope **Scope_L** fixes the processes in the inductive part of SVC: the secondary voltage of the SVC transformer, the firing pulses, the current, and the voltage of the reactor.

Let us begin with simulation. The breakers **Br1, Br2** are set in the state "Open"; their transition times are equal to 0.2 and 0.3 s, respectively. The switch **Br3** is set in the state "Open," and the *Transition time* is accepted as 30 s. After simulation, the plots of the voltage magnitude changing in the beginning and in the end of the line are obtained as shown in Figure 6.64. It is seen that the voltage at the line end

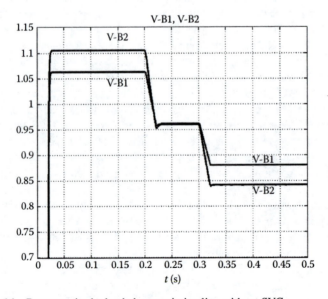

FIGURE 6.64 Processes in the loaded transmission line without SVC.

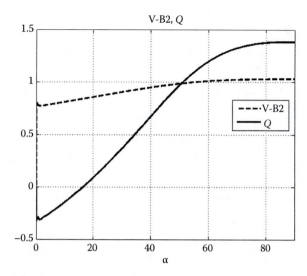

FIGURE 6.65 SVC characteristic (detailed model).

(point **B2**) varies in a wide range, which indicates that the line power is insufficient. For improving the situation, it is possible either to strengthen the line with the parallel wire or to use SVC that is made in the model under consideration.

Set the breaker **Br3** in the state "closed," and the simulation time and the transition times of the closed breakers **Br1** and **Br2** are equal to 2 s. Set **Switch** in the lower position (the angle α is determined by the block **alpha_ref**). The initial value is 0, and the final value is 90 in this block. **Rate Limiter** settles rather slow changing the actual value α. Run simulation. The changing of the line output voltage and the reactive power produced by SVC are shown in Figure 6.65. Transition from the inductive to the capacitive output when α = 17° and an increase in the load voltage as the capacitance power rises are seen.

Now repeat simulation with the simulation time of 0.8 s; the transition times of the open breakers **Br1** and **Br2** are 0.3 and 0.5 s, **Switch** in the upper position (the voltage feedback is closed). After simulation, the processes of voltage and reactive power changing depicted in Figure 6.66 are obtained. It is seen that, after transients, the voltage in the line end **V_B2** corresponds to the reference value with the good accuracy. It is useful to observe changing of the other quantities fixed by the scopes.

Use of the thyristor-switched capacitors (TSC) gives the opportunity to reduce reactor size; when it is necessary to lessen the voltage in the line, these capacitors are turned off by turns. The thyristors with a pulse control are used for capacitor switching. The total capacitive power is divided into n parts. The number of parts depends on many factors. In [4,6] the case $n = 3$ is considered. The control principle of such a circuit is considered also in [6]. It is noted in [6] that the reactive power of one capacitive part has be less than the reactive power of the reactor.

Model6_18 models such a unit. The electric power system is presented by the voltage source of 500 kV, **Three-Phase Series RL Branch** models, the impedance of the source and the transmission line with short-circuit power of 1000 MVA; the

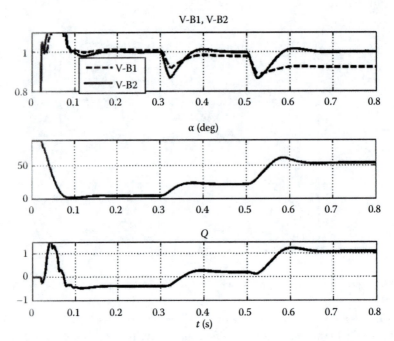

FIGURE 6.66 Processes in the loaded transmission line with the detailed SVC model.

load consists of the constant section 200 MW and the variable load whose power is determined by **Timer**; the maximum power of this load is 450 MW. The source voltage decreases as a sinusoid in the interval [1.6 s, 2.1 s] with the amplitude of 0.1 of the rated value. SVC is connected with the help of the transformer with parameters 500 kV/16 kV, 300 MVA. The SVC consists of the reactor block TSR power of 102 Mvar and three capacitor units TSC1, TSC2, and TSC3, each with a power of 86 Mvar. Small inductance in the capacitor circuit is useful to limit the surge current for possible malfunctions. Both the transformer and the units of SVC are connected in delta for eliminating the third harmonic currents (Figure 6.67). Therefore, the SVC reactive power can change from minus 102 Mvar (inductive) when $\alpha_{TSR} = 0°$ and turning-off TSC to plus 258 Mvar (capacitive) when $\alpha_{TSR} = 90°$ and turning-on TSCs. If the base power is accepted as $P_b = 100$ Mvar and taking into account that the transformer leakage inductance for this base power is 0.15/3 = 0.05 pu, by formula

$$B = \frac{Q*}{1 - 0.05Q*} \qquad (6.37)$$

where B is an SVC equivalent susceptance in pu referring to the primary transformer winding, $Q*$ is the reactive power in pu, and one can find that B changes from $-1.02/(1 + 1.02 \times 0.05) = -0.97$ to $2.58/(1 - 2.58 \times 0.05) = 2.96$.

The subsystem **SVC Controller** generates firing pulses for the thyristors. It consists of four units: **Discrete 3-phase Positive-Sequence Fundamental Value** described in Chapter 3, **Voltage Regulator**, the distribution unit that determines

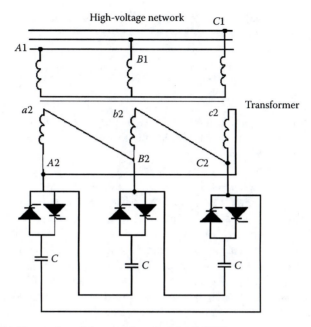

FIGURE 6.67 Connection of the capacitors in delta in SVC.

which blocks of SVC have to act in the present instant, and the subsystem with the blocks **Synchronized 6-pulse Generator**.

The output of the voltage controller defines the value B referred to the primary transformer winding. The distribution unit computes the control angle α_{TSR} and forms the signals for activation of the capacitor banks. Its diagram is shown in Figure 6.68. At first, the value B is referred to the second transformer winding by the inversion formula to (6.37). If $B < 0$, this value comes to the input of the amplifier **Amp** through the summer and the limiter. The **Amp** gain is equal to $1/B_{max}$ (in the case under

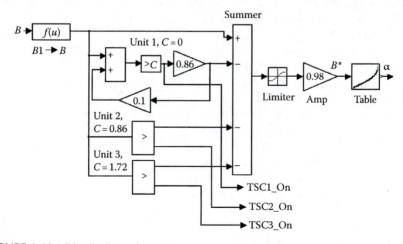

FIGURE 6.68 Distribution unit.

consideration $B_{max} = 1.02$), so that $B*$ changes from -1 to 0. The relationship between the equivalent SVC susceptance and the angle α is given as [6]:

$$B* = \frac{[\pi - 2\alpha - \sin(2\alpha)]}{\pi} \qquad (6.38)$$

The solution of the equation inverse to (6.38) is programmed in the table, at whose output the angle α_{TSR} is resulted. Data for the table are computed for model start (option *Callbacks/PreloadFcn*).

When B reaches zero level, **Unit1** switches on and gives the signal to turn on the first capacitor unit. Simultaneously, the summer output decreases by the value that is equal to the reactive power of this unit; this output gets negative and TSR is in the active range. If it turns out that activation of one capacitor unit is not sufficient, the signal B carries on to increase; when it reaches the value that is equal to the capacitor block power, the second capacitor unit is turned on and the summer output decreases by the same value, and so on. If turning on all the three blocks is not sufficient, and B remains positive, TCR has the angle $\alpha = 90°$; it is turned off in fact.

The output signals TSC goes in the subsystem **Pulse Generators** that contain 4 SPG. The first of them does not differ from the one used in the previous model; it controls TSR and works with the angle α_{TSR}. The other three are synchronized in antiphase to the first; that is, they are synchronized with the maximum value of the positive half-wave of phase voltage and not with that of the negative half-wave of phase voltage and act with the angle $\alpha \approx 0$ (is set $\alpha = 1°$).

The main scope **SVC** shows the voltage and the current of the phase A in the transformer primary winding, the reactive power, the reference and the actual values of the load voltage positive sequence, the angle α_{TSR}, and the number of the active TSC. Besides, the scope **Controller Signals** records a number of the additional quantities.

Begin with simulation. If simulation is carried out without SVC (breaker is opened) one can see that the load voltage is very oscillated (Figure 6.69). If the simulation is repeated with SVC, it is seen that the reference voltage is kept with sufficient accuracy. At the beginning, only TSR is active. As the load increases, TSC1 and afterwards TSC2 are getting active; TSC3 is active when the source voltage drops. In the end the dynamic load produces power, and all TSCs turn off. The plots of the transients are shown in Figure 6.69.

When this model is compared with the previous one, with the capacitor unit connected permanent, it is necessary to take into account that lessening of the reactor inductance is obtained at the expense of essential increasing of the thyristor numbers and of the control system complication; system reliability decreases. So, sending the firing pulse to TSC at the "wrong" instant leads to the large currents and overvoltages; the surge arrester has to be set for protection that makes the circuits more complicated. Simulation of this phenomenon is provided in the demonstrational model `power_svc_1tcr3tsc`. Thus, before choosing this or another variant, a careful technical-economical analysis has to be done.

Simulation of the long transmission line is considered in the conclusion of this section. There are often cases when the sites of production and consumption of

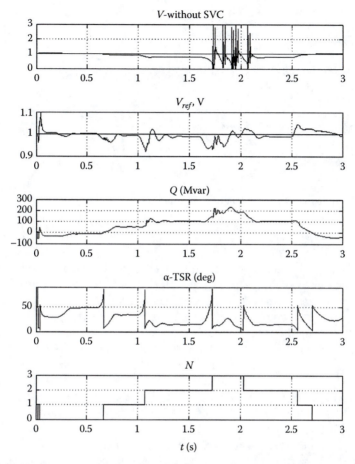

FIGURE 6.69 Processes of the load voltage stabilization in SVC with commutated capacitors.

electric energy are separated with a big distance—1000 km and more. Designing of long AC transmission line demands a lot of investigations. The line simplified model of 1200 km length is considered in model6_19. Its diagram is shown in Figure 6.70.

The equivalent generator SG simulates an electrical station, including the step-up transformers. The output voltage is 765 kV, and the rated power is 2200 MVA. The transmission line consists of four sections; each of them is 300 km in length. The units for compensation, measurement, and so on can be set between the sections. The points of measurement are designated as **B0–B5**. The line ends with the load of 2000 MW and the infinite power source with the voltage of 765 kV. Every section consists of a part of the transmission line with the parameters indicated in the figure and the capacitor for series compensation. Since the section inductive impedance is $Z_l = 0.93 \times 10^{-3} \times 314 \times 300 = 87.6\ \Omega$, and the impedance of the capacitors is $Z_e = 1/(91 \times 10^{-6} \times 314) = 35\ \Omega$, the degree of compensation is 35/87.6 = 40%. SVC is set in the line center. By (2.21)

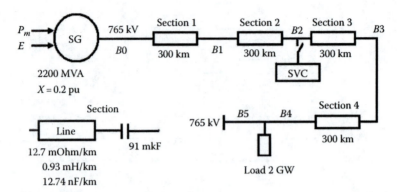

FIGURE 6.70 Structure of the long transmission line.

and (2.25), the wave impedance is $Z_c = (0.93/12.74)^{0.5} \times 10^3 = 270.2\ \Omega$ and the natural load is $NP = 765^2 \times 10^6/270.2 = 2.17\,\text{GW}$.

It is worth noting that, in practice, the capacitors for series compensation have to be protected from overvoltage. Such a protection is fulfilled with the help of varistors and lighting arresters. One can find the corresponding model in the demonstrational model Series-Compensated Transmission System.

The block **Simplified Synchronous Machine** in pu is used for SG simulation in model6_19, and the blocks **Distributed Parameters Line** are used for simulation of the transmission line sections. The section is arranged as the subsystem **Line**. SVC, mainly, is the same as in model6_17, but the capacitor unit and the filter are excluded, and the characteristic droop is added. Conversion of the controller output into angle α is fulfilled with the help of the table that carries out solution of the equation (6.38). Before the start of the model, the program SVC_Line that is written in the option *Callbacks/InitFcn* and computes the table content is fulfilled.

The scope **Generator** shows the load angle δ, the speed, and the power of SG. The scope **Voltages_In_Out** records the voltages in the beginning, in the end, and in the middle point of the line, and the scope **Voltages** fixes the amplitude of the positive sequence first harmonics of the same voltages. The scope **Power** writes the SG power and the powers on the buses **B4** and **B5**. The scope **Alpha** records the SVC firing angle, and the scope **SVC_Power**—SVC power.

The steady states are investigated, so, to exclude the electromechanical processes, it is accepted $H = \infty$ for SG. First, the line without compensating units is considered. For this aim, it is reasonable to make the copy of model6_19 under other name, and in this copy, in every subsystem **Line**, to remove the block with the capacitor. The breaker **Br1** is set in the position "Open." With the help of the Powergui option *Load Flow and Machine Initialization*, the SG initial values are set for the bus mode *P&V Generator*, 765,000 V, 2.3 10^9 W. It is seen after simulation by the scope **Voltages** that the amplitudes of the phase voltages at the line terminals are equal to ~625 kV (that corresponds to 625 × 1.73/1.42 = 765 kV for the rms phase-to-phase voltage), and in the line center the voltage amplitude is 508 kV; it is impermissible because the consumers can connect to the line all along the line. By the scope **Power**, the generated power is 2.3 GW, the power in the line end is 2.12 GW, and the power that is transferred to the infinitive power bus is 0.12 GW, so that the load power is 2 GW. The angle shift between input and output line voltages δ can be found by the scope **Voltages_In_Out**; it is equal to 95.4°. The power transferred by the line with the length d can be calculated by the formula [1]

$$P = \frac{V_1 V_2 \sin \delta}{Z_c \sin \lambda} \tag{6.39}$$

where
V_1, V_2 are the voltages at the beginning and at the end of the line
δ is the angle shift between them
λ is the wave line length that is equal to βd

In our case, with the use of (2.18), we can obtain the following: $\lambda = 314(0.93 \times 12.74 \times 10^{-12})^{0.5} \times 1200 = 1.3$, $P = 765^2 \times 10^6 \sin (95.4°)/(270.2 \sin (1.3)) = 2.24$ GW, which corresponds to the simulation result.

Return to model6_19, supposing that SVC is disconnected (**Br1** is in position "Open"). As discussed earlier, run simulation with the preliminary SG initialization. When SG power is 2.3 GW, the voltage in the line center increases essentially and is equal to 668 kV and the angle δ decrease to 43°; that is, it occurred as if equivalent to line shortening. When SG power is 1 GW, the voltage in the line center rises to 704 kV; therefore, it is necessary to take steps for the voltage limitation in the compensated line when there is load lessening. For the transferred power of 3 GW, the voltage in the line center is 610 kV.

Set the breaker Br1 in the position "close" and carry out initialization for the power of 2.2 GW.

The references $P_m = 1.02$, $E_{ref} = 0.985$ are obtained. Run simulation. The scope **Voltages** shows that the amplitudes of the phase voltages in the beginning, in the center, and in the end of the line are, respectively, 617, 642, and 625 kV. By the scope **Power**, the generated power is 2.13 GW, the power in the line end is 2 GW, and there is no power consumption from the infinite power source. By **Voltages_In_Out**, the angle δ is 41°, by **Alpha**, it is $\alpha = 25°$.

Repeat simulation using the power of 1 GW. One can find that the angle α decreases to 1° and the voltage in the line center increases to 650 kV. When the transferred power is equal to 3 GW, the angle $\alpha = 73.5°$ (SVC is turned off), the voltage in the line center is 610 kV, and the power in the line end is 2.76 GW, at which time 0.76 GW is transferred to the infinitive power bus.

Therefore, owing to SVC, the voltage deviations reduce to 0.4 of the previous value, from $(704 \text{ kV} - 610 \text{ kV}) = 94 \text{ kV}$ (15%) to $(650 \text{ kV} - 610 \text{ kV}) = 40 \text{ kV}$ (6.4%), during a change in the power transferred.

In order to decrease the voltage in the line with the series capacitors during load reduction, together with the considered devices, the reactors connected in parallel to the capacitors and controlled with the help of the thyristors can be used. These devices are called "Thyristor Controlled Series Capacitors—TCSC." There is the demonstrational model `power_tcsc.mdl` in SimPowerSystems with such devices, and these devices are not considered in this book.

6.7.2 STATCOM Simulation

6.7.2.1 Models of Standard STATCOM Systems

The Static Synchronous Compensator—STATCOM is intended for the voltage control in the point of its connection to the power system (PCC), absorbing or injecting a reactive power; this is a VSI. STATCOM has the following advantages compared to SVC: The power generated is proportional to the voltage value, but not to its square, as it is for SVC, because the output current can be controlled independent of the voltage; this feature is especially important during the faults; STATCOM that uses the force-commutated devices (switches) with rather large switching frequency can be used for filtration of the high harmonics in the grid; its speed of response is larger. Two STATCOM types are considered in SimPowerSystems: with the constant voltage V_{dc} in the DC link and with the altering V_{dc}. In the first case, STATCOM consists of VSI with IGBT and PWM, and VSI is connected to the power system by means of a transformer or a reactor; the control system regulates I_d and I_q currents of the VSI aligned alongside and perpendicular to the voltage space vector in PCC. At that point, I_d controller is subdued by the V_{dc} controller, as it takes place for VSI with an active front rectifier (model5_10), and I_q controller is subdued by the controller of the voltage in the network. The options are possible for STATCOM to operate either with constant I_q or with the constant reactive power.

When STATCOM operates with altering V_{dc}, the VSI uses GTO- or IGCT-thyristors; PWM is not used and the VSI output voltage changes in steps. The amplitude of this voltage is proportional to V_{dc}. In order to lessen the content of the harmonics in the output voltage and current, the multiphase schemes with several VSI are used. At that point, 4 three-level VSI with 4 transformers are used often; the transformers have the phase shifts in 7.5° about one another; by this way, 48-steps voltage with very little content of the harmonics is formed. For the power control, V_{dc} has to change. For this purpose, the firing angle α that is kept about zero in the steady state is shifted in the direction of phase lead or phase lag, forming by this way an active power flow that changes V_{dc}.

In the present time, a new subclass of STATCOM develops intensively, the so-called DSTATCOM (D—Distribution); these units are intended for operation in the utility distribution networks. They have lesser power (usually not more than some Mvar) and connect to the power network with relatively low voltage (3–35 kV). The control method with the constant V_{dc} and with PWM is used.

The block **Static Synchronous Compensator** is included in SimPowerSystems that simulates three-phase STATCOM of the first type in the Phasor mode. Since the VSI components and harmonic generation are not modeled, this block can be used as STATCOM of the second type for studying the transients.

The block picture is shown in Figure 6.71a. The block has two windows of the dialog box. In the window *Power data*, the following are specified consistently: the rated voltage (V) and frequency (Hz), the unit rated power S, the resistance and the inductance (in relationship to the rated quantities mentioned earlier) of the transformer and/or the reactor that connects VSI with the network, the initial unit current (usually accepted equal to zero), V_{dc} value (V), and the total capacitance in the

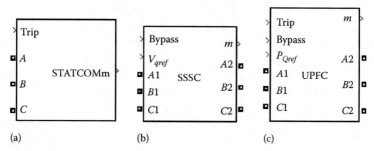

FIGURE 6.71 Pictures of the blocks (a) STATCOM, (b) SSSC, and (c) UPFC.

DC link (F). This quantity depends on the rated parameters and it is recommended to choose it by the relationship [4]

$$C = \frac{2St}{V_{dc}^2} \qquad (6.40)$$

where $t = 3$–4 ms.

The window *Control* gives the opportunity to select the unit operation mode: *Var Control* or *Voltage regulation*. Depending on the selected mode, the necessary parameters are specified, whose meanings are clear from the text in the window.

The block input *Trip* is provided for the blocking (the block is active when this signal is the logical 0). A total of 16 inner signals form the vector of the measured quantities at the output *m*, namely: 1, 2, 3 are the phase voltages (complex) in PCC, 4, 5, 6 are the input STATCOM currents (complex), 7 is the DC link voltage V_{dc} (V), 8 is the measured voltage of the positive sequence (pu), 9 is the reference voltage V_{ref} (pu), 10 is the reactive power Q (pu) ($Q > 0$ for the inductive power), 11 is the reactive power reference Q_{ref} (pu), 12 is the current I_d (pu), 13 is the current I_q (pu), 14 is the reference value I_{dref} (pu), 15 is the reference value I_{qref} (pu), 16 is the modulation factor, and $0 < m < 1$.

There is the demonstrational model power_statcom.mdl in SimPowerSystems that shows the use of this block. The power system chosen for simulation consists of two equivalent sources with power of 3 GVA and 2.5 GVA, and with the voltage of 500 kV that are connected by a transmission line 600 km in length. The source voltages are selected such that the power transferred by the line from the left to the right is 930 MW. STATCOM with power of ±100 Mvar is connected to the line in its middle point with the help of the transformer with the equivalent impedance of 0.22 pu for a power of 100 MVA.

Investigation of the transmission line operation with STATCOM can be made as follows. In the window *Control parameters* of the dialog box the *voltage regulation* mode with the external control of the reference voltage V_{ref} is selected. The block **Step Vref** defines a number of the values V_{ref}. The block **Fault** is turned off. Run simulation. By scope **VQ_STATCOM** one can observe voltage and reactive power changing. It is seen that the time constant of the voltage control loop is about 20 ms. Owing to the characteristic droop of 0.03, the measured voltage differs from V_{ref}; if *Droop* = 0, and the error will be equal to 0 too.

Comparison of STATCOM with SVC during the fault is provided in this model. If the subsystem **SVC Power System** is opened, one can see that it contains the same system, as the main model, but SVC is used instead of STATCOM. If the simulation is carried out when the block **Fault** is active (all the times are multiplied by 10 in the block **Step Vref**), one can compare the processes of voltage and power changing for both systems by the scope **SVC vs STATCOM**. It is seen that STATCOM generates more reactive power that leads to the voltage rapid recovery.

Because simulation in the Phasor mode does not give a complete understanding of the processes in the system, it is reasonable to have at one's disposal more detailed models that take into account the features of the unit implementation. The demonstrational model `power_statcom_gto48p.mdl` is included in SimPowerSystems, in which the second of the said STATCOM types is used, namely: using VSI without PWM, with reactive power control achieved by the change in DC link capacitor voltage. Describe this model in more detail. Its main circuits contain 4 three-level VSI with common capacitors in the DC link. **Zigzag Phase-Shifting Transformer**s are set at the VSI outputs. The transformers have the phase shift of +7.5° or −7.5° (Figure 6.72). The additional phase shift is provided by connection of the secondary

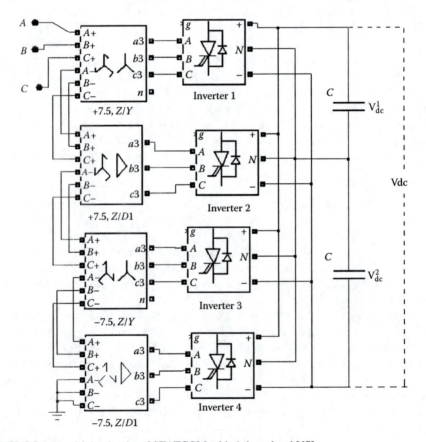

FIGURE 6.72 Main circuits of STATCOM with 4 three-level VSI.

windings of two transformers in delta; thus, the primary transformer voltages have the phase shifts as 7.5°, −22.5°, −7.5°, −37.5°. The transformer's primary windings are connected in series, and the modulating signals are shifted, respectively, by 0°, −30°, −15°, −45°, so that all the fundamental harmonics are in-phase, and the most of the high harmonics cancel out. We note that the shift of 30° between the transformers having star and delta connection leads to cancelling-out of the harmonics of the orders $5 + 12n$ and $7 + 12n$, and the shift by 15° between the first and the second groups leads to cancelling-out of the harmonics of the orders $11 + 24n$ and $13 + 24n$, so that harmonics of the orders 23, 25, 47, 49, and more remain.

The phase-to-phase voltages at the VSI outputs have a shape of the rectangular oscillations (Figure 6.73a); the resulting output voltage is shown in Figure 6.73b. If, instead of the rectangular voltage, the step-wise voltage is used, as shown in Figure 6.73c, which can be obtained by decreasing the period when the voltage $\pm V_{dc}/2$ is applied from 180° to $\sigma = 172.5°$, the 23rd and the 25th harmonics decrease essentially, and the 47th and the 49th harmonics remain most significant (Figure 6.73d). So, for the voltage in Figure 6.73b, THD = 7.6% with amplitudes of the 23rd and the 25th harmonics about 4%, for the voltage in Figure 6.73d—3.9% and 0.3%, respectively.

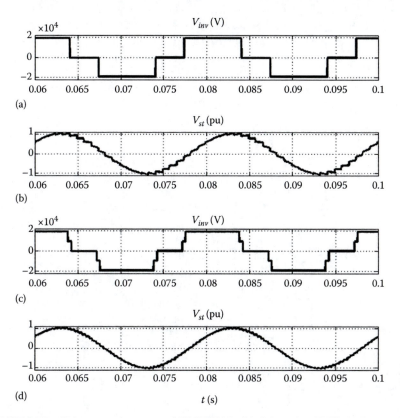

FIGURE 6.73 Voltages at the output of STATCOM with 4 three-level VSI. (a) Rectangular inverter voltage, (b) output voltage under $\sigma = 180°$, (c) stepwise inverter voltage, and (d) output voltage under $\sigma = 172.5°$.

STATCOM is controlled by the subsystem **STATCOM Controller** that uses the following relationships: If the circuit has the impedance X, the active and reactive powers transferred through it are

$$P = U_1 U_2 \frac{\sin \alpha}{X} \tag{6.41}$$

$$Q = \frac{U_1(U_1 - U_2 \cos \alpha)}{X} \tag{6.42}$$

where
  U_1, U_2 are the magnitudes of the voltage vectors at the input and at the output of the circuit
  α is the angle between vectors

In this case, U_1 is the voltage in the line at PCC and U_2 is the voltage generated by STATCOM. Therefore, for the reactive power control, it is reasonable to have the voltages U_1 and U_2 in-phase; to receive the capacitive reactive power, it must be $U_2 > U_1$; this control can be carried out only by changing the capacitor voltage because the voltage shape does not change. Since the active power is necessary to change the capacitor voltage, it can be realized by the shift of U_2 by some angle α about U_1. In addition, the control system has to balance capacitor voltages.

The structure of the control system is shown in Figure 6.74. The circuits for fabrication of the signals V_p, I_{sd}, I_{sq}, I_{sq}^*, θ and the voltage controller are the same as those

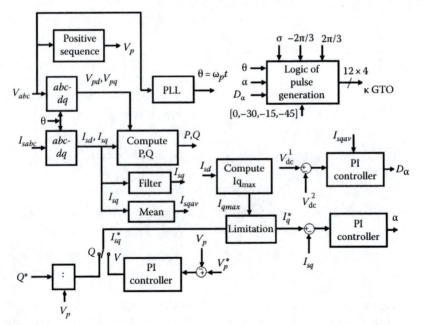

FIGURE 6.74 Diagram of the control system of STATCOM with 4 three-level VSI.

used in the previous model. The maximum value I_{qmax} that depends on the current I_{sd} is computed in the block **Compute I_{qmax}** from the equation

$$\sqrt{I_{sd}^2 + I_{q\,max}^2} < I_{max} \qquad (6.43)$$

The value I_{max} is accepted to be equal to 1.05 of the rated current. If $I_{sq}^* < I_{qmax}$, the reference current $I_q^* = I_{sq}^*$, otherwise $I_q^* = I_{qmax}$. The limitation of I_q^* changing rate is provided too. (In the model Equation 6.43 is solved in an indirect way, for some discrete points.) The output of the current I_{sq} controller defines the angle α of the shift of the STATCOM voltage fundamental harmonic relative to the network voltage.

Balancing of the capacitor voltages is carried out by changes in the time intervals, during which the voltages $V_{dc}/2$ and $-V_{dc}/2$ are applied. If, for instance, $V_{dc1} > V_{dc2}$ and STATCOM consumes the reactive power ($I_q > 0$), for voltage balancing it is necessary to increase the relative interval, and when $-V_{dc}/2$ is applied, and if the capacitive power is injected, this interval has to be decreased. PI controller of the voltage difference forms the angle D_α, taking into account the sign of the current average value I_{sqav} that, coming in the block of the pulse generating, changes the relative interval duration. Since for small values I_{sq} this scheme does not operate effectively, when $|I_{sqav}| < 0.1 \, D_\alpha = 0$.

The block for pulse generation that control GTO switching contains 12 circuits of the same type that are distinguished by the shift of the firing pulse generating instants by $-2\pi/3$ for the phase B and $2\pi/3$ for the phase C, and by the shift by $-30°$, $-15°$, $-45°$ for the inverter bridges 2, 3, and 4, respectively. The instants when the firing pulses are generated take into account the angles α and D_α and pulse duration σ (when $D_\alpha = 0$). The reader who wants to see how this structure is modeled can do it by opening corresponding subsystems.

The use of the described STATCOM model is shown in model6_20 (in addition to the demonstrational model power_statcom_gto48p from [4]). The circuits of electric power supply are the same, as in model6_17, and the source and load parameters are indicated in the model diagram. The breaker **Br1** closes at $t = 0.2\,s$, and **Br2**—at $t = 0.6\,s$. By scope **STATCOM** one can watch the inner STATCOM emf (V_{aSt}), its output voltage and current (V_a, I_a), its reactive power Q, the measured (V_{mes}) and reference (V_{ref}) voltages in PCC, and the capacitor voltage (V_{dc}). A number of the scopes are placed in the subsystem **Signals and Scopes**. By scope **V1** one can see the shape of the voltages at the outputs of the inverters, the scope **Ctrl_Sig** records the measured (V_{mes}) and the reference (V_{ref}) voltages (repeatedly), the reference and the actual values I_q, the STATCOM active and reactive powers (P, Q), and the angle α. The scope **Power P,L** fixes the voltages, the active and reactive powers of the network, and the load.

Run simulation with **Br3** open. It is seen by the scope **Power P,L** that the load voltage, at the end of each interval changing (i.e., at $t = 0.2$, 0.6, and 1 s), is equal to 1.16, 1.03, and 0.92, respectively; that is, when the load increases, an abrupt decrease in load voltage by 10%–12% takes place.

Repeat simulation with **Br3** kept closed, in the control mode *Var control* with *Fixed Qref* = 100 Mvar. One can see that the load voltages at the said instants are

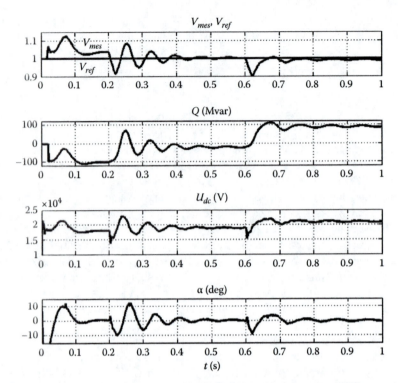

FIGURE 6.75 Processes in the system with STATCOM with 4 three-level VSI.

1.28, 1.13, 1; that is, the STATCOM power is sufficient for the voltage control but has to decrease with a reduction in load. If simulation is repeated in the *Voltage regulation* mode with the reference of 1 and with the droop of 0.01, the processes obtained are shown in Figure 6.75. It is seen that before the closing of **Br1**, the voltage is 1.03, that is, more than reference, and $Q = -100$ Mvar, which means that STATCOM inductive power is not sufficient for a reduction in voltage. After closing of **Br1** and afterwards **Br2**, the voltage reaches the reference value after transients with sufficient accuracy; the STATCOM power is −20 Mvar and 82 Mvar, respectively. The rise of Q is accompanied by an increasing in V_{dc}. It is seen that the quantities V_a, I_a are nearly sinusoidal.

6.7.2.2 DSTATCOM Simulation

In SimPowerSystems there are two demonstrational models with DSTATCOM and PWM: One of them uses the detailed STATCOM model, and another, the model for the average voltage values. The former will be of interest to us because the output voltage harmonics and the dynamic features are modeled. This demonstrational model is called `power_dstatcom_pwm`. The STATCOM main circuits are depicted in Figure 6.76, which consists of 2 two-level inverters with a common DC link. The output transformers are connected between the bridge phases of the same name and the bridge control signals are in antiphase. The inverter output voltages are filtered by *R-L-C* filters.

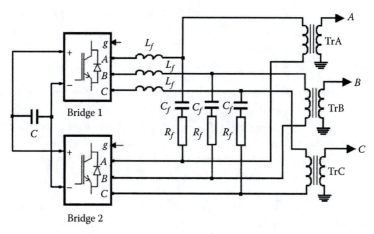

FIGURE 6.76 Main circuits of DSTATCOM.

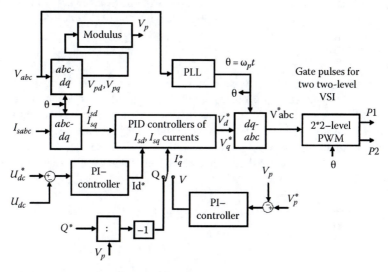

FIGURE 6.77 Control system of DSTATCOM.

The control system (Figure 6.77) contains PID controllers of the current I_{sd}, I_{sq} generated by DSTATCOM, which are aligned in the direction of the voltage space vector in PCC, and perpendicular to this direction, the block **PLL** is used for determination of the voltage vector angle θ. The controller outputs are transformed in the three-phase signal by the block **dq0_to_abc Transformation** that comes to the block **Discrete 3-phase PWM Generator**; the latter generates the antiphase firing pulses for both inverters.

The reference I_{sd}^* is formed at the capacitor voltage controller output; two possibilities are provided for fabrication of the reference I_{sq}^*: assignment of the reactive power Q^* and control of the voltage V_p in PCC. In the first case,

$$I_{sq}^* = \frac{Q}{V_p} \tag{6.44}$$

In the second case, I_{sq}^* is the output of the voltage V_p controller. DC link reference voltage is set equal to 2400 V. The reader is advised to acquaint him or herself in a more detailed manner both with the STATCOM diagram and its operation in the distribution network in the corresponding section of *Help*.

6.7.2.3 STATCOM with Cascaded H-Bridge Multilevel Inverter Simulation

There are a number of developments of the cascaded H-bridge multilevel inverters for STATCOM. Each individual bridge is powered from the capacitor (it means that the rectifier is removed from the diagrams depicted in Figure 4.38). Since STATCOM generates a reactive power, the average voltages across the charged capacitors do not have to change in ideal situations. However, owing to the losses, steps have to be taken to retain the capacitor voltages. The load connection to the source of limited power is simulated in model6_21. The source having the voltage of 6.6 kV supplies, with help of the transformer 6.3 kV/900 V, the active-inductive load that at $t = 2.1$ s changes stepwise from 500 kW, 500 kvar to 1000 kW, 1000 kvar. The voltage deviation at the transformer input without STATCOM (**Br2** is open) is shown in Figure 6.78 (the phase-to-phase base voltage is accepted to be equal to 6.6 kV).

STATCOM has three cells in each phase and the capacitor voltage is taken to be equal to 2 kV.

The structure of the control system is shown in Figure 6.79. PLL generates the signals sin (x), cos (x) where x is the angle position of the network voltage vector $\mathbf{V}_{abc}$ in the PCC. With the help of the inverse Clarke transformation, these signals are transformed in the signals U_a^*, U_b^*, U_c^*, which coincide by phase with corresponding phase voltages and their amplitudes are equal to 1. These signals, after adding the signals $U_{a(bc)dc}$ that control the capacitor voltages (see further), are multiplied by

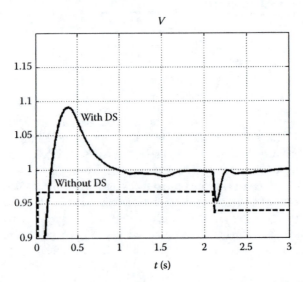

FIGURE 6.78 Voltage deviations in STATCOM with and without cascaded H-bridge multilevel inverter.

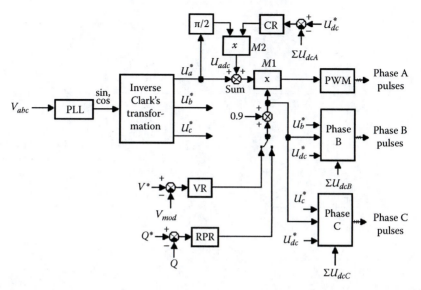

FIGURE 6.79 Control system of STATCOM with cascaded H-bridge multilevel inverter.

the modulation factor that is defined by the output of either the voltage controller VR or the reactive power controller RPR.

The control of the voltages across the capacitors is realized as follows. The reference value of the sum of three voltages and the sum of the capacitor voltages of the phases are compared at the input of CR controllers. The controller outputs modulate the signals that are shifted by 90° about U_a^*, U_b^*, U_c^*, forming the signals $U_{a(bc)dc}$; these signals are added to U_a^*, U_b^*, U_c^*, generating the control signals that are shifted about the phase voltages by the values that are proportional to the CR controller outputs, providing delivery of the active power to the capacitors. Transformation of the total signals into the sequence of the gate pulses is fulfilled by PWM that is similar to the one used in model5_13. The voltage of each of the capacitors can be controlled in the same way separately, if required.

A number of scopes are used for the observation of model functioning. The scope **V** shows the voltage magnitude in PCC (pu). The scopes **NETWORK**, **PowerS**, **PowerL** record the powers, the voltages, and the currents of the network, DSTATCOM, and load, respectively (the last one, only power). The scopes **UdcA**, **UdcB**, **UdcC** intended for the capacitor voltage control are set in the subsystem **STATCOM/Inverter**, and the scope **V_Ph** shows the inverter phase voltages.

Run simulation with the switch **Sw** in the subsystem **STATCOM** in the position V (voltage control). The voltage plot that comes out as a result of simulation is shown in Figure 6.78. The voltage deviation decreases essentially. One can see by scopes **NETWORK**, **PowerS**, **PowerL** how the powers are distributed between the parts of the scheme; it is seen specifically that the load active power is provided by the network and the reactive STATCOM power reaches 1.4 Mvar. By using Powergui, one can find that the second harmonic with the amplitude of 2.3% is perceptible in the network current, whereas the other harmonics are small. The scopes **Udca,b,c** show

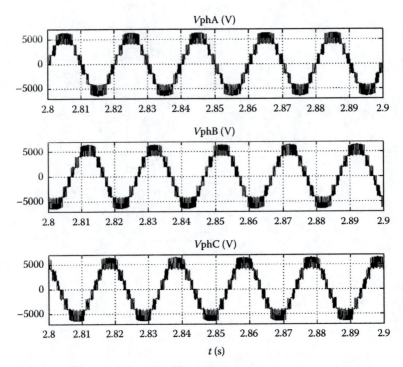

FIGURE 6.80 Phase voltages of the STATCOM inverter.

that the capacitor voltages are close to the reference of 2 kV, at which point these voltages oscillate with the frequency of 100 Hz and with a peak-to-peak value of 240 V. The inverter phase voltages are shown in Figure 6.80.

Repeat simulation with **Br1** closed and with the switch **Sw** in the position *Q*. The block **Step** defines *Q* changing from 0 to 0.5 at *t* = 1.5 s. The *Q* and *V* plots that are obtained as a result of simulation are shown in Figure 6.81. It is seen that power change is reproduced sufficiently rapidly.

Asymmetrical STATCOM is used in model6_22. Each phase has three cells having the same rated voltage U_{dc}. The transformer coupling is used between cells of one phase, as shown in Figure 4.40. Unlike this figure, the output voltages are accepted as *E*, 2*E*, and 6*E*. All nine cells are connected to the same capacitor (Figure 6.82). This way the 19th level of the output voltage can be obtained. The purpose of choosing such a voltage is to accomplish a smoothing voltage change with the help of PWM for the blocks having minimum power, that is, for the blocks *a*1, *b*1, *c*1. This is carried out by the following way (Figure 6.83) [7].

Suppose that the control signal U_{cn} changes from 0 to 9. While $0 \le U_{cn} \le 1$, the outputs of the blocks L1 and L2 are equal to zero, the signal $U_m = U_{cn}$, and the phase voltage *V* is defined by the output of the block *a*1 and changes from 0 to *E* proportional to U_{cn}. When U_{cn} becomes >1, the L1 output signal turns on the block *a*2 with voltage of 2*E*; at the same time, for this signal, which is equal to two sets $U_m = U_{cn} - 2 = -1$, the output *a*1 voltage is −*E*, and after switching-over *V* = 2*E* − *E* = *E*. During a subsequent increase in U_{cn} up to 3, U_m increases to 1, that is, *V* = 2*E* + *E* = 3*E*. When U_{cn} becomes >3, the

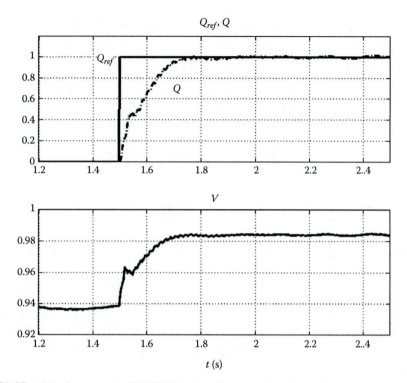

FIGURE 6.81   Processes in STATCOM with reference to the changes in reactive power.

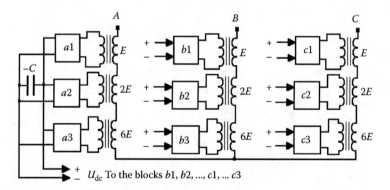

FIGURE 6.82   Diagram of the asymmetrical STATCOM.

signal that appears at the output of L2 turns on the block $a3$ with the voltage $6E$; at the same time, the signal at the output of L2 that is equal to six sets the signal equal to -3 at the output of Sum1; as a result, the signal that is equal to -2 appears at the output of L1 that puts the block $a2$ in the state with the output voltage $-2E$ and, hence, sets $U_m = 3 - 6 - (-2) = -1$; the $a1$ output voltage is $-E$ and after switching-over $V = 6E - E - 2E = 3E$. During a further increase in U_{cn} to 9, the voltage V is set subsequently $4E$, $5E$, and so on to $9E$, by the subsequent switching-over of the blocks $a1$ and $a2$, as was

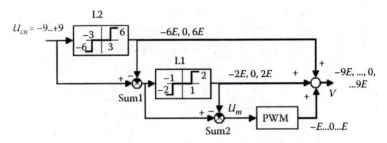

FIGURE 6.83 Control of the cascaded asymmetrical STATCOM.

described earlier. Analogously, the process of change of V takes place when U_{cn} changes from 0 to −9. Therefore, the block of the maximum power switches over with the minimum frequency. In Figure 6.84, where the control voltage and the output voltage of the each block are shown, the process of forming of the inverter output phase voltage V is depicted. It is seen that the latter is highly close to a sinusoid.

The subsystem **STATCOM** consists of the control system and the subsystem **INVERTER**. The latter in turn consists of three-phase subsystems that are united by the common capacitor and the output transformer; three blocks **Three-Phase Transformer 12 Terminals** are used for its model. The capacitor voltage is accepted to be equal to 4 kV, the transformer windings that are connected to the inverter

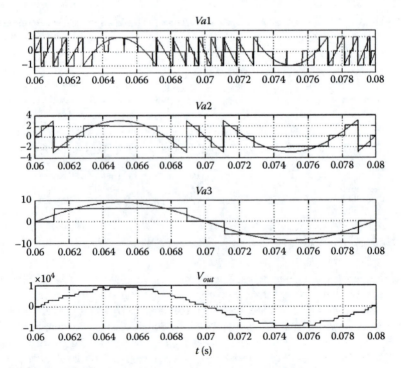

FIGURE 6.84 Voltage forming in the asymmetrical cascaded inverter.

switches have voltage rms value of 3.6 kV, and the output windings have the voltages of 8.5, 17, 51 kV for the blocks $a1$, $a2$, $a3$, respectively, so that the phase rated voltage is 76.5 kV; the total power is 24 MVA.

The subsystem of each phase consists of three single-phase (two arms) blocks **Universal Bridge** with GTO-thyristors and the control system that realizes the structure depicted in Figure 6.83. The block $a1$ switching frequency is 1 kHz; PWM is the same as seen in the previous model.

The STATCOM control system uses the previously described principle of the regulation of its current components I_d, I_q. The current controller output signals are transformed in the three-phase signal having the amplitude to nine units that are sent to the inverter phase voltage control inputs. The measures are taken to decrease the initial U_{dc} surge: at the beginning, the maximal signal at the inverter input is equal to 4.5 units, and only at $t = 0.25$ s rises smoothly to 9, besides, the reference for U_{dc} control appears also at $t = 0.25$ s.

The model considered simulates powering of the variable load from the source of the limited power with the voltage of 66 kV, and the source inner impedance takes into account the transmission line impedance. The load that is equal to 10 MW, 10 Mvar initially, increases twofold at $t = 1.5$ s. The scopes **NETWORK**, **PowerS**, **Load** show the powers, voltages, and currents of the network, STATCOM, and the load, respectively. The scope **Inverter** records the control signal and the inverter phase voltages, the scope **Volt**—U_{dc}, and the voltage amplitude V_s in PCC.

If the simulation is carried out with **Br2** open, V_s will change as shown in Figure 6.85a. The voltage differs a lot from the requested one (50 kV). If the simulation is repeated with **Br2** closed and with the switch I_q^* in the position "V," V_s will change as shown in Figure 6.85a too; it is seen that the voltage follows the reference with high accuracy in the steady state. The plots in Figure 6.85b show that the STATCOM active power is about zero and the reactive power rises to 15 Mvar. The capacitor voltage is retained with sufficient accuracy (Figure 6.85c). The shape of the inverter phase voltage is nearly sinusoidal (Figure 6.85d). The scope **Inverter** shows that the control signal does not exceed eight units; therefore, there is some margin for control.

One can repeat simulation with the switch I_q^* in the position "Q." The rated STATCOM power is accepted to be equal to 15 Mvar, and a step of such a reference is controlled in response for 0.5 s.

6.7.3 ACTIVE FILTER SIMULATION

An active filter is intended for decreasing of the higher harmonics in the grid current. It is a unit that injects in the grid such currents or voltages so that the grid current turns out to be free from the higher harmonics. Various technical decisions can be used for this purpose. Only a parallel active filter using a VSI is considered in the book; VSI connects to the grid with the help of a reactor in parallel to the nonlinear load that is a source of the higher harmonics. VSI forms the reactor current that, when added to the load current, produces the grid current without higher harmonics. The diagram of the filter connection is shown in Figure 6.86.

When the active filters are under investigation, it is usual to assume that the nonlinear load consists of the thyristor rectifier loaded on the resistance, the

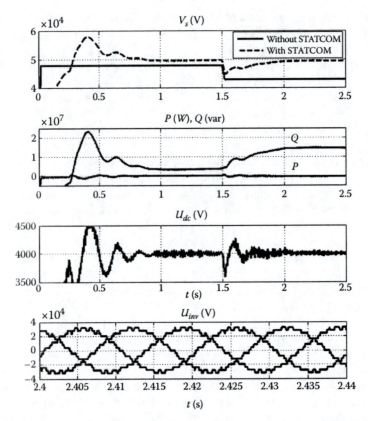

FIGURE 6.85 Processes in the system with the asymmetrical cascaded STATCOM.

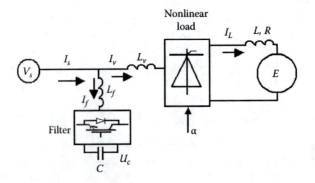

FIGURE 6.86 Diagram of the parallel active filter connection.

inductance, and the emf; the rectifier connects to the grid with the help of the reactor with the inductance L_v. In the control system, with the help of the filter, the higher harmonics of the current I_v are extracted that afterwards are reproduced, with an inverse sign, as the active filter output current I_f, so that the grid current I_s does not have higher harmonics (in ideal case).

The problem of filtration consists of two parts: forming of the signals—of the reference and the feedback—and the current control. At first, the first problem is considered.

Since it is lighter to pick out the variable component from a single-polar signal than to separate the high-frequency component from a bipolar signal, the load current is processed in the reference frame that rotates with the grid frequency. If the current is sinusoidal, the projection of the current space vector on the axis that rotates with the same rate is the constant quantity, so the load current high harmonics that are the references for the filter current controller can be found by subtracting the projected average value from the projected full value.

Another purpose for using the controller is to keep the capacitor C voltage equal to the reference value U_c. For this purpose, the same controller is used, as for the active front rectifier, whose output is added to the current reference that is aligned alongside the grid voltage space vector.

Since high-frequency current components have to be formed, the current controller speed response is of great significance. It was shown in [8] that the hysteresis controllers are most useful. Therefore, the hysteresis control of the filter currents I_{fd}, I_{fq} is used in model6_23 and described as follows. The control system uses the same principle that is applicable to IM DTC control (model5_8). In order to generate the current space vector components, the so-called virtual flux space vector is used instead of the voltage space vector that lags the voltage space vector by $\pi/2$; thus, the algorithm that is used in the previously mentioned model can be kept up, with the difference that the system dispense with zero states, so that only one table of the states **LOOKUP TABLE0** is used. In order to keep up the algorithm, the transformation abc_dq is not carried out by the standard block SimPowerSystems, as in the previous models, but is carried out by the following relationships:

$$I_d = I_\alpha \cos \theta + I_\beta \sin \theta$$

$$I_q = I_\alpha \sin \theta + I_\beta \cos \theta$$

(6.45)

Here, the component I_{fq} is perpendicular to the flux vector and, consequently, aligned alongside the voltage space vector and is "responsible" for the active power; the voltage controller output is added to the reference of this component. The difference $I_{fq}^* - I_{fq}$ comes to the input of the torque error, and the difference $I_{fd}^* - I_{fd}$ comes to the input of the flux error of the DTC system. The features of such a method of VSI current control are investigated in [9]. The blocks **Zero-Order Hold** model the computation delays. The controller block diagram is shown in Figure 6.87.

The described variant of the control system is the main one; it can be called "disturbance-compensating control" where disturbances are the load current higher harmonics. But there are other possibilities too. If I_{fd}^* is equal to the full current $-I_{vd}$, the load current reactive component will be compensated by the filter and the grid reactive power will be zero. If the current $-I_s$ is send to the input of the control system, instead of I_v, the feedback system results because the filter by the feedback principle will try to annihilate the grid current higher harmonics. Such a decision can be useful in some cases, when there is a lot of nonlinear loads placed far apart from one another and far from the point of grid entry.

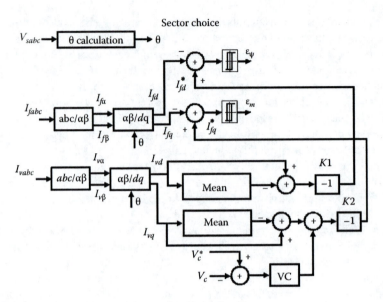

FIGURE 6.87   Block diagram of the active filter controller.

Compensation for all higher harmonics created by the load is supposed above, and the filter power has to be selected, taking into account the power of all higher harmonics. However, there can be situations when it is sufficient to only compensate for a part of the higher harmonics, in particular the lower-frequency ones; the higher-frequency harmonics can be filtered by the passive filters. Such a filtration is called a selective one. There are various methods of extracting the harmonics subjected to compensation, including the so-called synchronous method used as follows. If the Park transformation is used with the rotational frequency of the reference frame equal to $n\omega_s$ for the three-phase signal V that contains the fundamental harmonic with frequency of ω_s and a number of the higher harmonics with frequencies of $k\omega_s$, the output signal will contain the direct component that is equal to the amplitude of the harmonic $k = n$, and all remaining harmonics are transformed in the bipolar signals with the zero average values. If the direct components are picked out in the signals V_d and V_q obtained and, afterwards, the reverse Park's transformation is used with the rotational frequency of the reference frame equal to $n\omega_s$, the three-phase signal that contains only the harmonic of the order n appears at the output. If this signal, with the opposite sign, is taken as a reference for the current controller of the active filter inverter, the grid current will not contain this harmonic. Since the inverter has to keep up DC link capacitor voltage that is fulfilled by the active current control, the three-phase signal is added to the reference; this signal is in-phase with the grid voltage and the sign and the amplitude of this signal are controlled, depending on the capacitor voltage.

In model6_23, the thyristor rectifier with R-L-E load connects to the grid 380 V with help of the reactor with the inductance of 0.3 mH. The firing pulses are generated by the block **Synchronized 6-Pulse Generator**. Since the supplying voltage is distorted essentially, the reference voltages for this block are formed by the first-order filters with the time constant of 5.5 ms that give the phase shift of 60° for the frequency of 50 Hz; therefore, the diagram of connection of the reference and grid voltages changes. For instance, the output of the filter V_{ab}, with the opposite sign is sent to the input CA because $-V_{ab}$ leads V_{ca} by 60°. The load rated current is 150 A, which gives the load power of 60 kW. This current is obtained as $E = 400$ V, $\alpha = 29.5°$. When α changes, the value E has to change, in order to maintain the rated current. The filter inverter is connected in parallel to the load. Two options of its control system are provided, which are in the subsystems **Inv_Contr** and **Inv_Contr1**, in which the scopes are placed too. The first subsystem is intended for the filtration of all harmonics, the second one—for the selective filtration. There are two switches **SW1, SW2** in the subsystem **Filter_Contr/Inv_Contr** that select the control structure. If **SW1** is in the position L and **SW2** in the position F, the main filter variant is realized (disturbance-compensating control, harmonic compensation); if **SW2** is placed in position Q, the load reactive current is compensated; if **SW1** is in position S and **SW2** in position F, the grid higher harmonics are reduced by the feedback (at this point, the gains of K_1–K_3 have to be increased to 5–10); if in this situation we set **SW2** in position Q, the grid reactive current decreases nearly to zero. The capacitor voltage is 825 V. The signal average values are determined by the blocks **Mean Value**.

The scope **Currents** records the grid current, the input current of the thyristor rectifier, the filter current, and its rms value that is necessary for filter parameter choice. The scope **IL** fixes the rectified current, and the scope **Uc**, the voltage across the filter capacitor. The scope **Supply** shows the grid voltage and current (repeatedly).

The compensation for the fifth and the seventh harmonics actuated by the thyristor rectifier is carried out in the subsystem **Inv_Contr1**. Pay attention to the following fact. Fourier series for the phase A current is

$$Y_a = D\left(\sin x - \frac{1}{5}\sin 5x - \frac{1}{7}\sin 7x - \cdots\right), \quad x = \omega_s t, \ D \text{ is a factor,}$$

for the phase B

$$Y_b = D\left[\sin(x-120) - \frac{1}{5}\sin 5(x-120) - \frac{1}{7}\sin 7(x-120)\ldots\right]$$

$$= D\left[\sin(x-120) - \frac{1}{5}\sin(5x+120) - \frac{1}{7}\sin(7x-120)\ldots\right]$$

and for the phase C

$$Y_c = D\left[\sin(x+120) - \frac{1}{5}\sin 5(x+120) - \frac{1}{7}\sin 7(x+120)\ldots\right]$$

$$= D\left[\sin(x+120) - \frac{1}{5}\sin(5x-120) - \frac{1}{7}\sin(7x+120)\ldots\right].$$

It is seen that the fifth harmonic rotates in the opposite direction; therefore, the signal $-5\omega_s$ has to be sent to Park's transformation for the fifth harmonic.

The block PLL is used for determination of the reference frame angle position; this block is also used for the generation of phase voltage signals in the point of connection without distortions. The fifth and the seventh harmonics are extracted in the subsystems **5Harm** and **7Harm** according to the principle described earlier, and the direct components are picked out by the low-pass filters with the cutoff frequency of 1 Hz. The total signal to compensate is formed at the output of the summer **Sum**. The signals with the amplitudes equal to 1 that are in-phase with the corresponding phase voltages are generated by the blocks **FUa, FUb, FUc**. These signals are multiplied by the output of the capacitor voltage controller in the block **Product**; this way the capacitor voltage is kept constant by an effect on the active components of the fundamental harmonic of the filter phase currents. In the block **Sum1**, the resulting three-phase signal is added to the signal of the higher harmonics that are to be compensated, and thus, the reference for the phase current controllers is formed.

At first, simulation is carried out with **Switch** in the upper position.

Run simulation with the switches **SW1** in the position L and **SW2** in the position F (with $K_1 = K_2 = K_3 = 1$). The obtained plots of the grid current I_s, of the rectifier input current I_v and the filter current I_f are shown in Figure 6.88. It is seen that although I_v is nonsinusoidal essentially, the grid current is close to the sinusoid, THD = 5%. rms filter current is equal to 30 A. If Fourier transformation of the grid voltage and current with the same origin of a scale is performed, one can see that the phase displacement between their fundamental harmonics is 31.7°, which corresponds to

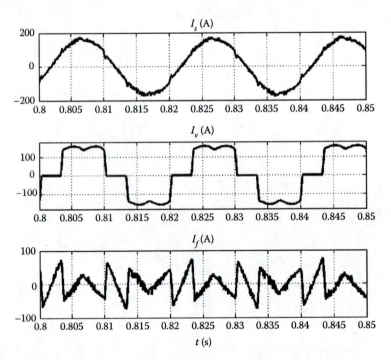

FIGURE 6.88 Processes in the active filter during load current measurement.

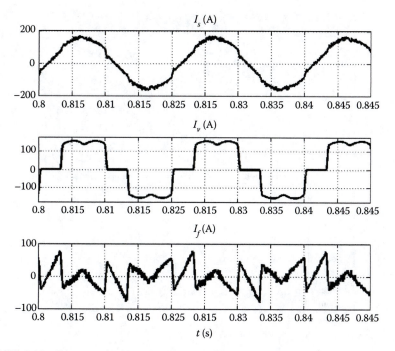

FIGURE 6.89 Processes in the active filter during grid current measurement.

an affirmation that the phase shift at the rectifier input is equal approximately to the firing angle. Repeat simulation with **SW2** in the position Q. The phase displacement reduces nearly to zero (0.4°), but filter rms current increases to 78 A and THD increases to 7.1%.

Repeat simulation with **SW1** in the position S and **SW2** in the position F, with the setting $K_1 = K_2 = K_3 = 10$. The plots obtained are shown in Figure 6.89; grid current THD = 6.5%, and the phase displacement between the fundamental harmonics of the grid voltage and current is 26.5°. If the simulation is carried out with **SW2** in the position Q, the said phase displacement decreases to 3°; that is, the reactive power consumed from the grid is nearly zero.

Run simulation with **Switch** in the lower position. The currents of the load and the grid are shown in Figure 6.90, and the content of higher harmonics is depicted in Figure 6.91. It is seen that the fifth and the seventh harmonics decrease by 10–15 times. The capacitor voltage U_c is equal to the reference one.

It is worth noting that a rise in the order of the harmonics compensated demands an increase in the speed of the computer.

6.7.4 STATIC SYNCHRONOUS SERIES COMPENSATOR SIMULATION

Static synchronous series compensator—SSSC is VSI with a capacitor in DC link that, with the help of a transformer, is connected in series with the transmission line (Figure 6.92). SSSC produces the nearly sinusoidal three-phase voltage V_q with the network frequency, whose phase is shifted by ±90° about the line current (it is shown

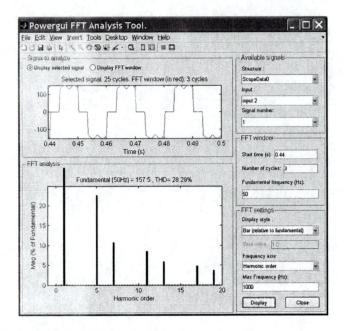

FIGURE 6.90   Spectrum of the load current.

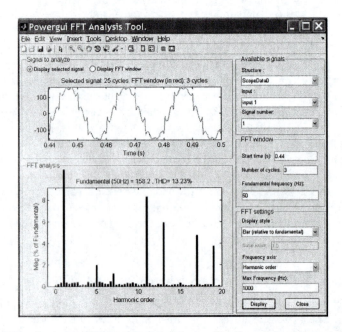

FIGURE 6.91   Spectrum of the grid current during filtration of the fifth and the seventh harmonics.

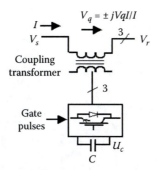

FIGURE 6.92 Diagram of SSSC connection.

in Figure 6.92. The maximum SSSC voltage does not exceed 10% of the network voltage usually. Therefore, SSSC executes the same functions as a series capacitor, but unlike it, SSSC voltage does not depend on the flowing current. It is worth noting that with the small line current, SSSC does not operate properly, because this current can be insufficient for producing the active power requested for SSSC losses compensation.

The SSSC control system regulates the amplitude and the phase (relatively the flowing current) of the SSSC voltage. Like a STATCOM, SSSC has two modifications: with the rectangular output voltage, which is proportional to the capacitor voltage U_c and, hence, with changeable U_c (without PWM), and with PWM and $U_c \approx$ constant. SSSC can operate in various modes: constant V_q, constant conductance X_q ($X_q = V_q/I$), and suppression of different type of oscillations.

The SSSC model in SimPowerSystems is intended for simulation in the Phasor mode, whose picture is shown in Figure 6.71b. The block is connected in series to the transmission line with the help of the terminals A1–A2, B1–B2, C1–C2. The reference output voltage is given to the input V_{qref}. When a logical 1 comes to the input *Bypass*, the breaker closes that is set in parallel to SSSC input–output. Both these inputs are seen when the corresponding mode of the external control is selected; 17 inner signals are collected at the output *m* as a vector, namely,1, 2, 3 are the input phase voltages in pu (complexes), 4 is the capacitor voltage U_c (V), 5, 6, 7 are the output phase voltages in pu (complexes), 8, 9, 10 are SSSC voltages (difference between 1, 2, 3 and 5, 6, 7), 11, 12, 13 are the currents through SSSC in pu, 14 is the reference voltage (pu), 15 is the SSSC measured voltage (pu), 16 is the measured current (pu), and 17 is the modulation factor, $0 < m < 1$.

The SSSC control system structure is shown in Figure 6.93. Its variables are aligned alongside the current space vector. The voltage V_q produced by SSSC is found as a difference between the input and output voltage components, which are perpendicular to the current space vector. The RVq controller sets this voltage equal to the reference one, and the feedforward signals that are proportional to the reference voltage and to the voltage drop across the SSSC inductance X_l are used for improving the dynamic characteristics. As a result, the reference for the component V_{q_inv} of the output VSI voltage forms. The reference for the component V_{d_inv} is formed by the capacitor voltage controller RV. Afterwards, both these voltages are transformed in the three-phase system that is a reference for generating of the VSI gate pulses.

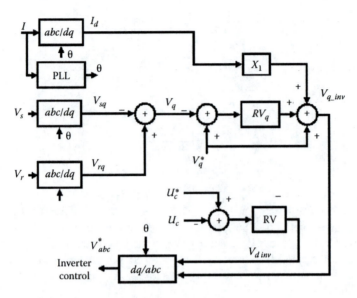

FIGURE 6.93 Block diagram of the SSSC control system.

The dialog box of SSSC has two widows: *Power data* and *Control parameters.* The transmission line voltage and frequency, the SSSC rated power and its maximum voltage, the active and the inductive SSSC impedances in pu, referring to the SSSC rated parameters (these impedances are defined by the SSSC output transformer and other inductances, if they exist), SSSC initial current, the rated value of DC link voltage, the capacitor capacitance, the control mode of the by-pass breaker, the mode of V_{qref} control (by inner or by external signal), the maximum rate of V_{qref} changing, and the controller gains are specified in the these windows.

The demonstrational model power_sssc.mdl with this block is included in SimPowerSystems. The electric power system simulated consists of two equivalent sources with power of 2.1 GVA and 1.4 GVA and voltage of 500 kV that are connected by two parallel transmission lines, with each with a length of about 300 km, and by a common transmission line with a length of 50 km. Some active loads are placed along the lines; the main load has the rated power of 2.2 GW that is modeled by the block **Three-Phase Dynamic Load**. The voltages of the sources are chosen such that the powers transmitted along the parallel lines from the first source to the load are 664 and 563 MW, and the power from the second source to the dynamic load is 990 MW. SSSC with power of ±100 Mvar connects to the input terminal of one of the parallel lines. SSSC has a DC link voltage of 40 kV and a DC link capacitor with capacitance of 375 μF. SSSC connects to the transmission line with the help of the transformer with the equivalent impedance of 0.16 pu at the power base of 100 MVA. The block of the ground fault is placed in the middle of the second parallel line. SSSC in this model is intended for dumping of the oscillations that arise after fault clearance, with the help of the device POD (Power Oscillation Damping) controller that operates, as a whole, as a differentiator in the frequency range 0.01–1 Hz; when power oscillations appear,

POD generates the signal that affect V_{qref} in such a way that these oscillations are dumped. A more thorough understanding of the SSSC function one be received by reading the *Help* section.

This demonstrational model offers further possibilities for studying the SSSC operation in various modes more thoroughly and gain a better understanding. At the same time, as it was noted repeatedly before, the Phasor simulation method does not reveal all the peculiarity of functioning and use of SSSC. Therefore, use of the detailed SSSC model that uses a three-level VSI with the switching frequency of 1050 Hz is shown in model6_24. Each of the capacitors has the capacitance of 2 mF, and the voltages across them are 4 kV. The processes of the initial charging and voltage balancing are not simulated. The source of 6310 V, through the inductive circuit that models the transmission line, feeds the load whose active and reactive powers can change from 5 MW, 5 Mvar to 10 MW, 10 Mvar. The blocks **V-I_S** and **V-I_R** measure the network and the load currents with the base voltage of 6300 V and the base power of 16.6 MVA.

The block **Three-Phase Transformer 12 Terminals** is used for connection of VSI in series to the network; the low-voltage windings are inserted at the network break and the high-voltage windings are connected in star and, with the help of the reactor with the inductance of 1 mH, connect to the VSI output. Closing of **Br1** brings SSSC out of the work. Inverter control (subsystem **Control**) is made according to the structure shown in Figure 6.93. If the switch **SW** is in the position I, the SSSC voltage reference V_{qref} is formed within the subsystem, in the position E—by the external controller, and in the considered model—by the voltage controller on the bus V-I_R. In the subsystem **Measurements** the scopes are gathered: **PowerS** fixes the power and the voltage on the bus **V-I_S PowerR** records the power, current, and amplitude of the positive sequence of the voltage on the bus **V-I_R**; and **Udc**, **modVq** shows the capacitor voltage and the value of SSSC output voltage component that is perpendicular to the current vector.

Run simulation with **Br1** closed. The plot of the voltage on the bus **V-I_R** is obtained (the upper plot V_r on Figure 6.94). It is seen that the bus voltage differs from the requested. Repeat simulation with **Br1** open and with **SW** in the position E. The rest of the plots in Figure 6.94 are obtained. It is seen that the voltage **Vr_SSSC** is much closer to the requested one; after transients the error is about zero. SSSC output changes from 0.035 to 0.06. The capacitor voltage is equal to the reference, and the network current is sinusoidal. By scopes **PowerS** and **PowerR** one can see that the difference between the reactive powers, that is, SSSC power, changes with the load increasing from 0.02 to 0.075.

6.7.5 UNIFIED POWER FLOW CONTROLLER SIMULATION

Unified power flow controller (UPFC) can be used to control, simultaneously or by choice, the active power P, the reactive power Q transmitted by the line, and the line voltage V_s. UPFC consists of STATCOM and SSSC with a common DC link (Figure 6.95). The former connects to the line by the parallel transformer PTR, and the latter by the series one STR. The main STATCOM function is to keep constant the DC link voltage U_{dc} and to control V_s by the reactive power

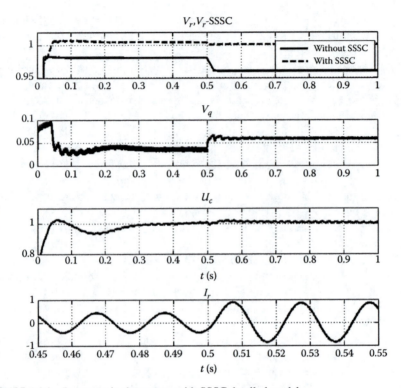

FIGURE 6.94 Processes in the system with SSSC detailed model.

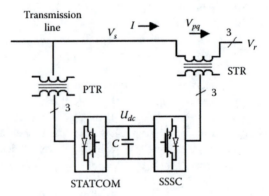

FIGURE 6.95 Block diagram of the UPFC.

flow control. Since U_{dc} is constant, independent of SSSC operation, the latter can form its voltage output space vector $\mathbf{V}_{pq}$ with any angle relative to the line current space vector, and the angle position ρ of the vector $\mathbf{V}_{pq}$ is counted from the vector $\mathbf{V}_s$. If the SSSC inner reactance and the difference between the voltage amplitudes in the points of SSSC connection and the line terminal are neglected, and these voltages are taken equal to V with the mutual angle δ (the so-called

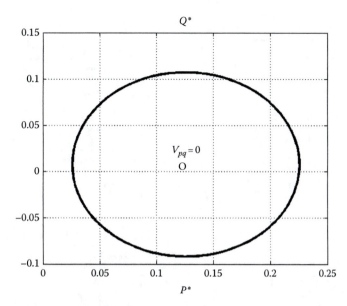

FIGURE 6.96 Idealized controllable region.

simplified model of the two-machine system), the active and the reactive powers transmitted by the line with the reactance X are defined as [6]

$$P = \frac{V^2}{X}\sin\delta - \frac{VV_{pq}}{X}\cos(\delta/2 + \rho) \tag{6.46}$$

$$Q = \frac{V^2}{X}(1 - \cos\delta) - \frac{VV_{pq}}{X}\sin(\delta/2 + \rho) \tag{6.47}$$

The dependence of the quantities $P^* = P\,X/V^2$, $Q^* = Q\,X/V^2$ with $V_{pq} = 0.1V$, $\delta = 7.22°$, $\rho = 0$–$360°$ is shown in Figure 6.96. Any values of the active and the reactive powers that are placed within this closed curve can be obtained. It is seen that a rather small change in SSSC voltage can increase the active power by 80% or generate or absorb the reactive power that is nearly equal to the transmitted active power.

The STATCOM control system in UPFC can be the same as shown in the separate STATCOM (Figures 6.74 and 6.77), but the SSSC control system differs. It is depicted in Figure 6.97. The values of active and reactive powers are calculated by (3.12) and (3.13) and come as a feedback to the inputs of the voltage V_d and V_q controllers, whose outputs are transformed into three-phase signals for the inverter control. There is also an opportunity to set the required V_d and V_q values directly, for instance, from the master controller.

The UPFC model in SimPowerSystems is intended for simulation in Phasor mode, whose picture is shown in Figure 6.71c. The block is connected in series to the transmission line with the help of the terminals A1–A2, B1–B2, C1–C2. The reference values of the SSSC active and reactive powers are sent to the input PQ_{ref}; the breaker that is connected in parallel to SSSC input-output closes when the logic 1 comes to the

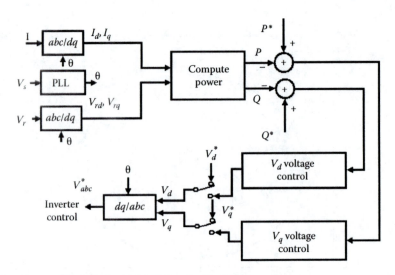

FIGURE 6.97   Block diagram of SSSC control in UPFC.

input *Bypass*. Both these inputs are seen when the corresponding operation modes are selected. When the logic 1 comes to the input *Trip*, SSSC is shunted and STATCOM is turned off; 34 inner signals are collected at the output *m* as a vector (three-phase voltages and currents are complexes in pu), namely,1, 2, 3 are the input phase voltages, 4, 5, 6 are the phase currents flowing in STATCOM, 7 is the DC link voltage (V), 8, 9, 10 are the phase output voltages, 11, 12, 13 are the SSSC voltages, 14, 15, 16 are the currents in STR, 17 is the positive sequence amplitude of the input voltage, 18 is the reference value of this voltage, 19 is the STATCOM reactive power, 20 is the reference of this quantity, 21 is the current I_d flowing in STATCOM, 22 is the current I_q flowing in STATCOM, 23 is I_d reference, 24 is I_q reference, 25 is the STATCOM modulation factor, $0 < m < 1$, 26 is the SSSC output active power, 27 is SSSC output reactive power, 28 is the reference for the active power, 29 is the reference for the reactive power, 30 is the SSSC V_d component, 31 is the SSSC V_q component, 32 is the SSSC V_{dref} component, 33 is the SSSC V_{qref} component, and 34 is the SSSC modulation factor, $0 < m < 1$.

The dialog box of UPFC has several windows for parameter specification that contain mainly the same data that for the STATCOM and SSSC used separately.

The demonstrational model power_upfc.mdl is included in SimPowerSystems that shows the use of this block. The simulated electric power system consists of two equivalent sources having the powers of 1 GVA and 1.2 GVA and the voltage of 230 kV and connected by two parallel transmission lines 65 and 100 km length; the second line has a voltage of 500 kV and connects to the first line with the help of the transformers. A load of 200 MW is placed in the middle of the second line; the rest of the power generated by the sources is transmitted to the external power system having the power of 15 GVA. The UPFC with a power of 100 MVA is also placed in the center of the second line. With the help of the option *Powergui/Load Flow and Machine Initialization*, the source active powers are set as 0.5 and 1 GW.

The purpose of UPFC is to decrease the load of **Tr2**. *Voltage regulation* is chosen for STATCOM and *Power flow control* for SSSC. It can be seen after simulation that the power distribution changes (it is shown on the model circuits with the blue numbers); the power transferred by **Tr2** decreases to 796 MW. A more detailed model description is given in the corresponding *Help* section.

SimPowerSystems contains also the detailed UPFC model `power_upfc_gto48p.mdl`. It is of great interest; therefore, it will be considered in detail. The model consists of STATCOM and SSSC with a common DC link. The capacitors of each unit that are connected in parallel in the UPFC mode can be disconnected, which gives the opportunity to use STATCOM and SSSC independently. The same circuits, as in model6_20, are used in each unit; the difference is the values of the transformer primary voltages: 125 kV for STATCOM and 12.5 kV for SSSC. The control system **UPFC Controller** contains three subsystems intended, respectively, for control of STATCOM, SSSC in UPFC, and SSSC operating independently. The latter is not considered because of control of the separated SSSC as described earlier. The STATCOM control system **SH_Control** is the same as in model6_20; that is, it contains the voltage controller. The SSSC control system **SE_Control** contains the blocks for forming of the reference currents I_d^*, I_q^*, PI controllers of these currents, whose outputs define the reference values V_d^*, V_q^* of the SSSC output voltage components; and the blocks for reproduction of these quantities. The vectors are aligned alongside the space vector of the voltage on the bus **B1**. The current reference values are found from (3.12), (3.13) with specified reference values of the active and reactive powers on the bus **B2** and with computed actual values of V_{d2}^*, V_{q2}^* of the voltage space vector on this bus.

The most complicated part of the control system are the blocks that generate the SSSC inverter gate pulses that provide the reference output voltage vector. Since PWM is not used and the output voltage has the rectangular shape with the amplitude to be proportional to the capacitor voltage V_{dc} (see model6_20) that changes, when STATCOM acts, the only way to the voltage control is to remove the rectangular sectors by a change in angle σ (see Section 6.7.2.1); at that point, the value σ depends on both the magnitude V^* of the required SSSC voltage and V_{dc}. Taking into account (4.3) and keeping in mind that the outputs of four transformers with the transformation ration of 1 are connected in series, we receive the SSSC output voltage amplitude as

$$V = 4\left(\frac{2}{\pi}\right)V_{dc}\sin\left(\frac{\sigma}{2}\right) \tag{6.48}$$

where, when $V = V^*$, the needed value σ is calculated from. As for realization of the requested relationship between components V_d^*, V_q^*, it is carried out by influencing the angle α (see model6_20). We will not give the details of controllers because of space limitations since most details are to be repeated from the previous examples.

The model adjustment is carried out in the window by clicking the button *UPFC GUI*. One of five operation modes can be selected: STATCOM with the reactive power control, STATCOM with the voltage control, SSSC with the specified voltage, STATCOM with the voltage control and SSSC with the specified voltage, UPFC. Only the last mode is of interest to us, and the mode parameters are given in last two lines of the window: The initial and the final values of the active and reactive powers are specified and also the time points at which they change. The reference voltage for STATCOM V_{ref} is specified in the third line of the window.

Three scopes are set for observing UPFC operation, which is initiated by clicking *Show Scopes*. The scope **STATCOM** with four axes shows: by the first axis—the sum of the inverter output voltages, the voltage on the bus B1, the STATCOM output current; by the second axis—the voltage across one of the capacitors; by the third axis—the actual and the reference values of the reactive power (the second signal is not actual in this mode); by the fourth axis—the actual and the reference voltages in PCC. The scope **SSSC** with six axes shows: by the first axis—the SSSC output voltage; by the second axis—the SSSC current; by the third axis—the actual and the reference values of the SSSC output voltage amplitude; by the fourth axis—the voltage across one of the capacitors; by the fifth axis—the active powers of the transmission lines L1, L2, L3; by the sixth axis—the reactive powers of these lines. The scope **UPFC** with four axes shows: by the first axis—the actual and the reference values of the active power; by the second axis—the same for the reactive power; the third and fifth axes repeat the last two axes of the previous scope. Besides, two displays are set with the purpose to show the voltages in PCC and the powers on the bus **B2**.

By clicking on the button *Show UPFC Controllable Region* that shows the controllable region, it can be seen that the initial power values (without UPFC) are equal to 870 MW and 60 Mvar and the limiting values are equal to 553–1225 MW and 295 Mvar minus 370 Mvar.

Run simulation. Note that with STATCOM reference voltage $V_{ref} = 1$, the active power changes from 8.7 to 10 pu with the base power of 100 MW at $t = 0.25$ s and the reactive power changes from −0.6 to 0.7 pu at $t = 0.5$ s. The first two axes of the scope **UPFC** show that the actual values of the powers follow the reference values with a delay of 0.1–0.15 s after the change in reference (Figure 6.98). It is seen that the STATCOM output voltage is nearly sinusoidal, whereas SSSC output voltage is much distorted.

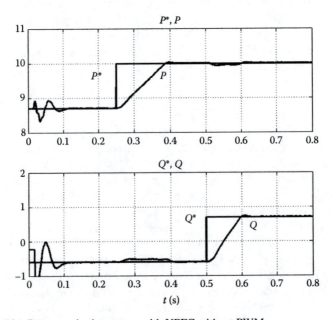

FIGURE 6.98 Processes in the system with UPFC without PWM.

UPFC with PWM, that is, with the constant value of the capacitor voltage V_{dc}, is used in model 6_25. The same circuits, as in model4_8, are used for STATCOM and SSSC: Each unit contains two similar blocks and each block consists of 2 three-level VSI and one three-phase transformer with 12 terminals. The windings of transformers that are connected to the transmission line connect in each phase in series and form a star-like connection and the terminals of each phase winding at VSI side are connected to the VSI terminals of the same names. The VSIs of the same block are controlled in antiphase, and the control of the VSIs of two blocks is shifted by 30°. The STATCOM control system **Statcom_Control** is the same as in Figure 6.77; that is, it contains the controllers of the transmission line voltage and the capacitor voltage. The difference is that 2 three-phase signals V_{abc1} and V_{abc2}, shifted one after the other by 30°, are formed at the outputs of the control system; these signals are shifted by ±15° relative to the voltage space vector in PCC that comes to the blocks of gate pulse generation for four VSI. This shift is accomplished with the help of the block **Synchron**. The unit for proportionaly decreasing the output signals on d and q axes, in case when the modulation factor could exceed 1, is provided.

The SSSC control system **SSSC_Control** is made according to the diagram in Figure 6.93 and contains the unit for generation of two three-phase signals shifted by 30° in the case of VSI control, analogous to the block **Statcom_Control**, its output is limited in the same manner. During power control, the feedback is closed by the switches **Switch1**, **Switch2** sometime after the start of simulation.

The purpose of UPFC is to reduce the power flow in the line 1 (e.g., keeping in mind the subsequent installation of the additional loads or carrying out repair works). The switch **On_S** turns on STATCOM, the switch **On_C**—SSSC. The constants **Pref** and **Qref** define the active and the reactive power in the power control mode. The scope **Lines** shows the voltages and powers in the points of the system, in which the measures B1–B5 are set. The displays that show the values of the powers in the same points are set to make observation lighter. Two scopes are set in addition in the subsystem **VPQ Measurements**. The scope **Statcom** records the instantaneous values of the STATCOM voltage and the current and the capacitor voltage; the scope **SSSC** records the V_{pq} components and the values of the power at the UPFC output.

Let us begin to work with the model. Carry out simulation with the switches **On_S** and **On_C** in the state "Off," The values of the powers indicated in red result at $t = 2$ s. Set both switches in the state "On" and the switches **SW1, SW2** in the subsystem **SSSC_Control** in the upper position. If the simulation is executed with the reference values V_{qref}, V_{dref} equal to (±1, 0), (0, ±1), (±0.7, ±0.7) (8 pairs total), the controllable region shown in Figure 6.99 will be obtained for $t = 2$ s.

Carry out simulation with the initial values in the blocks **Step Vqref** and **Vdref** of the subsystem **UPFC** equal to 0.7, the final values to −0.7, and a step time of 0.3 s. The obtained process is shown in Figure 6.100. It is seen that the power values are set quickly, but the damped oscillations have place. In practice, particular with regard to their elimination, the references do not change in steps; rather, they change smoothly.

Turn on the power controllers by setting the switches **SW1, SW2** in the subsystem **SSSC_Control** in the lower positions. The time of controller switching is set

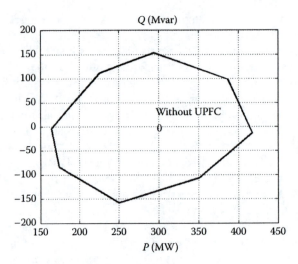

FIGURE 6.99 Controllable region for model6_25.

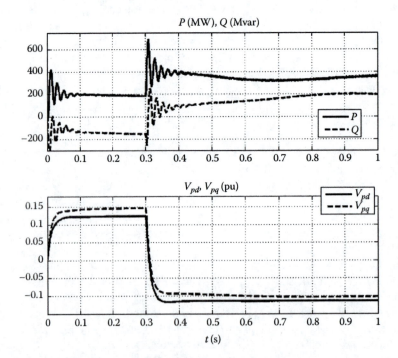

FIGURE 6.100 Processes in model6_25 for the output voltage reference.

equal to 0.1 s (switches **Switch1**, **Switch2**); the initial value of the active power is 290 MW, and it changes to 350 MW at $t = 1$ s. The reference of the reactive power is -30 Mvar. The processes plotted in Figure 6.101 are obtained. The values of the active power are indicated on the model by numbers appearing in blue. It is seen that power transferred by the first parallel line decrease by 46 MW. The scope **Statcom**

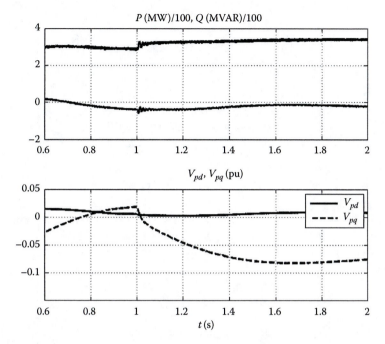

FIGURE 6.101　Processes in model6_25 for the output power reference.

shows that the capacitor voltage is equal to the reference. It is worth noting that simulation of this system runs slowly, so that the processes fixed do not reach the steady states a little.

6.7.6　PHASE-SHIFTING TRANSFORMER SIMULATION

It follows from (6.39) that the power flow in the transmission line can be controlled by changing the angle δ between the voltages in the beginning and in the end of the line. For this aim, in addition to the devices mentioned previously, the three-phase phase-shifting transformers can be used. The small power transformers are actually the IM with the wounded locked rotor, but usually they have a core with the three-phase windings; these windings are connected in a special way and have the taps; the phase shift between the input and output voltages are obtained by the switching-over of the taps. Such transformers have a variety of designs; simulation of one type of such a transformer is provided in SimPowerSystems. The transformer windings are connected in the closed hexagon (hence, the transformer is called Delta-Hexagonal). The primary W_1 and the secondary W_2 windings are connected as shown in Figure 6.102. All the windings have the same number of turns and are calculated for the voltage 2/3V where V is the rated phase-to-phase voltage of the transmission line. The secondary windings have taps, the voltages from these taps can come to input-output through switching devices; moreover, the switching devices connected with the primary (A, B, C) and the secondary (a, b, c) circuits move in the opposite directions symmetrically about the middle points of the windings (shown in figure). When the switching devices of both circuits are in the same

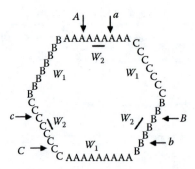

FIGURE 6.102   Scheme of winding connection of the phase-shifting transformer.

(middle) position, the phase shift between input and output voltages is zero. When the switching devices move in the opposite directions, the phase shift ψ appears, whose value is

$$\psi = 2 \arctan\left(\frac{-z}{1.73}\right), \quad z = \frac{N}{N_t} \tag{6.49}$$

where
  N is the switching position
  N_t is the number of the taps of each half-tapped winding (in SimPowerSystems $N_t = 10$, so that each winding has 21 taps total)

Therefore, $-60° \le \psi \le 60°$.

SimPowerSystems includes the transformer model that operates only in the Phasor mode: *Three-Phase OLTC Phase Shifting Transformer Delta-Hexagonal* (*Phasor Type*). However, in the demonstrational model `power_PSTdeltahex.mdl`, the complete model of such a transformer is given.

The model consists of three similar subsystems for the main circuits and the control subsystem. The former are connected according to the diagram in Figure 6.102. Each of these subsystems contains the model of the block **Multi_Winding Transformer**, whose circuits and parameters are selected properly, and of the block of switches. The latter consists of two circuits that contain the breakers connected to the winding terminals and the output switching-over circuit that connects the outputs of the circuits previously mentioned with the terminals of the primary and the secondary circuits, or in contrast, depending on the control signal *SwInv*. Eleven switches are set in each circuit, and the switches with the same numbers turn on simultaneously. The switches of both circuits connect with the middle point simultaneously.

When the pulse *Up* or *Down* comes to the control system input, the number of the requested tap increases or decreases by 1; this number appears at the output *TapNo* after *Tap selection time*. This number is compared with the possible positions of the tap switches, and when coincident with one of them, two switches with the same number are closed, one in the each circuit. If the signal *TapNo* = 1 during its increase or *TapNo* = −1 during its decrease, the signal **SwInv** is set in the state that determines which of the circuits forms the input voltage and which one forms the output voltage.

For the signal *SwTr*, the switching order is the same as the one followed for multi-winding transformer (Chapter 2).

The described model simulates operation of the delta-hexagonal transformer in a rather detailed manner, but there are other types of phase-shifting transformers; in order to use them the developer has to elaborate the model considered for use. At the same time, it is necessary to estimate quickly the effect of putting of such a transformer in the network. For this purpose, the model of IM with wounded rotor can be used. This idea is used in model6_26, which is a modification of model6_25. The UPFC model is replaced with the model of IM. IM with the wounded rotor has a power of 500 kVA and voltage of 230 kV. As a mechanical input, IM speed is used. The block **Control** generates the short pulses to this input, with frequency of 4/3 Hz and with area of 6.5°. At first, this area is positive and then the shift δ increases, beginning from $\delta = 0$, reaching the value $\delta = 39°$; at $t = 6$ s this area becomes negative, and δ decreases, reaching the value $\delta = -32°$. The simulation purpose is to show the influence of the displacement angle δ on the power distribution in the parallel lines. The scope **VP** records the angle δ, and the scope **Lines** fixes the powers on the buses **B1, B2, B3, B4** and the voltages on the buses **B1, B2**. The values of powers are shown by the blocks **Display** too. The curves of the powers together with the value δ are shown in Figure 6.103. It is seen how the powers in the parallel lines are redistributed (*P*1 and *P*2), with the changing δ; the power *P*3 that is transferred to the indefinite power network does not change perceptibly.

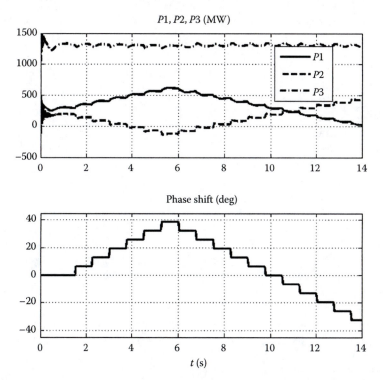

FIGURE 6.103 Processes that show operation of the phase-shifting transformer.

If the speed of simulation in this model is compared with the simulation speed that will occur when IM is replaced with the previously described ***Three-Phase OLTC Phase Shifting Transformer Delta-Hexagonal (Detailed Type)***, it can be seen the former requires much less simulation time and provides information about the actual shapes of voltages and currents in the system.

REFERENCES

1. Kundur, P., *Power System Stability and Control*, McGraw-Hill, New York, 1994.
2. Gross, C.A., *Power System Analysis*, John Wiley & Sons, New York, 1986.
3. Franklin, G.F., Powell, J.D., and Emami-Naeini, A., *Feedback Control of Dynamic Systems*, 3rd edn., Addison-Wesley Publishing Company, Reading, MA, 1994.
4. Mathworks, *SimPowerSystems™, User's Guide*, The MathWorks Inc., Natick, MA.
5. Singh, B., Murthy, S.S., and Gupta, S., STATCOM-based voltage regulator for self-excited induction generator feeding nonlinear loads, *IEEE Transactions on Industrial Electronics*, 53(5), 1437–1452, October 2006.
6. Hingorani, N.G. and Gyugyi, L., *Understanding FACTS; Concepts and Technology of Flexible AC Transmission Systems*, IEEE Press, New York, 2000.
7. Rodriguez, J., Bernet, S., Wu, B., Pontt, J., and Kouro, S., Multilevel voltage-source-converter topologies for industrial medium-voltage drives, *IEEE Transactions on Industrial Electronics*, 54(6), 2930–2945, December 2007.
8. Buso, S., Malesani, L., and Mattavelli, P., Comparison of current control techniques for active filter applications, *IEEE Transactions on Industrial Electronics*, 45(5), 722–729, October 1998.
9. Perelmuter, V.M., *Direct Control of the AC Motor Torque and Current (in Russian)*, Osnova, Kharkov, 2004.

List of the Models on CD

Model1_1: First example

Model1_2: Two generators supplying the remote load. Powergui demonstration

Model1_2a: Powergui demonstration, continuation

Model2_1: Three-phase programmable voltage source

Model2_2: Three-phase dynamic load

Model2_3: Mutual inductance

Model2_4: Saturated transformer with hysteresis

    Hysteresisa.mat: Hysteresis characteristic of the saturated transformer

Model2_5: Model of the three-phase regulated under load transformer

Model2_6: Transmission line with the regulated transformer

Model2_6_Phasor: Transmission line with the regulated transformer, simulation in Phasor mode

Model2_7: Rectifier with four-phase-shifting transformers

Model2_8: Energizing of a long line with DC source

Model2_9: Energizing of a long line with AC source

Model2_10: Pi section transmission line

Model2_11: Surge arrester using

    Model_use1: A simple model for investigation of the transmission line frequency characteristics

Model3_1: Employment of the three-phase sequence analyzer

Model3_2: Employment of the fast Fourier transformation—the first variant

Model3_2_1: Employment of the fast Fourier transformation—the second variant

Model3_3: Employment of the filters and the measuring blocks

Model3_4: Employment of the second-order filter

Model4_1: Supplying of the R-L-E load by the six-pulse thyristor rectifier

Model4_2: Four-quadrant thyristor rectifier without circulating currents

Model4_3: 12-pulse thyristor rectifier with series bridges

Model4_4: 12-pulse thyristor rectifier with parallel bridges

Model4_5: High-voltage DC transmission line, simple control

Model4_6: Two-level voltage-source inverter (VSI)

Model4_6a: Two-level VSI, averaging structure

Model4_7: Three-level VSI

Model4_8: Forming of the AC voltage with low harmonic content

Model4_9: Four-level inverter with "flying" capacitors and passive load

Model4_10: Simple model with Z-source inverter

Model4_11: Z-source inverter with voltage control

Model4_12: Resonance inverter

Model4_13: Modular multilevel inverter—starting charge

Model4_14: Modular multilevel inverter

Model4_15: Matrix converter

Model5_1: Two-quadrant chopper DC drive with 1 IGBT, hysteresis current controller and field weakening

Model5_2: Two-quadrant chopper DC drive with 1 IGBT, pulse-width current controller and field weakening

Model5_3: Two-quadrant chopper DC drive with 2 IGBT, hysteresis current controller and field weakening

Model5_4: Four-quadrant DC thyristor drive with field weakening and saturation

Model5_5: DC chopper drive with a series motor and saturation

    DC_contr.m: Model5_5 initialization

    dc_ser.mat: Result of carrying out of the program DC_contr.m

Model5_6: Direct IM start

Model5_7: AC4 drive with DTC

Model5_8: IM with two-level VSI and DTC—second version

Model5_9: IM AC3 with two-level VSI

Model5_10: IM drive with an active front-end rectifier

Model5_11: IM with three-level VSI and DTC

Model5_12: IM with three-level VSI and output LC filter

Model5_13: IM with CHB inverter without recuperation

Model5_14: IM with CHB inverter with recuperation

Model5_15: Four-level inverter with the "flying" capacitors and IM

Model5_16: IM with five-level HNPC inverter and vector control

Model5_17: IM with five-level HNPC inverter and DTC

    Tabl.m: Computation of the blocks Tabl for Model5_17

Model5_18: Resistance IM start

Model5_19: Scherbius cascade with DTC

Model5_20: IM with a current source inverter

Model5_21: IM soft start

Model5_22: IM model with six terminals

    IM_MODEL_6end: Parameter setting of the IM with six terminals

    Model5_23: Six-phase IM under supplying by stepwise voltage

IM_MODEL_6Ph: Parameter setting of the six-phase IM

Model5_24: Six-phase IM with two VSI

Model5_25: Not-regulated IM drive

Model5_26: IM reversal by phase sequence change

Model5_27: Load-commutated thyristor inverter with SM, with flux control

Model5_28: Load-commuted thyristor inverter with six-phase SM

    SM_MODEL_6ph: Parameter setting of six-phase SM

Model5_29: Load-commuted thyristor inverter with six-phase SM—second version

Model5_30: Cycloconverter with SM and three transformers

Model5_31: Cycloconverter with SM and multiwinding transformer

    SM_Param_2000.m Initialization of the model5_31

Model5_32: Simplified cycloconverter model with SM

Model5_33: VSI and SM

Model5_34: Power SM supplied by 4 three-level VSI

Model5_35: Power SM supplied by multi-sell cascade inverter

Model5_36: PMSM and DTC

  Contr_PMSM_DTC.m Initialization of the Model5_36

Model5_37: AC6 PMSM

Model5_38: Switch reluctance motor drive

Model5_39: Two-motor DC drive with elastic shifts

Model5_40: Two-motor DC drive with elastic shifts and backlashes

  LineParameters1.mat: Transmission line parameter computation (25 kV, 3 phase, 1 ground conductor)

  LineParameters2.mat: Two-circuit transmission line parameter computation (500 кB, 6 phase, 2 ground conductors)

  LineParameters3.mat: Two parallel transmission line (500 kV, 2*(3 phase, 1 ground conductor)) parameter computation

  LineParameters.mat: Transmission line parameter computation (735 kV, 3 phase, 2 ground conductors)

Model6_1: Two SG (simplified SG model) supplying a common load

Model6_2: Estimation of simplified SM parameters

Model6_3: Hydraulic turbine and generator supplying an isolated load

Model6_4: Two units hydraulic turbine and generator supplying a common load

Model6_5: Hydraulic turbine-generator and power source operating parallel

Model6_6: Two units hydraulic turbine—SG connected with a long line

Model6_7: Steam turbine and generator supplying an isolated load

Model6_8: Steam turbine—SG and power source operating parallel (multi-mass model)

Model6_9: Demonstration of the subsynchronous resonance

Model6_10: WG with squirrel-cage IM and power network operating parallel

Model6_11: WG with IM with wounded rotor and power network operating parallel

Model6_12: WG with SGPM and power grid operating parallel

Model6_13: WG with SGPM and diesel-generator operating parallel

Model6_14: Stand-alone WG model

Model6_15: Diesel-generator with squirrel-cage IM

Model6_16: Voltage stabilization with SVC (PHASOR model)

Model6_17: Voltage stabilization by SVC with nonregulated capacitor

Model6_18: Voltage stabilization by SVC with switched capacitors

Model6_19: Ultra-long transmission line with SVC

  SVC_Line.m Computation of the converter: demanded power—angle α for SVC

Model6_20: STATCOM with 4 three-level inverters

Model6_21: STATCOM with cascaded H-bridge multilevel inverter

Model6_22: Asymmetric STATCOM

Model6_23: Active filter

Model6_24: SSSC with three-level inverter

Model6_25: UPFC with three-level inverters and PWM

Model6_26: Power flow control by phase-shifting transformer

Index

A

Abc_to_dq0 transformation, 68, 76, 82, 280, 303
AC current source, 17–18
AC1 model, IM voltages in, 208
AC2 model
    block diagram, 204
    controller (parameter window), 205
AC3 model, 204, 208, 210, 217, 285, 304
    block diagram, 206
    controller (parameter window), 207
    vector control, 205
AC4 model, transient responses in, 201
AC5 model, 304
    parameter window, 286
    structure, 284–285
AC6 PMSM, 304–305
Active damping (AD), 214–215, 217
Active filter(s)
    controller, 400
    simulation, 397–403
Active front-end rectifier, and two-level VSI,
    209–211
Active & reactive power, 68–69
AC voltage source, 17–18
AD, see Active damping (AD)
All voltages and currents option, 90, 92
Asymmetrical cascaded inverter, 396
Asymmetrical (cascaded) STATCOM,
    394–396, 398

B

Bal-controller, 361
Ballast_Power, 361
Bal subsystem, 360
Bandpass filter, 26, 79–80, 83, 336
Bandstop filter, 79–80
Battery, 17–18, 22, 364
Bipolar HVDC electric power transmission
    system, 109
Bistable flip-flop, 77–78
Break algebraic loop in discrete saturation
    model, 40
Breaker, 6, 35, 40–41, 56, 58–60, 63, 165,
    254–256, 334
Brushless DC motor, see Trapezoidal PMSM
Bus selector, 196, 263, 299, 301
Bus type, 333–334, 337, 347
Bypass, 405, 410

C

Callbacks/InitFcn option, 382
Callbacks option, 141, 234
Callbacks/PreloadFcn option, 380
Cascaded asymmetrical STATCOM,
    see Asymmetrical (cascaded)
    STATCOM
Cascaded H-bridge multilevel (CHB) inverter
    IM and electric drives, 216–221
    power electrical drive, 291–297
    power electronics devices, 122–126
    STATCOM, 392–397
Characteristic (wave) impedance, 52
CHB inverter, see Cascaded H-bridge multilevel
    (CHB) inverter
Chopper control, DC drives with, 167–179
    armature circuit equation, 167
    armature motion equation, 168
    block mode, Simulink diagram of, 173
    converter, 177–178
    DC flux control, 172
    DCM current, 175
    DCM torque, 167
    dialog box, 169, 177
    four-quadrant chopper drive, 176
    generator mode, 169
    graphical representation, 168
    mode switch-over with contactors, 170
    motor mode, 169
    one-quadrant chopper drive, 176
    parameter window, 178
    pulse-width controller, 171
    relay current controller, 171
    speed regulation, 178
    static mode switch-over, 170
    torque regulation, 178
    transient responses, 174
    two-quadrant chopper drive, 176
Closed control system
    mains current, 295
    SM transients, 294
    SM voltages, 295
Compute RLC line parameters option,
    16, 318–320
Conductor construction, 320
Connection port, 59–60, 62
Continuous current mode, output voltage in, 92
Continuous DC drive model, 189
Control blocks, 77

abc_to_dq0 transformation, 82
bistable flip-flop, 77–78
continuous current mode, output voltage in, 92
discrete Butterworth filter, 82–83
discrete lead-lag, 82–83
discrete PI controller, 82–83
discrete PWM generator, 95, 98
discrete rate limiter, 82
discrete second-order variable-tuned
    filter, 82–83
discrete shift register, 82–83
discrete SV PWM generator, 95, 98, 100–101
discrete synchronized 6-pulse
    generator, 92–93
discrete synchronized 12-pulse
    generator, 92–93
discrete three-phase PWM generator,
    95, 98–100
discrete variable frequency mean value, 80
discrete variable transport delay, 82
discrete virtual PLL, 82–83
edge detector, 77–78
first-order filter, 77–79
monostable flip-flop, 77–78
one-phase PLL, 77, 80–82
on/off delay, 77
for power electronics, 92–101
pulse generators, 95
PWM generator, 95, 98–99
rate limiter, 82
sample & hold, 77
second-order filter, 77, 79, 83
space vector modulation, 97
synchronized 6-pulse generator, 92–93
synchronized 12-pulse generator, 92–95
three-phase PLL, 77, 82
three-phase programmable source, 77, 80
thyristors, 93
timer, 77–78
Controlled current source, 17–18, 90, 185, 251
Controlled voltage source, 17–18, 41, 90, 179,
    184–185, 283
Controller signals, 380
Control parameters option, 373, 385, 406
Converter(s), see Specific converters
Converters and DC bus option, 198–199
Coulomb friction value, 266
CSI, see Current source inverter (CSI)
C-type high-pass filter, 26–27, 29, 33
Current measurement, 65–66, 402–403
Current_Reg, 358
Currents and voltages, general equations for, 52
Current source inverter (CSI), 240–242
Cycloconverter simulation, 277
    controlled voltage source, 283
    control structure, 279–280
    electrical drive, transients in, 282
        main circuits, 278
        multiwinding transformer, 280
        regulators, 280–281, 284
        simplified cycloconverter model, 284
        SM and mains currents, 282–283
        SM_Param_2000 program, 281
        subsystems, 280

D

Data type of input reference vector option, 101
DC_contr.m program, 186–187
DC motors, see Direct current (DC) motors and
    drives
DC voltage source, 17–18
Dead zone, 141, 315
Delta-hexagonal transformer, 415, 417
Delta modulation, 146–147, 150
Detailed thyristor, 85–87
Device currents option, 90
Device voltages option, 90
Diesel and SG control, 360
Diesel-generator and SGPM, wind generation
    with, 359–363
Diesel–squirrel-cage IG, 369–371
Diode, 85–86, 89
Direct current (DC) motors and drives; see also
            Induction motors (IM); Switched
            reluctance motor (SRM); Synchronous
            motors (SM); Synchronous motor with
            permanent magnet (PMSM)
    with chopper control, 167–179
    saturation consideration, 179–187
    series motor main circuit, DC drive with, 186
    SimPowerSystems, electrical drives in, 187–189
    transient responses, 184
Direct torque control (DTC)
    and induction motor, 197, 199–200
    and PMSM, 303
    and three-level VSI, 211–214
    transient responses, 203
    and two-level VSI, 196–204
Discrete Butterworth filter, 82–83
Discrete control blocks, 82
Discrete gamma measurement, 111–113
Discrete HVDC controller, 111
Discrete lead-lag, 82–83
Discrete 3-phase PWM generator, 247, 391
Discrete PI controller, 82–83, 201, 267
Discrete PLL-driven fundamental value, 73, 76
Discrete 12-pulse HVDC control, 111
Discrete PWM generator, 95, 98, 115
Discrete rate limiter, 82
Discrete second-order variable-tuned filter, 82–83
Discrete shift register, 82–83
Discrete SV PWM generator, 95, 98, 100–101,
    116–117

Discrete synchronized 6-pulse generator, 92–93
Discrete synchronized 12-pulse generator, 92–93, 111
Discrete three-phase PLL-driven positive-sequence fundamental value, 73, 76
Discrete three-phase positive-sequence active & reactive power, 73–74
Discrete three-phase PWM generator, 95, 98–100, 119–120
Discrete three-phase total power, 73
Discrete variable frequency mean value, 73–74, 76, 80
Discrete variable transport delay, 82
Discrete virtual PLL, 82–83, 116
Discretize electrical model option, 3
Display wind turbine power characteristics option, 350
Distributed parameter line, 54–55, 111, 318, 320–321, 324, 347, 382
Distribution static synchronous compensator (DSTATCOM), 390–392
Disturbance-compensating control, 399
Double-fed induction generator, 351
Double pulsing option, 102
Double-tuned filter, 27–28, 31–32
Dq0-based active & reactive power, 68, 70
Dq0_to_abc transformation, 68
Droop characteristic realization, 329
DSTATCOM, *see* Distribution static synchronous compensator (DSTATCOM)
DTC, *see* Direct torque control (DTC)

E

Edge detector, 77–78
Edge:Rising option, 78
Electrical sources, 17
    AC current source, 17–18
    AC voltage source, 17–18
    battery, 17–18, 22
    controlled current source, 17–18
    controlled voltage source, 17–18
    DC voltage source, 17–18
    multimeter, 17
    three-phase programmable voltage source, 17–18
    three-phase source, 17–18, 20–21
Electric machine and electric drive simulation, *see Specific* motors and drives
Electric power production and transmission simulation
    diesel–squirrel-cage IG, 369–371
    FACTS simulation, 372
        active filter simulation, 397–403
        phase-shifting transformer, 415–418
        STATCOM simulation, 384–397
        static synchronous series compensator, 403–407
        static var compensator, 372–383
        UPFC, 407–415
    hydraulic-turbine generators, 326–337
    simplified SM model, 324–326
    steam turbine-synchronous generator, 338–348
    transmission line parameters, 317–324
    wind generation, 349
        with diesel-generator and SGPM, 359–363
        with induction generator, 349–356
        with SGPM, 356–359
        stand-alone WG, 363–368
Electronic PMSM, *see* Trapezoidal PMSM
Excitation, 336
Excitation controller, 182
Excitation system, 331
External control of PQ option, 24
External control of reference voltage option, 373
External mechanical torque option, 352
Extra library/control blocks, 344

F

FACTS, *see* Flexible AC Transmission Systems (FACTS)
Fast Fourier transformation (FFT), 73–75
FFT analysis option, 15, 20, 49, 102, 106, 117, 119–120, 126, 131, 138–139, 157, 163, 203, 215, 228, 281, 290
File/Model Properties/Callbacks/InitFcn option, 251
File/Model Properties/Callbacks/Model initialization function option, 3
File/Model Properties/Callbacks option, 227, 303, 329
Filter(s), *see Specific* filters
Firing angle reference, in 12-pulse converter, 105
First-order filter, 77–80, 141, 176, 183, 400
Five-level H-bridge neutral-point clamped (5L-HNPC) inverter, 223
    capacitor voltages, 228–229
    control code decoding, 233–234
    gate pulse forming, 226
    IM current, 235
    IM speed, 234
    load and supply currents, 228
    one phase, 224
    output voltages, 224
    phase and line load voltages, 228
    3-phase programmable source, 227
    PWM modulation, 225
    subsystem, 226–227, 231–232
    switching frequency, 234
    transient processes, 229, 235
    voltage space vector, 228, 230, 234

Flexible AC Transmission Systems (FACTS), 372
  active filter simulation, 397–403
  phase-shifting transformer, 415–418
  STATCOM simulation
    with CHB inverter, 392–397
    DSTATCOM simulation, 390–392
    standard STATCOM systems, 384–390
  static synchronous series compensator,
     403–407
  static var compensator, 372–383
  UPFC, 407–415
Flux animation tool option, 39
Flying capacitor inverter(s), 126–134, 221–223
Forced-commutated devices, 115
  currents, 119
  discrete PWM generator, 115
  discrete SV PWM generator, 116–117
  discrete three-phase PWM generator,
    119–120
  discrete virtual PLL, 116
  LC filter, 119
  multimeter, 118
  three-level bridge, 118
  three-level VSI, 119
    load voltage, 121
    output inverter voltages, 121
  three-phase programmable source, 116–117
  three-phase transformer 12 terminals, 120
  two-level VSI, 115
  universal bridge, 115, 118
  voltages, 116–117, 119
  VSI output voltage, controller of, 116
Form_Pulses subsystem, 267–268, 274
Fourier analyser, 325
Fourier blocks, 68–69, 72
Four-level inverter, *see* Flying capacitor
    inverter(s)
Four-quadrant chopper DC drive, 176
Frequency characteristics
  computation, 15, 30
  measurement, 58
Frequency measurement *vs.* impedance, 14–15
Fundamental and/or harmonic generation
    option, 19

G

Gate pulse generation
  bipolar modulation, 126
  flying capacitor inverter(s), 128
  single-polar modulation, 125
Generate report option, 16
Generator, 355–356, 358, 382
Generator buses, 11
Generic power system stabilizer, 335
Gen_Reg subsystem, 357–358
Geometric mean radius (GMR), 319–320

Ground block, 59–60, 62
Ground fault option, 61
Grounding transformer, 36, 47, 49–50
GTO, 3, 85–86, 91, 241, 389

H

Hall effect signal, 301
Harmonic generation option, 117
High-pass filter, 26–27, 29, 31–32, 78–80, 111,
    113, 335, 357
High-voltage direct current (HVDC) electric
    power transmission system, 108
  bipolar, 109
  currents, 114–115
  data acquisition, 113
  discrete gamma measurement, 111–113
  discrete HVDC controller, 111
  discrete 12-pulse HVDC control, 111
  discrete synchronized 12-pulse generator, 111
  distributed parameters line, 111
  HVDC discrete 12-pulse firing control, 111,
    113
  inverter, 110
  main circuits, 110
  master control, 112
  monopolar, 109
  PLL block, 111
  power-flow control, 109
  ramping unit, 113
  rectifier, 110
  simplified HVD, 113–114
  three-phase discrete PLL, 111
  voltages, 114–115
HVDC, *see* High-voltage direct current (HVDC)
    electric power transmission system
HVDC discrete 12-pulse firing control, 111, 113
Hydraulic turbine and governor, 328–331, 336,
    343–344
Hydraulic-turbine generators, 326–337

I

Ideal switch, 59–60, 63, 91–92, 161
Ideal switching device method, 5
IG, *see* Induction generator (IG)
IGBT, 85, 87
IGBT/Diode, 85, 87, 91
IM_Model_6term program, 246–247
IM motors, *see* Induction motors (IM), and
    electric drives
Impedance measurement, 29, 67
Impedances, and loads
  breaker, 35
  filter models, 27
  frequency characteristic computation, 30
  impedance measurement, 29

multimeter, 26
mutual impedance, 33
mutual inductance, 22, 33–36
parallel RLC branch, 22–23
parallel RLC load, 22–23
series RLC branch, 22–23
series RLC load, 22–23
three-phase breaker, 34–35
three-phase dynamic load, 22–26
three-phase harmonic filter, 22, 26
three-phase mutual inductance Z1–Z0, 22, 36
three-phase parallel RLC branch, 22–23
three-phase parallel RLC load, 22–23
three-phase programmable voltage source, 25
three-phase sequence analyzer, 26
three-phase series RLC branch, 22–23
three-phase series RLC load, 22–23
Impedance *vs.* frequency measurement, 14–15
IM soft-start, 242–244
Induction generator (IG)
 parameters, 352
 wind generation, 349–356
Induction motors (IM), and electric drives;
  see also Direct current (DC)
  motors and drives; Switched
  reluctance motor (SRM);
  Synchronous motors (SM);
  Synchronous motor with
  permanent magnet (PMSM)
 bus selector, 196
 CHB inverter, IM supplied from, 216–221
 with CSI, 240–242
 flying capacitor inverter(s), 221–223
 IM soft-start, 242–244
 IM voltages, equations of, 190
 5L-HNPC, 223–235
 line-fed IM, special operation modes of,
   254–256
 model description, 190–196
 with phase-wound rotor, 236–239
 SimPowerSystems, standard IM drives in,
   204–208
 six-phase IM model, 247–254
 with six terminals, 244–247
 squirrel-cage IM
  pu parameters, 195
  SI parameters, 195
 stator flux linkage, 190
 with three-level inverter and *L-C* filter,
   214–216
 three-level VSI and DTC, 211–214
 with two-level VSI
  and active front-end rectifier, 209–211
  and DTC, 196–204
 wounded-rotor IM
  pu parameters, 195
  SI parameters, 195

Initialize filter response option, 79
Integrator limited, 294
Interior magnet setting (IPM), 299, 301
Internal generation of modulating signal
   option, 99
Inverter control system, 205–206
Inverters, *see Specific* inverters
Inv_Reg subsystem, 358
IPM, *see* Interior magnet setting (IPM)

L

LC filter, 119, 161, 164, 214–216, 231, 365
5L-HNPC inverter, *see* Five-level H-bridge neutral-
   point clamped (5L-HNPC) inverter
Limiter, 182, 308
Linear transformer, 36–37
Line-fed IM, special operation modes of,
   254–256
Load buses, 11
Load-commutated converters
 six-phase SM, 274
 and synchronous motors, 264–270
Load flow and machine initialization option, 9,
   12, 332–333, 337, 343, 345, 382, 410
Logical switching unit, 103
Long transmission line, 380–382
Look under mask option, 42, 45, 196, 198, 328
Lookup table, 22, 179–180, 201, 212–213, 238
Lowpass filter, 78–80, 198, 331, 335, 402
LSU scheme, 103

M

Machines measurement demux, 196, 263
Magnetization curve, 180–181
Main circuit, measurement of, 65–66
 current measurement, 65
 impedance measurement, 67
 multimeter, 66
 three-phase V–I measurement, 65–67
 voltage measurement, 65
4-mass shaft block, 341
MATLAB®, 1, 65, 154, 186, 318
MATLAB Contr_PMSM_DTC, 302–303
Matrix converters (MC), 157–165
Mean value, 68, 72–73
Measurement option, 66, 90, 358
Measurement voltages option, 118
Mechanical coupling simulation, 311–312
 dead zone, 315
 mechanical shaft, 311, 313, 315
 reduction device, 311–313
 rotary kiln kinematics, 312
 single-mass system, 311, 314
 speed reducer, 311–312
 two-mass system, 314

Mechanical shaft, 311, 313, 315
Modular multilevel converters (MMC), 152–157
Monopolar HVDC electric power transmission
    system, 109
Monostable flip-flop, 77–78
MOSFET, 85–86, 91
Multi-band power system stabilizer, 336,
    343–344
Multimass simulation, adjustment for, 342–343
Multimass turbine, 345
Multimeter, 17, 26, 41, 55–56, 66, 71, 90, 103,
    106–107, 118, 141
Multiport switch, 336
Multiwinding transformer, 36–37, 41–43, 219,
    280, 416
Mutual impedance, 33, 71, 107, 318
Mutual inductance, 22, 33, 36
    dialog box, 34
    scheme, 35
    six windings, 34
    three-phase inductance, 34
    three windings, 34
    two windings, 34

N

Neutral block, 59–60, 62
Nominal π-model, 53, 59
Nonlinear turbine model, 328

O

Ode 23 tb method, 3, 5
OLTC regulating transformer, 42
    dialog box, 45
    main circuit switches, 44
    voltage regulator, 43, 46
One-phase PLL, 77, 80–82
One-quadrant chopper DC drive, 176
On/off delay, 77
Open control system
    mains current, 294
    SM transients, 293
    SM voltages, 294

P

Parallel active filter connection, 398
Parallel RLC branch, 22–23
Parallel RLC load, 22–23
Parameter calculation window, 322–324
Park's transformation, 68–69, 400
PCC, *see* Point of common coupling (PCC)
Permanent magnet option, 168
Permanent magnet synchronous option, 301
Phase current hysteresis controller, 207
Phase-locked loop (PLL) units, 76, 111

Phase-shifting transformer, 49, 126, 289,
    415–418; *see also* Zigzag phase-
    shifting transformer
Phase-wound rotor, 236–239
Phasor option, 4
Phasor simulation method, 3–4
PI section line, 54, 56, 59, 321
PLL, *see* Phase-locked loop (PLL) units
Plot filter response option, 79, 344
Plot magnetization curves option, 308
Plot selected measurements option, 66
π-model, 52–53
PMSM, *see* Synchronous motor with permanent
    magnet (PMSM)
POD, *see* Power oscillation damping (POD)
Point of common coupling (PCC), 355
Polyfit function, 186
Polyval function, 186
Position sensor, 308
Power
    generation, 382–383
    transfer, 355–356
Power circuit devices
    electrical sources, 17–22
    impedances and loads, 22–36
    miscellaneous, 59–63
    transformers, 36–50
    transmission line models, 51–59
Power data, 373, 384, 406
Power_dstatcom_pwm model, 390
Power electrical drive
    with CHB inverter, 291–297
    with three-level VSI, 287–291
Power electronics devices
    CHB inverter, 122–126
    control blocks for, 92–101
    flying capacitor inverter(s), 126–134
    forced-commutated devices, converters with,
        115–121
    HVDC electric power transmission system,
        108–115
    matrix converters, 157–165
    modular multilevel converters, 152–157
    power semiconductor devices, 85–92
    resonant inverters, 143–151
    semiconductor devices, 85
    thyristors, converter with, 101–108
    Z-source converters, 134–143
Power flow control, 410
Powergui, 7–16
    compute RLC line parameters, 318–319
    continuous mode, 12
    dialog box, 8
    discrete mode, 11
    FFT analysis, 15
    Fourier transformation computation, 16
    frequency-characteristic computation, 15

generate report, 16
generator buses, 11
hysteresis design tool, 16
impedance *vs.* frequency
        measurement, 14–15
initial states setting, 9
load buses, 11
load flow and machine
        initialization, 9, 12
load-flow computation, 13–14
main window, 8
reference bus, 11
simulation mode, 11
steady-state voltages and currents, 9, 12
update load flow, 12
window preferences, 7
Power oscillation damping (POD), 406
Power_PSS model, 347
Power_PSTdeltahex.mdl model, 416
Power semiconductor devices, 85–86
    controlled current source, 90
    controlled voltage source, 90
    detailed thyristor, 85, 87
    diode, 85–86, 89
    GTO, 85, 91
    ideal switch, 91–92
    IGBT, 85, 87
    IGBT/Diode, 85, 87, 91
    MOSFET, 85–86, 91
    multimeter, 90
    three-level bridge, 87–88, 91
    thyristor, 85–87, 89
    universal bridge, 87–89
    VSI, 90
Power_sssc.mdl model, 406
Power_statcom_gto48p.mdl model, 386
Power_statcom.mdl model, 385
Power_svc_ltcr3tsc model, 380
Power system stabilizers (PSS), 334
    diagram, 348
    system stability, influence on, 337
Power_tcsc.mdl model, 383
Power_termal model, 346
Power transformer, dialog box of, 183
Power_upfc_gto48p.mdl, 411
Power_upfc.mdl model, 410
Power_wind_dfig_det.mdl model, 354
Power_wind_dfig.mdl model, 354
Power_windgen.mdl model, 360
Power_wind_type_4_det.mdl model, 356
Preferences option, 7
Procedure to change the number of taps
        option, 45
PSS, *see* Power system stabilizers (PSS)
Pulse generators, 95, 380
PulseGen subsystem, 267–268, 274
Pulse-width current controller, 171

Pulse-width modulation (PWM), 92
    generator, 95, 98–99, 253
    inverter, 209
    rectifier, 209
P&V generator, 12, 333–334, 337, 347, 382
PV measurements, 337
PWM, *see* Pulse-width modulation (PWM)

R

Rate limiter, 82, 172, 182, 203, 218, 220, 304,
        357–358, 377
Reactance Xa of 1 m spacing, 320
Reactor_Currents, 290
Reduction device, 311–313
Reference bus, 11
Relay current controller, 171, 197
Reluctance motors
    sinusoidal, 305
    switched, 305–310
Resonant inverters, 143–151
Reversible DC drive, with flux weakening, 181
RMS block, 68, 72
Rotary kiln kinematics, 312
Round-rotor SM, 257, 263, 297

S

Saturable transformer, 36–38
Saturation parameters option, 261
Scherbius cascade scheme, 236–237, 239, 352
SCIG, *see* Squirrel-cage induction generators
        (SCIG)
Second-order filter, 77–80, 83, 141, 344
Semiconductor devices, *see* Power semiconductor
        devices
Sequence components, 70
Series-compensated transmission system
        model, 382
Series compensation scheme, 346
Series RLC branch, 22–23, 58, 161
Series RLC load, 22–23, 162
Servomotor model, 329
Set the initial capacitor voltage option, 22
Set the initial inductor current option, 22
SGPM, *see* Synchronous generator with
        permanent magnets (SGPM)
Show measurement port option, 63, 85
Show UPFC controllable region, 412
Sigma-delta modulation, 146–147, 151
Simplified cycloconverter model, 284
Simplified HVD, 113–114
Simplified SM model, 297–298
    bus selector, 299
    dialog box, 298
    dynamic torque, 298
    usage, 324–326

Simplified synchronous machine, 382
Simplify synchronous option, 299
SimPowerSystems
    AC current source, 17–18
    AC voltage source, 17–18
    battery, 17–18, 22
    breaker, 6
    characteristics, 1–7
    commands, 5–6
    controlled current source, 17–18
    controlled voltage source, 17–18
    controller diagram, 329
    DC electrical drives, 187–189
    DC voltage source, 17–18
    error report, 5
    generator buses, 11
    load buses, 11
    phasor simulation method, 3–4
    Powergui, 7–16
    reference bus, 11
    simple model scheme, 6
    standard IM drives in, 204–208
    three-phase programmable voltage source,
        17–18
    three-phase source, 17–18
    transformer model, 416
Simulate hysteresis option, 37, 41
Simulate saturation option, 261
Simulink®, 1, 4, 17, 68
    abc_to_dq0 transformation, 68, 76
    active & reactive power, 68–69
    discrete measurement blocks, 73
    discrete PLL-driven fundamental value,
        73, 76
    discrete three-phase PLL-driven positive-
        sequence fundamental value, 73, 76
    discrete three-phase positive-sequence active
        & reactive power, 73
    discrete three-phase total power, 73
    discrete variable frequency mean value,
        73–74, 76
    dq0-based active & reactive power, 68, 70
    dq0_to_abc transformation, 68
    FFT, 73–75
    Fourier blocks, 68–69, 72
    mean value, 68, 72–73
    multimeter, 71
    mutual inductance, 71
    Park's transformation, 68–69
    RMS, 68, 72
    sequence analyzer, 72
    sequence components, 70
    THD, 68, 72
    three-phase instantaneous active & reactive
        power, 68–70
    three-phase PLL-driven positive-sequence
        active & reactive power, 73, 76

    three-phase positive-sequence fundamental
        value, 73–74
    three-phase sequence analyzer,
        68, 70–71
    variable transport delay, 68
Single-mass simulation, 342
Single-mass system, 311, 314, 348
Single-phase transformer models, 37
Single-tuned filter, 27–30
Sinusoidal PMSM, 300
Sinusoidal reluctance motors, 305
Six-phase IM model, 247; *see also* Six-phase
        SM model
    arrangement, 249
    connection, 249
    controlled current source, 251
    control system, 252–253
    electrical drive
        control, 253
        processes in, 248, 254
    IM reverse, 256
    IM torque, 251
    polyphase motors, 247
    PWM generator, 253
    saturation, 253
    subsystem, 251
    torque oscillation, 252, 256
    voltage shape, 252
Six-phase SM model, 271; *see also* Six-phase
        IM model
    circuits, 276
    currents in, 277
    excitation current, 275
    Form_pulses, 274
    with load-commutated converter, 274
    phase currents, in motor and braking
        modes, 276
    SM torque, 272
    speed, 275
    stator voltage equations, 271
    subsystem, 273
    synchronized 12-pulse generator, 274
    torque, 275
    voltage in, 277
Six terminals, IM model with, 244–247
SM model, 256–264; *see also* Six-phase
        SM model
SM motors, *see* Synchronous motors (SM), and
        electric drives
SM_Param_2000 program, 281
SM voltage(s)
    with closed control system, 295
    with long cable connection, 295
    with open control system, 294
    spectrum, 296
Solver option, 7
Space vector modulation (SVM), 97

Specify impedance using short-circuit level option, 20
Speed controller
    of diesel generator, 361
    input circuits, 342
Speed reducer, 311–312
Speed Relay, 342
SPM, *see* Surface magnet setting (SPM)
Squirrel-cage IM
    pu parameters, 195
    SI parameters, 195
Squirrel-cage induction generators (SCIG), 350–351, 353; *see also* Diesel–squirrel-cage IG
SRM, *see* Switched reluctance motor (SRM)
SSR, *see* Subsynchronous resonance (SSR)
SSSC, *see* Static synchronous series compensator (SSSC)
SSSC_Control subsystem, 413
Stand-alone WG, 363–368
Standard IM drives, in SimPowerSystems, 204–208
Standard STATCOM systems, 384–390
STATCOM, *see* Static synchronous compensator (STATCOM)
Statcom_Control, 413
Static synchronous compensator (STATCOM), 372–385; *see also* Static var compensator (SVC)
    asymmetrical STATCOM, 395
    cascaded asymmetrical STATCOM, 396
    with CHB inverter, 392–397
    controller, 388
    control system, 393
    DSTATCOM simulation, 390–392
    phase voltages, 394
    standard STATCOM systems, 384–390
    with 4 three-level VSI, 386–388
    voltage deviations, 392
Static synchronous series compensator (SSSC), 385, 403–407
Static var compensator (SVC), 372–383; *see also* Static synchronous compensator (STATCOM)
Steady-state voltages and currents, 9, 12
Steam machine and governor, 340–341, 343–344
Steam turbine
    configuration, 338
    as controlled plant, 338
    control system, 339–340
    five-mass model, 340
    rotor, 339–340
Steam turbine-synchronous generator, 338–348
Step Vref block, 385
Subsynchronous resonance (SSR), 345–346
Surface magnet setting (SPM), 299, 301
Surge arrester, 59–62

SVC, *see* Static var compensator (SVC)
SVC modeled using positive-sequence option, 373
SVC_Power subsystem, 382
SVC power system, 386
Swing bus, 11–12, 333, 337
Switched PMSM, *see* Trapezoidal PMSM
Switched reluctance motor (SRM), 305, 307
    characteristics, 306
    converter, 308
    and current control, 310
    dialog box, 309
    electrical drive, 308
    limiter, 308
    merits, 306
    position sensor, 308
    rotor poles, 306
    and speed control, 310
    stator phase, 306
    structure, 307
    types, 308
Switching pattern option, 101
Symmetric component theory, 71
Synchronized 6-pulse generator, 92–93, 101–102, 182, 266, 274, 372, 375, 379, 400
Synchronized 12-pulse generator, 92–95, 104, 106, 274
Synchronous generator with permanent magnets (SGPM), 350, 369
    and diesel-generator, 359–363
    wind generation, 356–359
Synchronous machine pu fundamental, 359
Synchronous motors (SM), and electric drives, 259; *see also* Direct current (DC) motors and drives; Induction motors (IM); Switched reluctance motor (SRM); Synchronous motor with permanent magnet (PMSM)
    bus selector, 263
    cycloconverter simulation, 277–284
    DC and SM currents, 269
    dialog box, 260
    discrete PI controller, 267
    Flux_Est, 268
    flux linkages, differential equations for, 258
    Form_Pulses, 267–268
    and load-commutated converters, 264–270
    machines measurement demux, 263
    magnetization dependence, 261
    measuring block, 264
    PulseGen, 267–268, 274
    round-rotor SM, 257, 263, 297
    saturation parameters, 261
    simplified model, 297–299
    simulate saturation option, 261
    six-phase SM model, 271–277
    SM model, 256–264

standard dialog box, 262
subsystems, 266–267
synchronized 6-pulse generator, 266
three-phase series RLC branch, 266
three-phase V-I measurement, 266
transition to two-phase SM, 257
universal bridge, 266
and VSI
  AC5 model, 284–286
  CHB inverter, power electrical drive with,
    291–297
  electrical drive, transients in, 287
  standard model, 284–287
  three-level VSI, power electrical drive
    with, 287–291
Synchronous motor with permanent
    magnet (PMSM), 299–300; see
    also Direct current (DC) motors
    and drives; Induction motors
    (IM); Switched reluctance
    motor (SRM); Synchronous
    motors (SM)
  AC6 PMSM, 304–305
  bus selector, 301
  control principles, 301
  dialog box, 300
  and DTC, 303
  DTC_Reg, 301
  electrical part, 301
  feedback, 301
  flux, 301
  Hall effect signal, 301
  MATLAB Contr_PMSM_DTC, 302
  mechanical part, 301
  rate limiter, 304
  sinusoidal PMSM, 300
  torque, 299
  trapezoidal PMSM, 300
  voltage equations, 299

T

Tap selection time, 45
Tap transition time, 45
T/D ratio, 319–320
THD, see Total harmonic distortion (THD)
Three-level bridge, 87–88, 91, 118, 211–213
Three-level inverter and L-C filter, 214–216
Three-level VSI, see Voltage source
    inverter (VSI)
Three-phase breaker, 34–35, 59–60,
    254–255
Three-phase discrete PLL, 111
Three-phase dynamic load, 22–26, 406
Three-phase fault models, 59–60, 333,
    336–337, 346–347

Three-phase harmonic filter, 22, 26, 375
Three-phase instantaneous active & reactive
    power, 68–70
Three-phase measurement, 255
Three-phase mutual inductance Z1–Z0, 22, 36
Three-phase OLTC regulating transformer,
    see OLTC regulating
    transformer
Three-phase parallel RLC branch, 22–23
Three-phase parallel RLC load, 22–23
Three-phase PI section line, 46, 54, 56, 318,
    321, 348
Three-phase PLL, 77–78, 82
Three-phase PLL-driven positive-
    sequence active & reactive
    power, 73, 76
Three-phase positive-sequence fundamental
    value, 73–74, 141
Three-phase programmable source, 77, 80,
    116–117, 130–131, 138–139
Three-phase programmable voltage source,
    17–18, 25, 141, 255
Three-phase sequence analyzer, 26, 68,
    70–71, 156
Three-phase series RL branch, 377
Three-phase series RLC branch, 22–23, 266
Three-phase series RLC load, 22–23
Three-phase source, 17–18, 20–21, 163,
    333, 357
Three-phase transformer, 36, 47–48
Three-phase transformer inductance matrix
    type, 36, 47
Three-phase transformer 12 terminals, 36, 47,
    120, 396, 407
Three-phase V-I measurement, 65–67,
    209, 266
Three windings mutual inductance option, 33
Thyristor controlled series capacitors
    (TCSC), 383
Thyristors, 85–87, 89, 93, 101–108, 182,
    241, 266
Thyristor-switched capacitors (TSC),
    377–378, 380
Timer, 40, 44, 77–78, 103–104
Total harmonic distortion (THD), 68, 72
Tracking characteristic(s), 353–354
Transfer resistances, 45
Transformer(s), 36
  breaker, 40–41
  controlled voltage source, 41
  delta-hexagonal, 415, 417
  grounding, 36, 47, 49–50
  linear, 36–37
  minor hysteresis loop, 39–40
  multimeter, 41
  multiwinding, 36–37, 41–43, 219, 280, 416

OLTC regulating transformer, 42
  dialog box, 45
  main circuit switches, 44
  voltage regulator, 43, 46
phase-shifting, 49, 126, 289, 415–418
power, 183
saturable, 36–38
three-phase PI section line, 46
three-phase transformer, 36, 47–48
three-phase transformer inductance matrix
    type, 36, 47
three-phase transformer 12 terminals,
    36, 47, 120
timer, 40, 44
zigzag phase-shifting, 36, 47–48, 216, 386
Transmission line, 51
  breaker, 56, 58–59
  characteristic (wave) impedance, 52
  constructions, 319, 321
  currents and voltages, general
      equations for, 52
  distributed parameter line, 54–55
  frequency characteristics,
      measurement of, 58
  multimeter, 55–56
  nominal π-model, 53, 59
  parameters, 317–324
  PI section line, 54, 56, 59
  π-model, 52–53
  series RLC branch, 58
  three-phase PI section line, 54, 56
  transients in, 375
  without SVC, 376
Trapezoidal PMSM, 300
Trip, 352, 385, 410
TSC, see Thyristor-switched capacitors
    (TSC)
Turbine, 347, 352, 357
Turbine & regulators, 336, 348
Two-area system, 347
Two-level VSI, see Voltage source
    inverter (VSI)
Two-mass system, 314
Two-phase SM, 257
Two-quadrant chopper DC drive, 176
Type of mutual inductance option, 33

U

Unified power flow controller (UPFC),
    385, 407
  block diagram, 408
  controller, 411
  model adjustment, 411–412
  power controllers, 413–415
  purpose, 410–411

with PWM, 413
scopes, 412
in SimPowerSystems, 409
SSSC control, block diagram of, 410
STATCOM control system, 409
without PWM, 412
Unit simulation, see
    Diesel–squirrel-cage IG
Universal bridge, 87–89, 101, 115, 118,
    182, 186, 237, 240–241,
    266, 397
Update load flow option, 12, 332–333, 347
UPFC, see Unified power flow controller
    (UPFC)
Use a label option, 66

V

Var control mode, 373, 385, 389
Variable inductance model, 185
Variable resistor model, 236
Variable transport delay, 68
Vector control, 210
Voltage measurement, 65
Voltage regulation mode, 373, 385,
    390, 410
Voltage regulator, 45, 378–379
Voltages, 382–383
Voltages_In_Out, 382–383
Voltage source inverter (VSI), 90, 369
  and active front-end rectifier, 209–211
  and DTC, 196–204, 211–214
  output voltage, controller of, 116
  and synchronous motors
    AC5 model, 284–286
    CHB inverter, power electrical drive
        with, 291–297
    electrical drive, transients in, 287
    standard model, 284–287
    three-level VSI, power electrical drive
        with, 287–291
  three-level, 119
    and DTC, 211–214
    load voltage, 121
    and output filter, 214
    output inverter voltages, 121
    power electrical drive, 287–291
    voltage space vectors, 211
    without capacitance filter, 216
  two-level, 115
    and active front-end rectifier,
        209–211
    and DTC, 196–204
  voltage space vectors, 198
VPQ measurements, 413
VSI, see Voltage source inverter (VSI)

W

Wind generation (WG), 349–350
    characteristics, 351
    dialog box, 351
    with diesel-generator and SGPM, 359–363
    with induction generator, 349–356
    SCIG, transients with, 353
    with SGPM, 356–359
    stand-alone WG, 363–368
    wind turbine, power of, 349
    with wounded rotor, 353, 355
Wind_Syst subsystem, 360
Wind turbine, 140, 349–350, 352, 357
Wounded rotor

induction motors
    pu parameters, 195
    SI parameters, 195
    wind generation, 353, 355
Wound-rotor induction generators (WRIG),
        350–351

Z

Zero-order hold model, 399
Zigzag phase-shifting transformer, 36, 47–48,
        216, 386; *see also* Phase-shifting
        transformer
Zoom around hysteresis option, 39
Z-source converters, 134–143